ARCHITECTURE DESIGN FOR SOFT ERRORS

ARCHITECTURE DESIGN FOR SOFT ERRORS

Shubu Mukherjee

AMSTERDAM • BOSTON • HEIDELBERG • LONDON
NEW YORK • OXFORD • PARIS • SAN DIEGO
SAN FRANCISCO • SINGAPORE • SYDNEY • TOKYO

Morgan Kaufmann Publishers is an imprint of Elsevier

Acquisitions Editor	Charles Glaser
Publishing Services Manager	George Morrison
Project Manager	Murthy Karthikeyan
Editorial Assistant	Matthew Cater
Cover Design	Alisa Andreola
Compositor	diacriTech
Cover Printer	Phoenix Color, Inc.
Interior Printer	Sheridan Books

Morgan Kaufmann Publishers is an imprint of Elsevier.
30 Corporate Drive, Suite 400, Burlington, MA 01803, USA

This book is printed on acid-free paper. ∞

Library of Congress Cataloging-in-Publication Data
Mukherjee, Shubu.
   Architecture design for soft errors/Shubu Mukherjee.
       p. cm.
   Includes index.
   ISBN 978-0-12-369529-1
1. Integrated circuits. 2. Integrated circuits—Effect of radiation on. 3. Computer architecture.
4. System design. I. Title.
   TK7874.M86143 2008
   621.3815–dc22

                                                                              2007048527

British Library Cataloguing in Publication Data
A catalogue record for this book is available from the British Library

ISBN: 978-0-12-369529-1

For information on all Morgan Kaufmann publications,
visit our Web site at *www.mkp.com* or *www.books.elsevier.com*

Printed and bound in the United Kingdom

Transferred to Digital Printing, 2011

To my wife Mimi, my daughter Rianna, and my son Ryone

and

In remembrance of my late father Ardhendu S. Mukherjee

Contents

Foreword

I am delighted to see this new book on architectural design for soft errors by Dr. Shubu Mukherjee. The metrics used by architects for processor and chipset design are changing to include reliability as a first-class consideration during design. Dr. Mukherjee brings his extensive first-hand knowledge of this field to make this book an enlightening source for understanding the cause of this change, interpreting its impact, and understanding the techniques that can be used to ameliorate the impact.

For decades, the principal metric used by microprocessor and chipset architects has been performance. As dictated by Moore's law, the base technology has provided an exponentially increasing number of transistors. Architects have been constantly seeking the best organizations to use this increasing number of transistors to improve performance.

Moore's law is, however, not without its dark side. For example, as we have moved from generation to generation, the power consumed by each transistor has not fallen in direct proportion to its size, so both the total power consumed by each chip and the power density have been increasing rapidly. A few years ago, it became vogue to observe, given current trends, that in a few generations the temperature on a chip would be hotter than that on the surface of the sun. Thus, over the last few years, in addition to their concerns about improving performance, architects have had to deal with using and managing power effectively.

Even more recently, another complicating consequence of Moore's law has risen in significance: reliability. The transistors in a microprocessor are, of course, used to create logic circuits, where one or more transistors are used to represent a logic bit with the binary values of either 0 or 1. Unfortunately, a variety of phenomena, such as from radioactive decay or cosmic rays, can cause the binary value held by a transistor to change. Chapters 1 and 2 contain an excellent treatment of these device- and circuit-level effects.

Since a change in a bit, which is often called a bit flip, can result in an erroneous calculation, the increasing number of transistors provided by Moore's law has a direct impact on the reliability of a chip. For example, if we assume (as is roughly projected over the next few process generations) that the reliability of each individual transistor is approximately unchanged across generations, then a doubling of the number of transistors might naively be expected to double the error rates of the chips. The situation is, however, not nearly so simple, as a single erroneous bit value may not result in a user-visible error.

The fact that not every bit flip will result in a user-visible error is an interesting phenomenon. Thus, for example, a bit flip in a prediction structure, like a branch predictor, can never have an effect on the correctness of the computation, while a bit flip in the current program counter will almost certainly result in an erroneous calculation. Many other structures will fall in between these extremes, where a bit flip will sometimes result in an error and other times not. Since every structure can behave differently, the question arises of how is each structure affected by bit flips and overall how significant a problem are these bit flips? Since the late 1990s that has been a focus of Dr. Mukherjee's research.

By late 2001 or early 2002, Dr. Mukherjee had already convinced himself that the reliability of microprocessors was about to become a critical issue for microarchitects to take into consideration in their designs. Along with Professor Steve Reinhardt from the University of Michigan, he had already researched and published techniques for coping with reliability issues, such as by doing duplicate computations and by comparing the results in a multithreaded processor. It was around that time, however, that he came into my office discouraged because he was unable to convince the developers of a future microprocessor that they needed to consider reliability as a first-class design metric along with performance and power.

At that time, techniques existed and were used to analyze the reliability of a design. These techniques were used late in the design process to validate that a design had achieved its reliability goals. Unfortunately, the techniques required the existence of essentially the entire logic of the design. Therefore, they could not be used either to guide designs on the reliability consequences of a design decision or for early projections of the ultimate reliability of the design. The consequence was that while opinions were rife, there was little quantitative evidence to base reliability decisions on early in the design process.

The lack of a quantitative approach to the analysis of a potentially important architectural design metric reminded me of an analogous situation from my early days at Digital Equipment Corporation (DEC). In the early 1980s when I was starting my career at DEC, performance was the principal design metric. Yet, most performance analysis was done by benchmarking the system after it was fully designed and operational. Performance considerations during the design process were largely a matter of opinion.

One of my most vivid recollections of the range of opinions (and their accuracy) concerned the matter of caches. At that time, the benefits of (or even the necessity for) caches were being hotly debated. I recall attending two design meetings. At the first meeting, a highly respected senior engineer proposed for a next-generation machine that if the team would just let him design a cache that was twice the size of the cache of the VAX-11/780, he would promise a machine with twice the performance of the 11/780. At another meeting, a comparably senior and highly respected engineer stated that we needed to eliminate all caches since "bad" reference patterns to a cache would result in a performance worse than that with no cache at all. Neither had any data to support his opinion.

My advisor, Professor Ed Davidson at the University of Illinois, had instilled in me the need for quantitatively analyzing systems to make good design decisions. Thus, much of the early part of my career was spent developing techniques and tools for quantitatively analyzing and predicting the performance of design ideas (both mine and others') early in the design process. It was then that I had the good fortune to work with people like Professor Doug Clark, who also helped me promulgate what he called the "Iron Law of Performance" that related the instructions in a program, the cycles used by the average instruction, and the processor's frequency to the performance of the system. So, it was during this time that I generated measurements and analyses that demonstrated that both senior engineers' opinions were wrong: neither any amount of reduction of memory reference time could double the performance nor "bad" patterns could happen negating all benefits of the cache.

Thus, in the early 2000s, we seemed to be in the same position with respect to reliability as we had been with respect to performance in the early 1980s. There was an abundance of divergent qualitative opinions, and it was difficult to get the level of design commitment that would be necessary to address the issue. So, in what seemed a recapitulation of the earlier days of my career, I worked with Dr. Mukherjee and the team he built to develop a quantitative approach to reliability. The result was, in part, a methodology to estimate reliability early in the design process and is included in Chapters 3 and 4 of this book.

With this methodology in hand, Dr. Mukherjee started to have success at convincing people, at all levels, of the exact extent of the problem and how effective were the design alternatives being proposed to remediate it. In one meeting in particular, after Dr. Mukherjee presented the case for concerns about reliability, an executive noted that although people had been coming to him for years predicting reliability problems, this was the first time he had heard a compelling analysis of the magnitude of the situation.

The lack of good analysis methodologies resulting in a less-than-optimal engineering is ironically illustrated in an anecdote about Dr. Mukherjee himself. Prior to the development of an adequate analysis methodology, an opinion had formed that a particular structure in a design contributed significantly to the reliability of the processor and needed to be protected. Then, Dr. Mukherjee and other members of the design team invented a very clever technique to protect the structure. Later, after we developed the applicable analysis methodology, we found that the structure was actually intrinsically very reliable and the protection was overkill.

Now that we have good analysis methodologies that can be used early in the design cycle, including in particular those developed by Dr. Mukherjee, one can practice good engineering by focusing remediation efforts on those parts of the design where the cost-benefit ratio is the best. An especially important aspect of this is that one can also consider techniques that strive to meet a reliability goal rather than strive to simply achieve perfect (or near-perfect) reliability. Chapters 5, 6, and 7 present a comprehensive overview of many hardware-based techniques for improving processor reliability, and Chapter 8 does the same for software-based techniques. Many of these error protection schemes have existed for decades, but what makes this book particularly attractive is that Dr. Mukherjee describes these techniques in the light of the new quantitative analysis outlined in Chapters 3 and 4.

Processor architects are now coming to appreciate the issues and opportunities associated with the architectural reliability of microprocessors and chipsets. For example, not long ago Dr. Mukherjee made a presentation of a portion of our quantitative analysis methodology at an internal conference. After the presentation, an attendee of the conference came up to me and said that he had really expected to hate the presentation but had in fact found it to be particularly compelling and enlightening. I trust that you will find reading this book equally compelling and enlightening and a great guide to the architectural ramifications of soft errors.

Dr. Joel S. Emer
Intel Fellow
Director of Microarchitecture Research, Intel Corporation

Preface

As kids many of us were fascinated by black holes and solar flares in deep space. Little did we know that particles from deep space could affect computing systems on the earth, causing blue screens and incorrect bank balances. Complementary metal oxide semiconductor (CMOS) technology has shrunk to a point where radiation from deep space and packaging materials has started causing such malfunction at an increasing rate. These radiation-induced errors are termed "soft" since the state of one or more bits in a silicon chip could flip temporarily without damaging the hardware. As there are no appropriate shielding materials to protect against cosmic rays, the design community is striving to find process, circuit, architectural, and software solutions to mitigate the effects of soft errors.

This book describes architectural techniques to tackle the soft error problem. Computer architecture has long coped with various types of faults, including faults induced by radiation. For example, error correction codes are commonly used in memory systems. High-end systems have often used redundant copies of hardware to detect faults and recover from errors. Many of these solutions have, however, been prohibitively expensive and difficult to justify in the mainstream commodity computing market.

The necessity to find cheaper reliability solutions has driven a whole new class of quantitative analysis of soft errors and corresponding solutions that mitigate their effects. This book covers the new methodologies for quantitative analysis of soft errors and novel cost-effective architectural techniques to mitigate their effects. This book also reevaluates traditional architectural solutions in the context of the new quantitative analysis.

These methodologies and techniques are covered in Chapters 3–7. Chapters 3 and 4 discuss how to quantify the architectural impact of soft errors. Chapter 5 describes error coding techniques in a way that is understandable by practitioners and without covering number theory in detail. Chapter 6 discusses how redundant computation streams can be used to detect faults by comparing outputs of the two streams. Chapter 7 discusses how to recover from an error once a fault is detected.

To provide readers with a better grasp of the broader problem definition and solution space, this book also delves into the physics of soft errors and reviews current circuit and software mitigation techniques. In my experience, it is impossible to become the so-called soft error or reliability architect without a fundamental grasp of the entire area, which spans device physics (Chapter 1), circuits (Chapter 2), and software (Chapter 8). Part of the motivation behind adding these chapters had grown out of my frustration at some of the students working on architecture design for soft errors not knowing why a bit flips due to a neutron strike or how a radiation-hardened circuit works.

Researching material for this book had been a lot of fun. I spent many hours reading and rereading articles that I was already familiar with. This helped me gain a better understanding of the area that I am already supposed to be an expert in. Based on the research I did on this book, I even filed a patent that enhances a basic circuit solution to protect against soft errors. I also realized that there is no other comprehensive book like this one in the area of architecture design for soft errors. There are bits and pieces of material available in different books and research papers. Putting all the material together in one book was definitely challenging but in the end, has been very rewarding.

I have put emphasis on the definition of terms used in this book. For example, I distinguish between a *fault* and an *error* and have stuck to these terminologies wherever possible. I have tried to define in a better way many terms that have been in use for ages in the classical fault tolerance literature. For example, the terms *fault, errors*, and *mean time to failure* (MTTF) are related to a domain or a boundary and are not "absolute" terms. Identifying the silent data corruption (SDC) MTTF and detected unrecoverable error (DUE) MTTF domains is important to design appropriate protection at different layers of the hardware and software stacks. In this book, I extensively use the acronyms SDC and DUE, which have been adopted by the large part of industry today. I was one of those who coined these acronyms within Intel Corporation and defined these terms precisely for appropriate use.

I expect that the concepts I define in this book will continue to persist for several years to come. A number of reliability challenges have arisen in CMOS. Soft error is just one of them. Others include process-related cell instability, process variation, and wearout causing frequency degradation and other errors. Among these areas, architecture design for soft errors is probably the most evolved area and hence ready to be captured in a book. The other areas are evolving rapidly, so one can expect books on these in the next several years. I also expect that the concepts from this book will be used in the other areas of architecture design for reliability.

I have tried to define the concepts in this book using first principles as much as possible. I do, however, believe that concepts and designs without implementations leave incomplete understanding of the concepts themselves. Hence, wherever possible I have defined the concepts in the context of specific implementations. I have also added simulation numbers—borrowed from research papers—wherever appropriate to define the basic concepts themselves.

In some cases, I have defined certain concepts in greater detail than others. It was important to spend more time describing concepts that are used as the basis of other proliferations. In some other cases, particularly for certain commercial systems, the publicly available description and evaluation of the systems are not as extensive. Hence, in some of the cases, the description may not be as extensive as I would have liked.

How to Use This Book

I see this book being used in four ways: by industry practitioners to estimate soft error rates of their parts and identify techniques to mitigate them, by researchers investigating soft errors, by graduate students learning about the area, and by advanced undergraduates curious about fault-tolerant machines. To use this book, one requires a background in basic computer architectural concepts, such as pipelines and caches. This book can also be used by industrial design managers requiring a basic introduction to soft errors.

There are a number of different ways this book could be read or used in a course. Here I outline a few possibilities:

- Complete course on architecture design for soft errors covering the entire book.
- Short course on architecture design for soft errors, including Chapters 1, 3, 5, 6, and 7.
- Reference book on classical fault-tolerant machines, including Chapters 6 and 7 only.
- Reference book on circuit course on reliability, including Chapters 1 and 2 only.
- Reference book on software fault tolerance, including Chapters 1 and 8 only.

At the end of each chapter, I have provided a summary of the chapter. I hope this will help readers maintain the continuity if they decide to skip the chapter. The summary should also be helpful for students taking courses that cover only part of the book.

Acknowledgements

Writing a book takes a lot of time, energy, and passion. Finding the time to write a book with a full-time job and "full-time" family is very difficult. In many ways, writing this book had become one of our family projects. I want to thank my loving wife, Mimi Mukherjee, and my two children, Rianna and Ryone, for letting me work on this book on many evenings and weekends. A special thanks to Mimi for having the confidence that I will indeed finish writing on this book. Thanks to my brother's family, Dipu, Anindita, Nishant, and Maya, for their constant support to finish this book and letting me work on it during our joint vacation.

This is the only book I have written, and I have often asked myself what prompted me to write a book. Perhaps, my late father, Ardhendu S. Mukherjee, who was a professor in genetics and had written a number of books himself, was my inspiration. Since I was 5 years old, my mother, Sati Mukherjee, who founded her own school, had taught me how learning can be fun. Perhaps the urge to convey how much fun learning can be inspired me to write this book.

I learned to read and write in elementary through high school. But writing a technical document in a way that is understandable and clear takes a lot of skill. By no means do I claim to be the best writer. But whatever little I can write, I ascribe that to my Ph.D. advisor, Prof. Mark D. Hill. I still joke about how Mark made me revise our first joint paper seven times before he called it a first draft! Besides Mark, my coadvisors, Prof. James Larus and Prof. David Wood, helped me significantly in my writing skills. I remember how Jim had edited a draft of my paper and cut it down to half the original size without changing the meaning of a single sentence. From David, I learned how to express concepts in a simple and a structured manner.

After leaving graduate school, I worked in Digital Equipment Corporation for 10 days, in Compaq for 3 years, and in Intel Corporation for 6 years. Throughout this work life, I was and still am very fortunate to have worked with Dr. Joel Emer. Joel had revolutionized computer architecture design by introducing the notion of quantitative analysis, which is part and parcel of every high-end microprocessor design effort today. I had worked closely with Joel on architecture design for reliability and particularly on the quantitative analysis of soft errors. Joel also has an uncanny ability to express concepts in a very simple form. I hope that part of that has rubbed off on me and on this book. I also thank Joel for writing the foreword for this book.

Besides Joel Emer, I had also worked closely with Dr. Steve Reinhardt on soft errors. Although Steve and I had been to graduate school together, our collaboration on reliability started after graduate school at the 1999 International Symposium on Computer Architecture (ISCA), when we discussed the basic ideas of Redundant Multithreading, which I cover in this book. Steve was also intimately involved in the vulnerability analysis of soft errors. My work with Steve had helped shape many of the concepts in this book.

I have had lively discussions on soft errors with many other colleagues, senior technologists, friends, and managers. This list includes (but is in no way limited to) Vinod Ambrose, David August, Arijit Biswas, Frank Binns, Wayne Burleson, Dan Casaletto, Robert Cohn, John Crawford, Morgan Dempsey, Phil Emma, Tryggve Fossum, Sudhanva Gurumurthi, Glenn Hinton, John Holm, Chris Hotchkiss, Tanay Karnik, Jon Lueker, Geoff Lowney, Jose Maiz, Pinder Matharu, Thanos Papathanasiou, Steve Pawlowski, Mike Powell, Steve Raasch, Paul Racunas, George Reis, Paul Ryan, Norbert Seifert, Vilas Sridharan, T. N. Vijaykumar, Chris Weaver, Theo Yigzaw, and Victor Zia.

I would also like to thank the following people for providing prompt reviews of different parts of the manuscript: Nidhi Aggarwal, Vinod Ambrose, Hisashige Ando, Wendy Bartlett, Tom Bissett, Arijit Biswas, Wayne Burleson, Sudhanva Guru-murthi, Mark Hill, James Hoe, Peter Hazucha, Will Hasenplaugh, Tanay Karnik, Jerry Li, Ishwar Parulkar, George Reis, Ronny Ronen, Pia Sanda, Premkishore Shiv-akumar, Norbert Seifert, Jeff Somers, and Nick Wang. They helped correct many errors in the manuscript.

Finally, I thank Denise Penrose and Chuck Glaser from Morgan Kaufmann for agreeing to publish this book. Denise sought me out at the 2004 ISCA in Munich and followed up quickly thereafter to sign the contract for the book.

I sincerely hope that the readers will enjoy this book. That will certainly be worth the 2 years of my personal and family time I have put into creating this book.

Shubu Mukherjee

Introduction

1.1 Overview

In the past few decades, the exponential growth in the number of transistors per chip has brought tremendous progress in the performance and functionality of semiconductor devices and, in particular, microprocessors. In 1965, Intel Corporation's cofounder, Gordon Moore, predicted that the number of transistors per chip will double every 18–24 months. The first Intel microprocessor with 2200 transistors was developed in 1971, 24 years after the invention of the transistor by John Bardeen, Walter Brattain, and William Shockley in Bell Labs. Thirty-five years later, in 2006, Intel announced its first billion-transistor Itanium® microprocessor—codenamed Montecito—with approximately 1.72 billion transistors. This exponential growth in the number of transistors—popularly known as Moore's law—has fueled the growth of the semiconductor industry for the past four decades.

Each succeeding technology generation has, however, introduced new obstacles to maintaining this exponential growth rate in the number of transistors per chip. Packing more and more transistors on a chip requires printing ever-smaller features. This led the industry to change lithography—the technology used to print circuits onto computer chips—multiple times. The performance of off-chip dynamic random access memories (DRAM) compared to microprocessors started slowing down resulting in the "memory wall" problem. This led to faster DRAM technologies, as well as to adoption of higher level architectural solutions, such as prefetching and multithreading, which allow a microprocessor to tolerate longer latency memory operations. Recently, the power dissipation of semiconductor chips started reaching astronomical proportions, signaling the arrival of the "power wall." This caused manufacturers to pay special attention to reducing power dissipation via innovation in process technology as well as in architecture and circuit design. In this series of challenges, transient faults from alpha particles and neutrons are next in line. Some refer to this as the "soft error wall."

Radiation-induced transient faults arise from energetic particles, such as alpha particles from packaging material and neutrons from the atmosphere, generating electron–hole pairs (directly or indirectly) as they pass through a semiconductor device. Transistor source and diffusion nodes can collect these charges. A sufficient amount of accumulated charge may invert the state of a logic device, such as a latch, static random access memory (SRAM) cell, or gate, thereby introducing a logical fault into the circuit's operation. Because this type of fault does not reflect a permanent malfunction of the device, it is termed *soft* or *transient*.

This book describes architectural techniques to tackle the soft error problem. Computer architecture has long coped with various types of faults, including faults induced by radiation. For example, error correction codes (ECC) are commonly used in memory systems. High-end systems have often used redundant copies of hardware to detect faults and recover from errors. Many of these solutions have, however, been prohibitively expensive and difficult to justify in the mainstream commodity computing market.

The necessity to find cheaper reliability solutions has driven a whole new class of quantitative analysis of soft errors and corresponding solutions that mitigate their effects. This book covers the new methodologies for quantitative analysis of soft errors and novel cost-effective architectural techniques to mitigate them. This book also reevaluates traditional architectural solutions in the context of the new quantitative analysis. To provide readers with a better grasp of the broader problem definition and solution space, this book also delves into the physics of soft errors and reviews current circuit and software mitigation techniques.

Specifically, this chapter provides a general introduction to and necessary background for radiation-induced soft errors, which is the topic of this book. The chapter reviews basic terminologies, such as faults and errors, and dependability models and describes basic types of permanent and transient faults encountered in silicon chips. Readers not interested in a broad overview of permanent faults could skip that section. The chapter will go into the details of the physics of how alpha particles and neutrons cause a transient fault. Finally, this chapter reviews architectural models of soft errors and corresponding trends in soft error rates (SERs).

1.1.1 Evidence of Soft Errors

The first report on soft errors due to alpha particle contamination in computer chips was from Intel Corporation in 1978. Intel was unable to deliver its chips to AT&T, which had contracted to use Intel components to convert its switching system from mechanical relays to integrated circuits. Eventually, Intel's May and Woods traced the problem to their chip packaging modules. These packaging modules got contaminated with uranium from an old uranium mine located upstream on Colorado's Green River from the new ceramic factory that made these modules. In their 1979 landmark paper, May and Woods [15] described Intel's problem with alpha particle contamination. The authors introduced the key concept of *Qcrit* or "critical charge," which must be overcome by the accumulated charge generated by the particle strike to introduce the fault into the circuit's operation. Subsequently, IBM Corporation faced a similar problem of radioactive contamination in its chips from 1986 to 1987. Eventually, IBM traced the problem to a distant chemical plant, which used a radioactive contaminant to clean the bottles that stored an acid required in the chip manufacturing process.

The first report on soft errors due to cosmic radiation in computer chips came in 1984 but remained within IBM Corporation [30]. In 1979, Ziegler and Lanford predicted the occurrence of soft errors due to cosmic radiation at terrestrial sites and aircraft altitudes [29]. Because it was difficult to isolate errors specifically from cosmic radiation, Ziegler and Lanford's prediction was treated with skepticism. Then, the duo postulated that such errors would increase with altitude, thereby providing a unique signature for soft errors due to cosmic radiation. IBM validated this hypothesis from the data gathered from its computer repair logs. Subsequently, in 1996, Normand reported a number of incidents of cosmic ray strikes by studying error logs of several large computer systems [17].

In 1995, Baumann et al. [4] observed a new kind of soft errors caused by boron-10 isotopes, which were activated by low-energy atmospheric neutrons. This discovery prompted the removal of boro-phospho-silicate glass (BPSG) and boron-10 isotopes from the manufacturing process, thereby solving this specific problem.

Historical data on soft errors in commercial systems are, however, hard to come by. This is partly because it is hard to trace back an error to an alpha or cosmic ray strike and partly because companies are uncomfortable revealing problems with their equipment. Only a few incidents have been reported so far. In 2000, Sun Microsystems observed this phenomenon in their UltraSPARC-II-based servers, where the error protection scheme implemented was insufficient to handle soft errors occurring in the SRAM chips in the systems. In 2004, Cypress semiconductor reported a number of incidents arising due to soft errors [30]. In one incident, a single soft error crashed an interleaved system farm. In another incident, a single soft error brought a billion-dollar automotive factory to halt every month. In 2005, Hewlett-Packard acknowledged that a large installed base of a 2048-CPU server system in Los Alamos National Laboratory—located at about 7000 feet above sea level—crashed frequently because of cosmic ray strikes to its parity-protected cache tag array [16].

1.1.2 Types of Soft Errors

The cost of recovery from a soft error depends on the specific nature of the error arising from the particle strike. Soft errors can either result in a *silent data corruption* (SDC) or *detected unrecoverable error* (DUE). Corrupted data that go unnoticed by the user are benign and excluded from the SDC category. But corrupted data that eventually result in a visible error that the user cares about cause an SDC event. In contrast, a DUE event is one in which the computer system detects the soft error and potentially crashes the system but avoids corruption of any data the user cares about. An SDC event can also crash a computer system, besides causing data corruption. However, it is often hard, if not impossible, to trace back where the SDC event originally occurred. Subtleties in these definitions are discussed later in this chapter. Besides SDC and DUE, a third category of benign errors exists. These are corrected errors that may be reported back to the operating system (OS). Because the system recovers from the effect of the errors, these are usually not a cause of concern. Nevertheless, many vendors use the reported rate of correctable errors as an early warning that a system may have an impending hardware problem.

Typically, an SDC event is perceived as significantly more harmful than a DUE event. An SDC event causes loss of data, whereas a DUE event's damage is limited to unavailability of a system. Nevertheless, there are various categories of machines that guarantee high reliability for SDC, DUE, or both. For example, the classical mainframe systems with triple-modular redundancy (TMR) offer both high degree of data integrity (hence, low SDC) and high availability (hence, low DUE). In contrast, web servers could often offer high availability by failing over to a spare standby system but may not offer high data integrity.

To guarantee a certain level of reliable operation, companies have SDC and DUE budgets for their silicon chips. If you ask a typical customer about how many errors he or she expects in his or her computer system, the response is usually zero. The reality is, though, computer systems do encounter soft errors that result in SDC and DUE events. A computer vendor tries to ensure that the number of SDC and DUE events encountered by its systems is low enough compared to other errors arising from software bugs, manufacturing defects, part wearout, stress-induced errors, etc.

Because the rate of occurrence of other errors differs in different market segments, vendors often have SDC and DUE budgets for different market segments. For example, software in desktop systems is expected to crash more often than that of high-end server systems, where after an initial maturity period, the number of software bugs goes down dramatically [27]. Consequently, the rate of SDC and DUE events needs to be significantly lower in high-end server systems, as opposed to computer systems sitting in homes and on desktops. Additionally, hundreds and thousands of server systems are deployed in a typical data center today. Hence, the rate of occurrence of these events is magnified 100 to 1000 times when viewed as an aggregate. This additional consideration further drives down the SDC and DUE budgets set by a vendor for the server machines.

1.1.3 Cost-Effective Solutions to Mitigate the Impact of Soft Errors

Meeting the SDC and DUE budgets for commercial microprocessor chips, chipsets, and computer memories without sacrificing performance or power has become a daunting task. A typical commercial microprocessor consists of tens of millions of circuit elements, such as SRAM (random access memory) cells; clocked memory elements, such as latches and flip-flops; and logic elements, such as NAND and NOR gates. The mean time to failure (MTTF) of such an individual circuit element could be as high as a billion years. However, with hundreds of millions of these elements on the chip, the overall MTTF of a single microprocessor chip could easily come down to a few years. Further, when individual chips are combined to form a large shared-memory system, the overall MTTF can come down to a few months. In large data centers—using thousands of these systems—the MTTF of the overall cluster can come down to weeks or even days.

Commercial microprocessors typically use several flavors of fault detection and ECC to protect these circuit elements. The die area overheads of these gate- or transistor-level detection and correction techniques could range roughly between 2% to greater than 100%. This extra area devoted to error protection could have otherwise been used to offer higher performance or better functionality. Often, these detection and correction codes would add extra cycles in a microprocessor pipeline and consume extra power, thereby further sacrificing performance. Hence, microprocessor designers judiciously choose the error protection techniques to meet the SDC and DUE budgets without unnecessarily sacrificing die area, performance, or even power.

In contrast, mainframe-class solutions, such as TMR, run identical copies of the same program on three microprocessors to detect and correct any errors. While this approach can dramatically reduce the SDC and DUE, it comes with greater than 200% overhead in die area and a commensurate increase in power. This solution is deemed an overkill in the commercial microprocessor market. In summary, gate- or transistor-level protection, such as fault detection and ECC, can limit the incurred overhead but may not provide adequate error coverage, whereas mainframe-class solutions can certainly provide adequate coverage but at a very high cost (Figure 1.1).

The key to successful design of a highly reliable, yet competitive, microprocessor or chipset is a systematic analysis and modeling of its SER. Then, designers can choose cost-effective protection mechanisms that can help bring down the

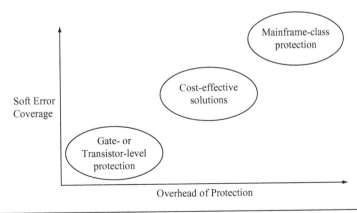

FIGURE 1.1 Range of soft error protection schemes.

SER within the prescribed budget. Often this process is iterated several times till designers are happy with the predicted SER. This book describes the current state-of-the-art in soft error modeling, measurement, detection, and correction mechanisms.

This chapter reviews basic definitions of faults, errors, and metrics, and dependability models. Then, it shows how these definitions and metrics apply to both permanent and transient faults. The discussion on permanent faults will place radiation-induced transient faults in a broader context, covering various silicon reliability problems.

1.2   Faults

User-visible *errors,* such as soft errors, are a manifestation of underlying *faults* in a computer system. Faults in hardware structures or software modules could arise from defects, imperfections, or interactions with the external environment. Examples of faults include manufacturing defects in a silicon chip, software bugs, or bit flips caused by cosmic ray strikes.

Typically, faults are classified into three broad categories—permanent, intermittent, and transient. The names of the faults reflect their nature. Permanent faults remain for indefinite periods till corrective action is taken. Oxide wearout, which can lead to a transistor malfunction in a silicon chip, is an example of a permanent fault. Intermittent faults appear, disappear, and then reappear and are often early indicators of impending permanent faults. Partial oxide wearout may cause intermittent faults initially. Finally, transient faults are those that appear and disappear. Bit flips or gate malfunction from an alpha particle or a neutron strike is an example of a transient fault and is the subject of this book.

Faults in a computer system can occur directly in a user application, thereby eventually giving rise to a user-visible error. Alternatively, it can appear in any abstraction layer underneath the user application. In a computer system, the abstraction layers can be classified into six broad categories (Figure 1.2)—user application, OS, firmware, architecture, circuits, and process technology. Software bugs are faults arising in applications, OSs, or firmware. Design faults can arise in architecture or circuits. Defects, imperfections, or bit flips from particle strikes are examples of faults in the process technology or the underlying silicon chip.

A fault in a particular layer may not show up as a user-visible error. This is because of two reasons. First, a fault may be masked in an intermediate layer. A defective transistor—perhaps arising from oxide wearout—may affect performance but may not affect correct operation of an architecture. This could happen, for example, if the transistor is part of a branch predictor. Modern architectures typically use a branch predictor to accelerate performance but have the ability to recover from a branch misprediction.

Second, any of the layers may be partially or fully designed to tolerate faults. For example, special circuits—radiation-hardened cells—can detect and recover

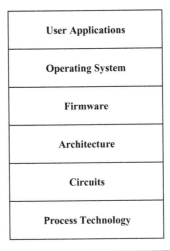

| User Applications |
| Operating System |
| Firmware |
| Architecture |
| Circuits |
| Process Technology |

FIGURE 1.2   Abstraction layers in a computer system.

from faults in transistors. Similarly, each abstraction layer, shown in Figure 1.2, can be designed to tolerate faults arising in lower layers. If a fault is tolerated at a particular layer, then the fault is avoided at the layer above it.

The next section discusses how faults are related to errors.

1.3   Errors

Errors are manifestation of *faults*. Faults are necessary to cause an error, but not all faults show up as errors. Figure 1.3 shows that a fault within a particular scope may not show up as an error outside the scope if the fault is either masked or tolerated. The notion of an error (and units to characterize or measure it) is fundamentally tied to the notion of a *scope*. When a fault is detected in a specific scope, it becomes an error in that scope. Similarly, when an error is corrected in a given a scope, its effect usually does not propagate outside the scope. This book tries to use the terms *fault detection* and *error correction* as consistently as possible. Since an error can propagate and be detected again in a different scope, it is also acceptable to use the term *error detection* (as opposed to fault detection).

Three examples are considered here. The first one is a fault in a branch predictor. No fault in a branch predictor will cause a user-visible error. Hence, there is no scope outside which a branch predictor fault would show up as an error. In contrast, a fault in a cache cell can potentially lead to a user-visible error. If the cache cell is protected with ECC, then a fault is an error within the scope of the ECC logic. Outside the scope of this logic where our typical observation point would be, the fault gets tolerated and never causes an error. Consider a third scenario in which three complete microprocessors vote on the correct output. If the output of one of the processors is incorrect, then the voting logic assumes that the other two are

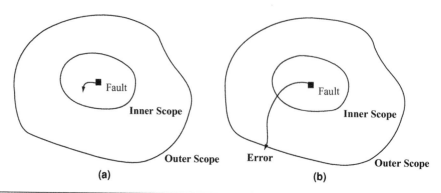

FIGURE 1.3 (a) Fault within the inner scope masked and not visible outside the inner scope. (b) Fault propagated outside the outer scope and visible as an error.

correct, thereby correcting any internal fault. In this case, the scope is the entire microprocessor. A fault within the microprocessor will never show up outside the voting logic.

In traditional fault-tolerance literature, a third term—*failures*—is used besides faults and errors. Failure is defined as a system malfunction that causes the system to not meet its correctness, performance, or other guarantees. A failure is, however, simply a special case of an error showing up at a boundary where it becomes visible to the user. This could be an SDC event, such as a change in the bank account, which the user sees. This could also be a detected error (or DUE) caught by the system but not corrected and may lead to temporary unavailability of the system itself. For example, an ATM machine could be unavailable temporarily due to a system reboot caused by a radiation-induced bit flip in the hardware. Alternatively, a disk could be considered to have failed if its performance degrades by 1000x, even if it continues to return correct data.

Like faults, errors can be classified as permanent, intermittent, or transient. As the names indicate, a permanent fault causes a permanent or hard error, an intermittent fault causes an intermittent error, and a transient fault causes a transient or soft error. Hard errors can cause both infant mortality and lifetime reliability problems and are typically characterized by the classic bathtub curve, shown in Figure 1.4. Initially, the error rate is typically high because of either bugs in the system or latent hardware defects. Beyond the infant mortality phase, a system typically works properly until the end of its useful lifetime is reached. Then, the wearout accelerates causing significantly higher error rates. The silicon industry typically uses a technique called *burn-in* to move the starting use point of a chip to the beginning of the useful lifetime period shown in Figure 1.4. Burn-in removes any chips that fail initially, thereby leaving parts that can last through the useful lifetime period. Further, the silicon industry designs technology parameters, such as oxide thickness, to guarantee that most chips last a minimal lifetime period.

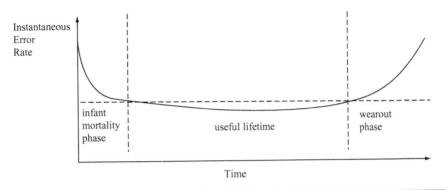

FIGURE 1.4 Bathtub curve showing the relationship between failure rate, infant mortality, useful lifetime, and wearout phase.

1.4 Metrics

Time to failure (TTF) expresses fault and error rates, even though the term TTF refers specifically to failures. As the name suggests, TTF is the time to a fault or an error, as the case may be. For example, if an error occurs after 3 years of operation, then the TTF of that system for that instance is 3 years. Similarly, MTTF expresses the mean time elapsed between two faults or errors. Thus, if a system gets an error every 3 years, then that system's MTTF is 3 years. Sometimes reliability models use median time to failure (MeTTF), instead of MTTF, such as in Black's equation for electromigration (EM)–related errors (see Electromigration, p. 15).

Under certain assumptions (e.g., an exponential TTF, see Reliability, p. 12), the MTTF of various components comprising a system can be combined to obtain the MTTF of the whole system. For example, if a system is composed of two components, each with an MTTF of 6 years, then the MTTF of the whole system is

$$MTTF_{system} = \cfrac{1}{\cfrac{1}{MTTF_{component\ 1}} + \cfrac{1}{MTTF_{component\ 2}}} = \frac{1}{\frac{1}{6} + \frac{1}{6}} = 3$$

More generally,

$$MTTF_{system} = \cfrac{1}{\sum\limits_{i=0}^{n} \frac{1}{MTTF_i}}$$

Although the term MTTF is fairly easy to understand, computing the MTTF of a whole system from individual component MTTFs is a little cumbersome, as expressed by the above equations. Hence, engineers often prefer the term *failure in time* (FIT), which is additive.

One FIT represents an error in a billion (10^9) hours. Thus, if a system is composed of two components, each having an error rate of 10 FIT, then the system has a total error rate of 20 FIT. The summation assumes that the errors in each component are independent.

The error rate of a component or a system is often referred to as its FIT rate. Thus, the FIT rate equation of a system is

$$FIT\ rate_{system} = \sum_{i=0}^{n} FIT\ rate_i$$

As may be evident by now, FIT rate and MTTF of a component are inversely related under certain conditions (e.g., exponentially distributed TTF):

$$MTTF\ in\ years = \frac{10^9}{FIT\ rate \times 24\ hours \times 365\ days}$$

Thus, an MTTF of 1000 years roughly translates into a FIT rate of 114 FIT.

■ EXAMPLE

A silicon chip consists of a billion transistors, each with a FIT rate of 0.00001 FIT. What will be the MTTF of a system composed of 100 such chips?

SOLUTION The FIT rate of each chip $= 10^9 \times 0.00001$ FIT $= 10^4$ FIT. The FIT rate of 100 such chips $= 100 \times 10^4 = 10^6$ FIT. Then, the MTTF of a system with 100 such chips $= 10^9/(10^6 \times 24) \sim 40$ days.

■ EXAMPLE

What is the MTTF of a computer's memory system that has 16 gigabytes of memory? Assume FIT per bit is 0.00001 FIT.

SOLUTION The FIT rate of the memory system $= 16 \times 2^{30} \times 8 \times 0.00001 = 1\,374\,390$ FIT. This translates into an MTTF of $10^9/(1\,374\,390 \times 24) \sim 30$ days.

Besides MTTF, two terms—*mean time to repair* (MTTR) and *mean time between failures* (MTBF)—are commonly used in the fault-tolerance literature. MTTR represents the mean time needed to repair an error once it is detected. MTBF

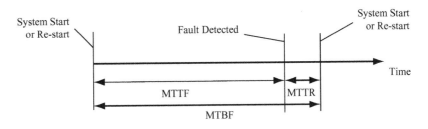

FIGURE 1.5   Relationship between MTTF, MTTR, and MTBF.

represents the average time between the occurrences of two errors. As Figure 1.5 shows, MTBF = MTTF + MTTR. Typically, MTTR ≪ MTTF. The next section examines how these terms are used to express various concepts in reliable computing.

Recently, Weaver et al. [26] introduced the term *mean instructions to failure* (MITF). MITF captures the average number of instructions committed in a microprocessor between two errors. Similarly, Reis et al. [19] introduced the term *mean work to failure* (MWTF) to capture the average amount of work between two errors. The latter is useful for comparing the reliability for different workloads. Unlike MTTF, both MITF and MWTF try to capture the amount of work done till an error is experienced. Hence, MITF and MWTF are often useful in doing trade-off studies between performance and error rate.

The definitions of MTTF and FIT rate have one subtlety that may not be obvious. Both terms are related to a particular scope (as explained in the last section). Consider a bit with ECC, which can correct an error in the single bit. The MTTF(bit) is significantly lower than the MTTF(bit + ECC). Conversely, the FIT rate(bit) is significantly greater than the FIT rate(bit + ECC). In both cases, it is the MTTF that is affected and not the MTBF. Vendors, however, sometimes incorrectly report MTBF numbers for the components they are selling, if they add error correction to the component.

All the above metrics can be applied separately for SDC or DUE. Thus, one can talk about SDC MTTF or SDC FIT. Similarly, one can express DUE MTTF or DUE FIT. Usually, the total SER is expressed as the sum of SDC FIT and DUE FIT.

1.5   Dependability Models

Reliability and availability are two attributes typically used to characterize the behavior of a system experiencing faults. This section discusses mathematical models to describe these attributes and the foundation behind the metrics discussed in the last section. This section will also discuss other miscellaneous related models used to characterize systems experiencing faults.

1.5.1 Reliability

The reliability $R(t)$ of a system is the probability that the system does not experience a user-visible error in the time interval $(0, t]$. In other words, $R(t) = P(T > t)$, where T is a random variable denoting the lifetime of a system. If a population of N_0 similar systems is considered, then $R(t)$ is the fraction of the systems that survive beyond time t. If N_t is the number of systems that have survived until time t and $E(t)$ is the number of systems that experienced errors in the interval $(0, t]$, then

$$R(t) = \frac{N_t}{N_0} = \frac{N_0 - E(t)}{N_0} = 1 - \frac{E(t)}{N_0}$$

Differentiating this equation, one gets

$$\frac{dR(t)}{dt} = -\frac{\frac{dE(t)}{dt}}{N_0}$$

The instantaneous error rate or hazard rate $h(t)$—graphed in Figure 1.4—is defined as the probability that a system experiences an error in the time interval Δt, given that it has survived till time t. Intuitively, $h(t)$ is the probability of an error in the time interval $(t, t + \Delta t]$.

$$h(t) = P(t < T \leq t + \Delta t | (T > t)) = \frac{\frac{dE(t)}{dt}}{N_t} = \frac{\frac{dE(t)}{dt}}{\frac{N_t}{N_0}} = \frac{\frac{dR(t)}{dt}}{R(t)}$$

Rewriting,

$$\frac{dR(t)}{dt} = -h(t)R(t)$$

The general solution to this differential equation is

$$R(t) = e^{-\int h(t)dt}$$

If one assumes that $h(t)$ has a constant value of λ (e.g., during the useful lifetime phase in Figure 1.4), then

$$R(t) = e^{-\lambda t}$$

This exponential relationship between reliability and time is known as the *exponential failure law*, which is commonly used in soft error analysis. The expectation of $R(t)$ is the MTTF and is equal to λ.

The exponential failure law lets one sum FIT rates of individual transistors or bits in a silicon chip. If it is assumed that a chip has n bits, where the ith bit has a constant and independent hazard rate of h_i, then, $R(t)$ of the whole chip can be expressed as

$$R(t) = \prod_{i=0}^{n-1} R_i(t) = \prod_{i=0}^{n-1} e^{-h_i t} = e^{-\left(\sum_{i=0}^{n-1} h_i\right)t}$$

Thus, the reliability function of the chip is also exponentially distributed with a constant FIT rate, which is the sum of the FIT rates of individual bits.

The exponential failure law is extremely important for soft error analysis because it allows one to compute the FIT rate of a system by summing the FIT rates of individual components in the system. The exponential failure law requires that the instantaneous SER in a given period of time is constant. This assumption is reasonable for soft error analysis because alpha particles and neutrons introduce faults in random bits in computer chips. However, not all errors follow the exponential failure law (e.g., wearout in Figure 1.4). The Weibull or log-normal distributions could be used in cases that have a time-varying failure rate function [18].

1.5.2 Availability

Availability is the probability that a system is functioning correctly at a particular instant of time. Unlike reliability, which is defined over a time interval, availability is defined at an instant of time. Availability is also commonly expressed as

$$\text{Availability} = \frac{\text{system uptime}}{\text{system uptime} + \text{system downtime}} = \frac{\text{MTTF}}{\text{MTTF} + \text{MTTR}} = \frac{\text{MTTF}}{\text{MTBF}}$$

Thus, availability can be increased either by increasing MTTF or by decreasing MTTR.

Often, the term five 9s or six 9s is used to describe the availability of a system. The term five 9s indicates that a system is available 99.999% of the time, which translates to a downtime of about 5 minutes per year. Similarly, the term six 9s indicates that a system is available 99.9999% of the time, which denotes a system downtime of about 32 seconds per year. In general, n 9s indicate two 9s before the decimal point and $(n - 2)$ 9s after the decimal point, if expressed in percentage.

▪ E X A M P L E

If the MTTR of a system is 30 minutes, how many crashes can it sustain per year and still maintain a five 9s uptime? What is the MTTF in this case?

S O L U T I O N A five 9s uptime denotes a total downtime of about 5 hours per year. Hence, the number of system crashes allowed for this system per year is $(5 \times 60/30) = 10$. The MTTF is (1 year–5 hours)/10 = 876 hours.

1.5.3 Miscellaneous Models

Three other models, namely maitainability, safety, and performability, are often used to describe systems experiencing faults. Maintainability is the probability that a failed system will be restored to its original functioning state within a specific period of time. Maintainability can be modeled as an exponential repair law, a concept very similar to the exponential failure law.

Safety is the probability that a system will either function correctly or fail in a "safe" manner that causes no harm to other related systems. Thus, unlike reliability, safety modeling incorporates a "fail-stop" behavior. Fail-stop implies that when a fault occurs, the system stops operating, thereby preventing the effect of the fault to propagate any further.

Finally, performability of a system is the probability that the system will perform at or above some performance level at a specific point of time [10]. Unlike reliability, which relates to correct functionality of all components, performability measures the probability that a subset of functions will be performed correctly. Graceful degradation, which is a system's ability to perform at a lower level of performance in the face of faults, can be expressed in terms of a performability measure.

These models are added here for completeness and will not be used in the rest of this book. The next few sections discuss how the reliability and availability models apply to both permanent and transient faults.

1.6 Permanent Faults in Complementary Metal Oxide Semiconductor Technology

Dependability models, such as reliability and availability, can characterize both permanent and transient faults. This section examines several types of permanent faults experienced by complementary metal oxide semiconductor (CMOS) transistors. The next section discusses transient fault models for CMOS transistors. This section reviews basic types of permanent faults to give the reader a broad understanding of the current silicon reliability problems, although radiation-induced transient faults are the focus of this book.

Permanent faults in CMOS devices can be classified as either extrinsic or intrinsic faults. Extrinsic faults are caused by manufacturing defects, such as contaminants in silicon devices. Extrinsic faults result in infant mortality, and the fault rate usually decreases over time (Figure 1.4). Typically, a process called burn-in, in which silicon chips are tested at elevated temperatures and voltages, is used to accelerate the manifestation of extrinsic faults. The defect rate is expressed in defective parts per million.

In contrast, intrinsic faults arise from wearout of materials, such as silicon dioxide, used in making CMOS transistors. In Figure 1.4, the intrinsic fault rate corresponds to the wearout phase and typically increases with time. Several architecture researchers are examining how to extend the useful lifetime of a transistor device by delaying the onset of the wearout phase and decreasing the use of the device itself.

This section briefly reviews intrinsic fault models affecting the lifetime reliability of a silicon device. Specifically, this section examines metal and oxide failure modes. These fault models are discussed in greater detail in Segura and Hawkins' book [23].

1.6.1 Metal Failure Modes

This section discusses the two key metal failure modes, namely EM and metal stress voiding (MSV).

Electromigration

EM is a failure mechanism that causes voids in metal lines or interconnects in semiconductor devices (Figure 1.6). Often, these metal atoms from the voided region create an extruding bulge on the metal line itself.

EM is caused by electron flow and exacerbated by rise in temperature. As electrons move through metal lines, they collide with the metal atoms. If these collisions transfer sufficient momentum to the metal atoms, then these atoms may get displaced in the direction of the electron flow. The depleted region becomes the void, and the region accumulating these atoms forms the extrusion.

Black's law is commonly used to predict the MeTTF of a group of aluminum interconnects. This law was derived empirically. It applies to a group of metal interconnects and cannot be used to predict the TTF of an individual interconnect wire. Black's law states that

$$\text{MeTTF}_{\text{EM}} = \frac{A_0}{j_e^2} e^{\frac{E_a}{kT}}$$

where A_0 is a constant dependent on technology, j_e is electron current density (A/cm^2), T is the temperature (K), E_a is the activation energy (eV) for EM failure, and k is the Boltzmann constant. As technology shrinks, the current density usually increases, so designers need to work harder to keep the current density at acceptable levels to prevent excessive EM. Nevertheless, the exponential temperature term has a more acute effect on MeTTF than current density.

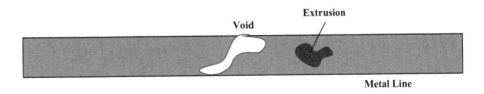

FIGURE 1.6 Void in a metal line from EM.

■ E X A M P L E

Use Black's equation to estimate relative average lifetimes of two identical parts. A metal line in part 1 runs at 70°C with a maximum current density of $1\,MA/cm^2$. A similar metal line in part 2 runs at 100°C with a maximum current density of $2\,MA/cm^2$. Use $E_a = 0.8\,eV$ and $k = 86.17\mu\,eV/K$.

S O L U T I O N   Plugging in the numbers

$$\frac{MeTTF_{part\,1}}{MeTTF_{part\,2}} = \frac{2^2}{1^2} \times e^{\frac{E_a}{k}\left[\frac{1}{(273+70)} - \frac{1}{(273+100)}\right]} = 35$$

Hence, product 1 will last 35 times longer than product 2.

An additional phenomenon called the Blech effect dictates whether EM will occur. Ilan Blech demonstrated that the product of the maximum metal line length (l_{max}) below which EM will not occur and the current density (j_e) is a constant for a given technology.

Metal Stress Voiding

MSV causes voids in metal lines due to different thermal expansion rates of metal lines and the passivation material they bond to. This can happen during the fabrication process itself. When deposited metal reaches 400°C or higher for a passivation step, the metal expands and tightly bonds to the passivation material. But when cooled to room temperature, enormous tensile stress appears in the metal due to the differences in the thermal coefficient of expansion of the two materials. If the stress is large enough, then it can pull a line apart. The void can show up immediately or years later.

The MTTF due to MSV is given by

$$MTTF_{MSV} = \frac{B_0}{(T_0 - T)^n} e^{\frac{E_b}{kT}},$$

where T is the temperature, T_0 is the temperature at which the metal was deposited, B_0, n, and E_b are material-dependent constants, and k is the Boltzmann constant. For copper, $n = 2.5$ and $E_b = 0.9$. The higher the operating temperature, lower is the term $(T_0 - T)$ and higher the $MTTF_{MSV}$. Interestingly, however, the exponential term drops rapidly with a rise in the operating temperature and usually has the more dominant effect.

In general, copper is more resistive to EM and MSV than aluminum. Copper has replaced aluminum for metal lines in the high-end semiconductor industry. Copper, however, can cause severe contamination in the fab and therefore needs a more carefully controlled process.

1.6.2 Gate Oxide Failure Modes

Gate oxide reliability has become an increasing concern in the design of high-performance silicon chips. Gate oxide consists of thin noncrystalline and amorphous silicon dioxide (SiO_2). In a bulk CMOS transistor device (Figure 1.7), the gate oxide electrically isolates the polysilicon gate from the underlying semiconductor crystalline structure known as the substrate or bulk of the device. The substrate can be constructed from either p-type silicon for n-type metal oxide semiconductor (nMOS) transistors or n-type silicon for p-type metal oxide semiconductor (pMOS) transistors. The source and drain are also made from crystalline silicon but implanted with dopants of polarity opposite to that of the substrate. Thus, for example, an nMOS source and drain would be doped with an n-type dopant.

The gate is the control terminal, whereas the source provides electrons or hole carriers that are collected by the drain. When the gate terminal voltage of an nMOS (pMOS) transistor is increased (decreased) sufficiently, the vertical electric field attracts minority carriers (electrons in nMOS and holes in pMOS) toward the gate. The gate oxide insulation stops these carriers causing them to accumulate at the gate oxide interface. This creates the conducting channel between the source and drain, thereby turning on the transistor.

The switching speed of a CMOS transistor—going from off to on or the reverse—is a function of the gate oxide thickness (for a given gate oxide). As transistors shrink in size with every technology generation, the supply voltage is reduced to maintain the overall power consumption of a chip. Supply voltage reduction, in turn, can reduce the switching speed. To increase the switching speed, the gate oxide thickness is correspondingly reduced. Gate oxide thicknesses, for example, have decreased from $750\,\text{Å}$ from the 1970s to $15\,\text{Å}$ in the 2000s, where $1\,\text{Å} = 1\,\text{angstrom} = 10^{-10}\text{m}$. SiO_2 molecules are $3.5\,\text{Å}$ in diameter, so gate oxide thicknesses rapidly approach molecular dimensions. Oxides with such a low thickness—less than $30\,\text{Å}$—are referred to as ultrathin oxides.

Reducing the oxide thickness further has become challenging since the oxide layer runs out of atoms. Further, a thinner gate oxide increases oxide leakage. Hence, the industry is researching into what is known as high-k materials, such as hafnium dioxide (HfO_2), zirconium dioxide (ZrO_2), and titanium dioxide (TiO_2),

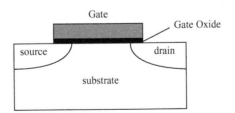

FIGURE 1.7 Physical structure of a bulk CMOS transistor. The substrate is also referred to as bulk.

which have a dielectric constant or "k" above 3.9, the "k" of silicon dioxide. These high-k oxides are thicker than SiO_2. Besides, they reduce the oxide leakage when the transistor is off without affecting the transistor's performance when it is on.

This section discusses three oxide failure mechanisms—wearout in ultrathin oxides, hot carrier injection (HCI), and negative bias temperature instability (NBTI).

Gate Oxide Wearout

Ultrathin oxide breakdown causes a sudden discontinuous increase in conductance, often accompanied by an increased current noise. This causes a reduction in the "on" current of the transistor. Gradual oxide breakdown may initially lead to intermittent faults but may eventually prevent the transistor from functioning correctly, thereby causing a permanent fault in the device.

The breakdown is caused by gradual buildup of electron traps, which are oxide defects produced by missing oxygen atoms. Such electron traps can exist from the point of oxide creation, or they can be created when the SiO_2–SiO_2 bonds are broken by energetic particles, such as electrons, holes, or radiations. The precise point at which the breakdown occurs is statistically distributed, so only statistical averages can be predicted. The breakdown occurs when a statistical distribution of these traps is vertically aligned and allows a thermally damaging current to flow through the oxide. This is known as the *percolation* model of wearout and breakdown.

The time-to-breakdown (T_{bd}) for gate oxide could be expressed as

$$T_{bd} = Ce^{\gamma\left(\alpha t_{ox} + \frac{E_a}{kT_j} - V_G\right)},$$

where C is a constant, t_{ox} is the gate oxide thickness, T_j is the average junction temperature, E_a is the activation energy, V_G is the gate voltage, and γ and α are technology-dependent constants. Thus, T_{bd} decreases with decreasing oxide thickness but increases with decreasing V_G.

The T_{bd} model is still an area of active research. Please refer to Strathis [22] for an in-depth discussion of this subject.

Hot Carrier Injection

HCI results in a degradation of the maximum operating frequency of the silicon chip. HCI arises from impact ionization when electrons in the channel strike the silicon atoms around the drain–substrate interface. This could happen from one of several conditions, such as higher power supply, short channel lengths, poor oxide interface, or accidental overvoltage in the power rails.

The ionization produces electron–hole pairs in the drain. Some of these carriers enter the substrate, thereby increasing the substrate current I_{sub}. A small fraction of carriers created from the ionization may have sufficient energy (3.1 eV for electrons and 4.6 eV for holes) to cross the oxide barrier and enter the oxide to cause damage. Because these carriers have a high mean equivalent temperature, they are referred to as "hot" carriers. Interestingly, however, HCI becomes worse as ambient temperature decreases because of a corresponding increase in carrier mobility.

■ EXAMPLE

Compute the mean equivalent temperature of an electron with energy of 3.1 eV. Assume that the thermal energy follows the Boltzmann distribution: $E_t = kT/q$, where E_t is thermal energy, T is the temperature (K), and k is the Boltzmann constant $= 1.38 \times 10^{-23}$ J/K, and $q = 1.6 \times 10^{-19}$ C.

SOLUTION Rearranging the terms of the equation, the mean equivalent temperature $T = E_t q/k = 3.1 \times 1.6 \times 10^{-19}/(1.38 \times 10^{-23}) \sim 36\,000$ K.

Typically, the drain saturation current (I_{Dsat}) degradation is used to measure HCI degradation because I_{Dsat} is one of the key transistor parameters that most closely approximates the impact on circuit speed and because HCI-related damage occurs only during normal operation when the transistor is in saturation. Oxide damage due to HCI raises the threshold voltage of an nMOS transistor causing I_{Dsat} to degrade.

Frequency guardbanding is a typical measure adopted by the silicon chip industry to cope with HCI-related degradation. The expected lifetime of a silicon chip is often between 5 and 15 years. Usually, the frequency degradation during the expected lifetime is between 1% and 10%. Hence, the chips are rated at a few percentage points below what they actually run at. This reduction in the frequency is called the frequency guardband.

Transistor lifetime degradation (τ) due to HCI (e.g., 3% reduction in threshold voltage) is specified as

$$\tau = \text{Constant} \frac{\frac{W}{I_D}}{\left(\frac{I_{sub}}{I_D}\right)^3},$$

where W is the transistor width, I_D is the drain current, and I_{sub} is the substrate current. The I_D and I_{sub} parameters are typically estimated for the use condition of the chip (e.g., power on).

Negative Bias Temperature Instability

Like HCI, NBTI causes degradation of maximum frequency. Unlike HCI, which affects both nMOS and pMOS devices, NBTI only affects short-channel pMOS transistors (hence the term "negative bias"). The "hydrogen-release" model provides the most popular explanation for this effect. Under stress (e.g., high temperature), highly energetic holes bombard the channel–oxide interface, electrochemically react with the oxide interface, and release hydrogen atoms by breaking the silicon–hydrogen bonds. These free hydrogen atoms combine with oxygen or nitrogen atoms to create positively charged traps at the oxide–channel interface. This causes a reduction in mobility of holes and a shift in the pMOS threshold voltage in the more negative direction. These effects cause the transistor drive current to degrade, thereby slowing down the transistor device. The term "instability" refers to the variation of threshold voltage with time. Researchers are actively looking into models that can predict how NBTI will manifest in future process generations.

1.7   Radiation-Induced Transient Faults in CMOS Transistors

Transient faults in semiconductor devices can be induced by a variety of sources, such as transistor variability, thermal cycling, erratic fluctuations of minimum voltage at which a circuit is functional, and radiation external to the chip. Transistor variability arises due to the random dopant fluctuations, use of subwavelength lithography, and high heat flux across the silicon die [5]. Thermal cycling can be caused by repeated stress from temperature fluctuations. Erratic fluctuations in the minimum voltage of a circuit can be caused by gate oxide soft breakdown in combination with high gate leakage [1].

This book focuses on radiation-induced transient faults. There are two sources of radiation-induced faults—alpha particles from packaging and neutrons from the atmosphere. This section discusses the nature of these particles and how they introduce errors in silicon chips. The permanent faults described earlier in this chapter and the transient faults discussed in the last paragraph can mostly be taken care of before a chip is shipped. In contrast, a radiation-induced transient fault is typically addressed in the field with appropriate fault detection and error correction circuitry.

1.7.1   The Alpha Particle

An alpha particle consists of two protons and two neutrons bound together into a particle that is identical to a helium nucleus. Alpha particles are emitted by radioactive nuclei, such as uranium or radium, in a process known as alpha decay. This sometimes leaves the nucleus in an excited state, with the emission of a gamma ray removing the excess energy.

The alpha particles typically have kinetic energies of a few MeV, which is lower than those of typical neutrons that affect CMOS chips. Nevertheless, alpha particles can affect semiconductor devices because they deposit a dense track of charge and create electron–hole pairs as they pass through the substrate. Details of the interaction of alpha particles and neutrons with semiconductor devices are described below.

Alpha particles can arise from radioactive impurities used in chip packaging, such as in the solder balls or contamination of semiconductor processing materials. It is very difficult to eliminate alpha particles completely from the chip packaging materials. Small amounts of epoxy or nonradioactive lead can, however, significantly reduce a chip's sensitivity to alpha particles by providing a protective shield against such radiation. Even then, chips are still exposed to very small amounts of alpha radiation. Consequently, chips need fault detection and error correction techniques within the semiconductor chip itself to protect against alpha radiation.

1.7.2 The Neutron

The neutron is one of the subatomic particles that make up an atom. Atoms are considered to be the basic building blocks of matter and consist of three types of subatomic particles: protons, neutrons, and electrons. Protons and neutrons reside inside an atom's dense center. A proton has a mass of about 1.67×10^{-27} kg. A neutron is only slightly heavier than a proton. An electron is about 2000 times lighter than both a proton and a neutron. A proton is positively charged, a neutron is neutral, and an electron is negatively charged. An atom consists of an equal number of protons and electrons and hence it is neutral itself.

Figure 1.8 shows the two dominant models of an atom. In the Bohr model (Figure 1.8a), electrons circle around the nucleus at different levels or orbitals much like planets circle the sun. Electrons can exist at definite energy levels, but can move from one energy state to another. Electrons release energy as electromagnetic radiation when they change state. The Bohr model explains the mechanics of the simplest atoms, like hydrogen. Figure 1.8b shows the wave model of an atom in which electrons form a cloud around the nucleus instead of orbiting around the nucleus, as in the Bohr model. This is based on quantum theory. Recently, the *string* theory has tried explaining the structure of an atom as particles on a string.

The neutrons that cause soft errors in CMOS circuits arise when atoms break apart into protons, electrons, neutrons. The half-life of a neutron is about

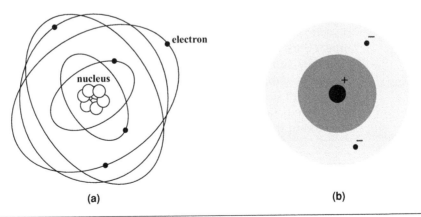

FIGURE 1.8   (a) Bohr model of an atom. (b) Wave model of an atom.

10–11 minutes,[1] unlike a proton whose half-life is about 10^{32} years. Thus, protons can persist for long durations before decaying and constitute the majority of the *primary cosmic rays* that bombard the earth's outer atmosphere. When these protons and associated particles hit atmospheric atoms, they create a shower of secondary particles, which constitute the *secondary cosmic rays*. The particles that ultimately hit the earth's surface are known as *terrestrial cosmic rays*. The rest of the section describes these different kinds of cosmic rays.

Computers used in space routinely encounter primary and secondary cosmic rays. In contrast, computers at the earth's surface need only deal with terrestrial cosmic rays, which are easier to protect against compared to primary and secondary cosmic rays. This book focuses on architecture design for soft errors encountered by computers used closer to the earth's surface (and not in space).

Primary Cosmic Rays

Primary cosmic rays consist of two types of particles: galactic particles and solar particles. Galactic particles are believed to arise from supernova explosions, stellar flares, and other cosmic activities. They consist of about 92% protons, 6% alpha particles, and 2% heavier atomic nuclei. Galactic particles typically have energies above 1 GeV.[2] The highest energy recorded so far for a galactic particle is 3×10^{20} eV. As a reference point, the energy of a 10^{20} eV particle is the same as that of a baseball thrown at 50 miles per hour [7]. These particles have a flux of about 36 000 particles/cm^2-hour (compared to about 14 particles/cm^2-hour that hit the earth's surface).

As the name suggests solar particles arise from the sun. Solar particles have significantly less energy than galactic particles. Typically, a particle needs about 1 GeV of energy or more to penetrate the earth's atmosphere and reach sea level, unless the particle traverses down directly into the earth's magnetic poles. Whether solar particles have sufficient energy or flux to penetrate the earth's atmosphere depends on the solar cycle.

The solar cycle—also referred to as the sunspot cycle—has a period of about 11 years. Sunspots are dark regions on the sun with strong magnetic fields. They appear dark because they are a few thousand degrees cooler than their surroundings. Few sunspots appear during the solar minimum when the luminosity of the sun is stable and quite uniform. In contrast, at the peak of the solar cycle, hundreds of sunspots appear on the sun. This is accompanied by sudden, violent, and unpredictable bursts of radiation from solar flares. The last solar maximum was around the years 2000–2001.

---

[1] The half-life for a given particle is the time for half the radioactive nuclei in any sample to undergo radioactive decay.

[2] 1 eV is a unit of energy equal to the work done by an electron accelerated through a potential difference of 1 V.

Interestingly, the sea-level neutron flux is minimum during the solar maximum and maximum during the solar minimum. During the solar maximum, the number of solar particles does indeed increase by a million-fold and exceeds that of the galactic particles. This large number of solar particles creates an additional magnetic field around the earth. This field increases the shielding against intragalactic cosmic rays. The net effect is that the number of sea-level neutrons decreases by 30% during the solar maximum compared to that during the solar minimum.

Overall, neutrons from galactic particles are still the dominant source of neutrons on the earth's surface. This conclusion is further supported by the fact that flux of terrestrial cosmic rays varies by less than 2% between day and night.

Both the flux and energy of neutrons determine the SER experienced by CMOS chips. To the first order for a given CMOS circuit, the SER from cosmic rays is proportional to the neutron flux. The energy of these particles also makes a difference, but the relationship is a little more complex and will be explained later in the chapter.

Secondary Cosmic Rays

Secondary cosmic rays are produced in the earth's atmosphere when primary cosmic rays collide with atmospheric atoms. This interaction produces a cascade of secondary particles, such as pions, muons, neutrons. Pions and muons decay spontaneously because their mean lifetimes are in nanoseconds and microseconds, respectively. Neutrons have a mean lifetime of 10–11 minutes, so they survive longer. But most of these neutrons lose energy and are lost from the cascade. Nevertheless, they collide again with atmospheric atoms and create new showers and further cascades.

The flux of these secondary particles varies with altitude. The flux of secondary particles is relatively small in the outer atmosphere where the atmosphere is not as thick. The flux continues to increase as the altitude drops and peaks at around 15 km (also known as the Pfotzer point). The density of secondary particles continues to decrease thereafter till sea level.

Figure 1.9 shows the variation of neutron flux with altitude. This variation in flux is given by the following equation [30], where H is the altitude in kilometers:

$$\text{Flux increase over sea level} = e^{\left(\frac{119.685 \times H - 4.585 \times H^2}{136}\right)}$$

This equation is a rough approximation of how the neutron flux varies with altitude. A more detailed calculation can be found in the joint electron device engineering council (JEDEC) standard [13].[3]

---

[3]The Web site http://www.seutest.com provides a calculator to compute the neutron flux at a location based on a variety of parameters, such as latitude, longitude, altitude, and geometric rigidity (discussed later in this section).

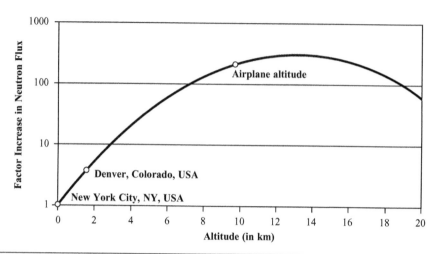

FIGURE 1.9 Variation of neutron flux with altitude. In Denver, Colorado, for example, the neutron flux is about 3.5 times higher than that in New York City, which is at sea level.

EXAMPLE

Airplanes typically fly at an altitude of 10 km. What is the increase in neutron flux over the sea level at this altitude?

SOLUTION The equation given above shows that the neutron flux increases by 228 times over sea level at an altitude of 10 km at which airplanes typically fly. This implies that the SER of CMOS chips operating in airplanes will be 228 times higher than what the same chips experience at sea level.

Please note that the SER experienced by CMOS chips will be slightly higher than indicated by the neutron flux equation given above. This is because of the presence of other particles, such as pions and muons, besides neutrons.

Terrestrial Cosmic Rays

Terrestrial cosmic rays refer to those cosmic particles that finally hit the earth's surface. Terrestrial cosmic rays primarily arise from the cosmic ray cascades and consist of fewer than 1% primary particles.

Both the distribution of energy and flux of neutrons determine the SER experienced by CMOS chips. Figure 1.10 shows the distribution of neutron flux with energy based on a recent measurement by Gordon et al. [9]. Typically, only neutrons

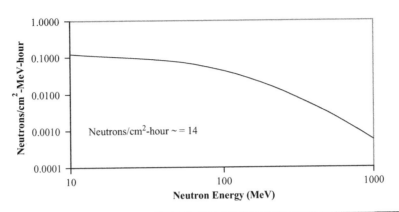

FIGURE 1.10   Terrestrial differential neutron flux (neutrons/cm²-MeV-hour) plotted as a function of the neutron energy. The area under the curve gives a total neutron flux of about 14 neutrons/cm²-Mev-hour.

above 10 MeV affect CMOS chips today.[4] To the first order, the SER is proportional to the neutron flux. Integrating the area under the curve in Figure 1.10 gives a total neutron flux of about 14 neutrons per cm²-hour. Earlier measurements by Ziegler in 1996, however, had suggested about 20 neutrons per cm²-hour [28]. The JEDEC standard [13] has agreed to use the recent data of Gordon et al. instead of Ziegler's, possibly because the more recent measurements are more accurate.

The neutron flux varies not only with altitude but also with the location on the earth. The earth's magnetic field can bend both primary and secondary cosmic particles and reflect them back into space. The minimum momentum necessary for a normally incident particle to overcome the earth's magnetic intensity and reach sea level is called the "geomagnetic rigidity (GR)" of a terrestrial site. The higher the GR of a site, the lower is the terrestrial cosmic ray flux. The GR at New York City is 2 GeV and hence only particles with energies above 2 GeV can cause a terrestrial cascade.

Typically, the magnetic field only causes the neutron flux to vary across any two terrestrial sites by only about a factor of 2x. The GR is highest near the equator (about 17 GeV) and hence the neutron flux is lowest there. In contrast, the GR is lowest near the north and south poles (about 1 GeV), where the neutron flux is the highest. The city of Kolkata in India, for example, is close to sea level and has one of the highest GR (15.67 GeV). Nevertheless, its neutron flux is only about half of that

---

[4]Terrestrial neutrons have two other peaks below the one between 10 and 1000 MeV. The other two peaks appear roughly between 0.01 and 0.02 eV (also called thermal neutrons) and between 0.01 and 5 MeV. These neutrons could also cause transient faults. Further, thermal neutrons may also arise in semiconductor processes that use the boron-10 isotope.

of New York City, although its GR is eight times higher than that of New York City. The JEDEC standard [13] describes how to map a given terrestrial site to its GR and compute the corresponding neutron flux at that location. In summary, three factors influence the neutron flux at a terrestrial site: the solar cycle, its altitude, and its latitude and longitude, which determines its GR.

The neutron flux discussed so far is without any shielding and measured in open air. The flux seen within a concrete building can be somewhat lower. For example, a building with 3-feet concrete walls could see a 30% reduction in SER. Unfortunately, unlike for alpha particles, there is no known shielding for these atmospheric cosmic rays except 10–15 feet of concrete. Consequently, semiconductor chips deep inside the basement of a building are less affected by atmospheric neutrons compared to those directly exposed to the atmosphere (e.g., next to a glass window). Since it is impractical to ship a silicon chip with a 10-feet concrete slab, silicon chip manufacturers look for other ways to reduce the error rate introduced by these neutrons.

1.7.3 Interaction of Alpha Particles and Neutrons with Silicon Crystals

Alpha particles and neutrons slightly differ in their interactions with silicon crystals. Charged alpha particles interact directly with electrons. In contrast, neutrons interact with silicon via inelastic or elastic collisions. Inelastic collisions cause the incoming neutrons to lose their identity and create secondary particles, whereas elastic collisions preserve the identity of the incoming particles. Experimental results show that inelastic collisions cause the majority of the soft errors due to neutrons [21], hence inelastic collisions will be the focus of this section.

Stopping Power

When an alpha particle penetrates a silicon crystal, it causes strong field perturbations, thereby creating electron–hole pairs in the bulk or substrate of a transistor (Figure 1.11). The electric field near the p–n junction—the interface between the

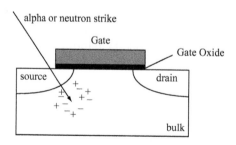

FIGURE 1.11   Interaction of an alpha particle or a neutron with silicon crystal.

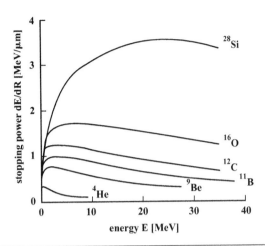

FIGURE 1.12 Stopping power of different particles in silicon. Reprinted with permission from Karnik et al. [12]. Copyright © 2004 IEEE.

bulk and diffusion—can be high enough to prevent the electron–hole pairs from recombining. Then, the excess carriers could be swept into the diffusion regions and eventually to the device contacts, thereby registering an incorrect signal.

Stopping power is one of the key concepts necessary to explain the interaction of alpha particles with silicon. Stopping power is defined as the energy lost per unit track length, which measures the energy exchanged between an incoming particle and electrons in a medium. This is same as the linear energy transfer (LET), assuming all the energy absorbed by the medium is utilized for the production of electron–hole pairs. The maximum stopping power is referred to as the Bragg peak. Figure 1.12 shows the stopping power of different particles in a silicon crystal.

Stopping power quantifies the energy released from the interaction between alpha particles and silicon crystals, which in turn can generate electron–hole pairs. About 3.6 eV of energy is required to create one such pair. For example, an alpha particle (^{4}He) with a kinetic energy of 10 MeV has a stopping power of about 100 keV/μm (Figure 1.12) and hence can roughly generate about 2.8×10^4 electron–hole pairs/μ m.[5] The charge on an electron is 1.6×10^{-19}C, so this generates roughly a charge as high as 4.5 fC/μm.[6] Whether the generated charge can actually cause a malfunction or a bit flip depends on two other factors, namely, charge collection efficiency and critical charge of the circuit, which are covered later in this section.

---

[5] In reality, the stopping power varies over its penetration distance (as expressed in μm) [21].
[6] 1 fC = 1 femto Coulomb = 10^{-15} C.

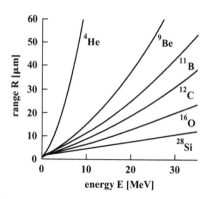

FIGURE 1.13 Penetration range in silicon of different particles as a function of energy. Reprinted with permission from Karnik et al. [12]. Copyright © 2004 IEEE.

■ E X A M P L E

Compute the amount of charge necessary to flip a memory cell with a capacitance of $2\,fF/\mu m$ and a supply voltage of 1.2V.[7]

S O L U T I O N Total charge in the cell = capacitance × voltage = $2 \times 1.2 = 2.4\,fC/\mu m$. It should be noted that a 10 MeV alpha particle could flip this cell.

■ E X A M P L E

Can a single photon of visible light carrying 2 eV cause an upset?

S O L U T I O N The minimum energy needed to generate an electron–hole pair is 3.6 eV. So, it is highly unlikely that a photon will generate an electron–hole pair. Even if it did generate a single electron–hole pair, the generated charge would correspond to the charge on a single electron or 0.00016 fC, which is several orders of magnitude less than 1–10 fC of charge stored by devices in current technologies. Hence, a single photon cannot cause an upset.

Neutrons do not directly cause a transient fault because they do not directly create electron–hole pairs in silicon crystals (hence their stopping power is zero). Instead, these particles collide with the nuclei in the semiconductor resulting in the emission of secondary nuclear fragments. These fragments could consist of particles such as pions, protons, neutrons, deuteron, tritons, alpha particles, and other

---

[7]fF = femto Farad, unit of measure for capacitance.

heavy nuclei, such as magnesium, oxygen, and carbon. These secondary fragments can cause ionization tracks that can produce a sufficient number of electron–hole pairs to cause a transient fault in the device. The probability of a collision that produces these secondary fragments, however, is extremely small. Consequently, about 10^5 times greater number of neutrons is necessary than alpha particles to produce the same number of transient faults in a semiconductor device.

To understand the interaction, consider the following example of an inelastic collision provided by Tang [24] in which a 200 MeV neutron interacts with ^{28}Si. This can produce the following interaction:

$$n + {}^{28}Si \rightarrow 2p + 2n + {}^{25}Mg^*,$$

where n is a neutron, p is a proton, and ^{25}Mg* is an excited compound nucleus, which deexcites as

$$^{25}Mg^* \rightarrow n + 3\,{}^4He + {}^{12}C$$

This creates one neutron, three alpha particles (^{4}He), and a residual nucleus ^{12}C. The ^{12}C nucleus has the smallest kinetic energy of all these particles but the highest stopping power estimated at 1.25 MeV/μm (Figure 1.12) with a maximum penetration range of 3 μm (Figure 1.13) around its Bragg peak. This can generate about 3.5×10^5 electron–hole pairs with a total charge of about 55.7 fC. As shown in an earlier example for the alpha particle, this charge is often sufficient to cause a transient fault. The high stopping power of these ions also explain why neutrons produce an intense current pulse with a small width, whereas alpha particles produce a shorter but wider current pulse.

Besides neutrons, other particles, such as pions and muons, exist in the terrestrial cosmic rays. However, neither pions nor muons are a significant threat to semiconductor devices. The number of pions is negligible compared to neutrons and therefore pions can cause far less upsets than neutrons. The kinetic energy of muons is usually very high, and muons do interact directly with electrons. Nevertheless, typically, muons do not create a sufficiently dense electron–hole trail to cause an upset. Ziegler and Puchner [30] predict that these particles can cause soft errors worth only a few FIT.

Critical Charge (Qcrit)

Stopping power explains why and how many electron–hole pairs may be generated by an alpha or a neutron strike, but it does not explain if the circuit will actually malfunction. The charge accumulation needs to cross a certain threshold before an SRAM cell, for example, will flip the charge stored in the cell. This minimum charge necessary to cause a circuit malfunction is termed as the critical charge of the circuit and represented as *Qcrit*. Typically, Qcrit is estimated in circuit models by repeatedly injecting different current pulses through the circuit till the circuit malfunctions.

Hazucha and Svensson [11] proposed the following model to predict neutron-induced circuit SER:

$$\text{Circuit SER} = \text{Constant} \times \text{Flux} \times \text{Area} \times e^{-\frac{Q_{crit}}{Q_{coll}}}$$

Constant is a constant parameter dependent on the process technology and circuit design style, Flux is the flux of neutrons at the specific location, Area is the area of the circuit sensitive to soft errors, and Qcoll is the charge collection efficiency (ratio of collected charge and generated charge per unit volume). It should be noted that the SER is a linear function of the neutron flux, as well as the area of the circuit. The parameter Qcoll depends strongly on doping and Vcc and is directly related to the stopping power. The greater is the stopping power, the greater is Qcoll. Qcoll can be derived empirically using either accelerated neutron tests or device physics models, whereas Qcrit is derived using circuit simulators. Although Hazucha and Svensson formulated this equation for neutrons, it can also be used to predict the SER of alpha particles. In reality, the SER equations used in industrial models can be far more complicated with a number of other terms to characterize the specific technology generation.

The Hazucha and Svensson equation does explain the basic trends in SERs over process technology. With every process generation, the area of the same circuit goes down, so this should reduce the effective SER encountered by a circuit scaled down from one process generation to the next. However, Qcrit also decreases because the voltage of the circuit typically goes down across process generations. At present, for latches and logic, this effect appears to cancel each other out, resulting in roughly a constant circuit SER across generations. However, if Qcrit is sufficiently low, such as seen in SRAM devices, which are usually 5–10 times smaller than latches in the same technology, then the impact of the area begins to dominate. This is usually referred to as the saturation effect, where the SER for a circuit decreases with process generations. Interestingly, however, the circuit is highly vulnerable to soft errors in the saturation region. In the extreme case, as Qcrit approaches zero, almost any amount of charge generated by alpha or neutron strikes will result in a transient fault.

Chapter 2 discusses in greater detail how to compute Qcrit and map it into a circuit-level SER.

1.8   Architectural Fault Models for Alpha Particle and Neutron Strikes

Microprocessors and other silicon chips can be considered to have different levels of abstractions: transistors that create circuits, circuits that create logic gates and storage devices, and finally the gates and storage devices themselves. The last few sections describe how a transistor collects charge from an alpha particle or a

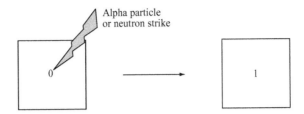

FIGURE 1.14 Strike on a storage device can flip the bit stored from zero to one.

neutron strike. When this charge is sufficient to overwhelm a circuit, then it may malfunction. Logically, at the gate or cell level, this malfunction appears as a bit flip. For storage devices, the concept is simple: when a bit residing in a storage cell flips, a transient fault is said to have occurred (Figure 1.14).

For logic devices, however, a change in the value of the input node feeding a gate or output node coming out of a gate does not necessarily mean a transient fault has actually occurred. Only when this transient fault propagates to a forward latch or storage cell does one say a transient fault has occurred. Chapter 2 discusses transient faults in logic devices in greater detail.

An alpha particle or a neutron strike can, however, cause bit flips in multiple storage or logic gates. There are two types of multibit faults: spatial and temporal. Spatial multibit faults arise when a single neutron can cause flips in multiple contiguous cells. In today's technology, such multibit faults primarily arise only in SRAM and DRAM cells because latches and clocked logic devices are significantly larger than these memory cells. Temporal multibit errors occur when two different neutron or alpha particles strike two different bits. Typically, these are related to error detection codes and are discussed in Chapter 5.

Maiz et al. [14] computed the probability of a spatial multibit error in 130- and 90-nm process technology for Intel SRAM cells based on experiments done under an accelerated neutron beam (Figure 1.15).[8] The probability of an error in three or more contiguous bits is still quite low, but the double-bit error rate could be as high as 1–5% of the single-bit error rate. Such double-bit errors could arise because of not only the small size of transistors but also the aggressive layout optimizations of memory cells. As process technology continues to shrink, this effect will get worse and is likely to increase the number of spatial double-bit errors. Fortunately, current processors and chipsets can use interleaved error detection and correction codes to tackle such errors (see Chapter 5). For further details on multibit errors, please refer to recent analysis of multibit errors in CMOS technology by Seifert et al. [20].

---

[8]Chapter 2 discusses accelerated neutron tests.

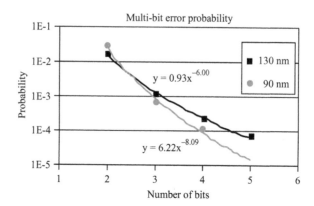

Multi-bit error probability

$y = 0.93x^{-6.00}$

$y = 6.22x^{-8.09}$

- ■ 130 nm
- ● 90 nm

Probability

Number of bits

FIGURE 1.15   Probability of a multibit error compared to a single-bit error in a sample of SRAM cells. Reprinted with permission from Maiz et al. [14]. Copyright © 2003 IEEE.

1.9   Silent Data Corruption and Detected Unrecoverable Error

For an alpha particle or a neutron to cause a soft error, the strike must flip the state of a bit. Whether the bit flip eventually affects the final outcome of a program (Figure 1.3) depends on whether the error propagates without getting masked, whether there is error detection, and whether there is error detection and correction. Architecturally, the error detection and correction mechanisms create two categories of errors: SDC and DUE. Much of the industry has embraced this model because of two reasons. First, different market segments care to a different degree about SDC versus DUE. Second, this allows semiconductor manufacturers to specify what the error rates of their chips are.

The rest of this section explains these definitions, the subtleties around the definitions, and soft error budgets vendors typically create for their silicon chips.

1.9.1   Basic Definitions: SDC and DUE

Figure 1.16 illustrates the possible outcomes of a single-bit fault. Outcomes labeled 1–3 indicate nonerror conditions. The most insidious form of error is SDC (outcome 4), where a fault induces the system to generate erroneous outputs. SDC can be expressed as both FIT and MTTF. To avoid SDC, designers often use basic error detection mechanisms, such as parity.

The ability to detect a fault but not correct it avoids generating incorrect outputs, but cannot recover when an error occurs. In other words, simple error detection does not reduce the overall error rate but does provide fail-stop behavior and thereby avoids any data corruption. Errors in this category are called DUE. Like

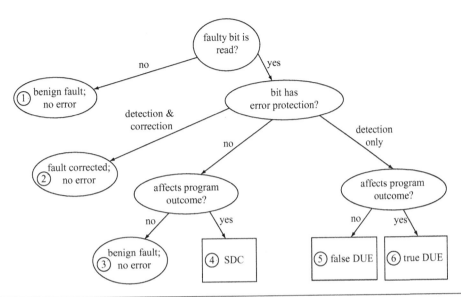

**FIGURE 1.16 Classification of the possible outcomes of a faulty bit in a micro-processor. Reprinted with permission from Weaver et al. [26]. Copyright ©
2004 IEEE.**

SDC, DUE is also expressed in both FIT and MTTF. Currently, much of the industry
specifies SERs in terms of SDC and DUE numbers.

DUE events are further subdivided according to whether the detected fault
would have affected the final outcome of the execution. Such benign detected faults
are called false DUE events (outcome 5 of Figure 1.16) and others true DUE events
(outcome 6). A conservative system that signals all detected faults as processor
failures will unnecessarily raise the DUE rate by failing on false DUE events. Alternatively, if the processor can identify false DUE events (e.g., the fault corrupted
only the result of a wrong-path instruction), then it can suppress the error signal.

DUE events can also be divided into process-kill and system-kill categories (not
shown in Figure 1.16). In some cases, such as a parity error on an architectural
register, an OS can isolate the error to a specific process or set of processes. The
OS can then kill the affected process or processes but leave the rest of the system
running. Such a DUE event is called process-kill DUE. The remaining DUE events
fall into the system-kill DUE category, as the only recourse is to bring down the
entire system.

EXAMPLE

A silicon chip had an initial SDC of 1000 FIT purely from soft errors. The vendor
decided to add parity to every bit that contributed to the 1000 FIT SDC. The

vendor also estimated that the false DUE per bit will cause a DUE increase of 20%. What is the resulting DUE of the chip? Assume only single-bit faults.

SOLUTION Adding parity to the chip converts SDC to DUE. So, the total DUE of the chip would be $(1 + 0.2) \times 1000 = 1200 \, \text{FIT}$.

There are four subtleties in the definitions of SDC and DUE. First, inherent in the definition of a DUE event is the idea that it is a fail-stop. That is, on detecting the error, the computer system prevents propagation of its effect beyond the point at which it has been detected. Typically, a computer system can reboot—either automatically or manually—and return to normal function after a DUE event. That may not necessarily be true for an SDC event.

Second, because a DUE event is caused by an error detected by the computer system, it is often possible to trace back to the point where the error occurred (see Chapter 5). A computer system will typically or optionally log such an event either in hardware or in software, allowing system diagnosis. In contrast, it is usually very hard to trace back and identify the origin of an SDC event in a computer system. Logs may be missing if the system crashes due to an SDC event before the computer has a chance to log the error.

Third, a system crash may not necessarily be a DUE event. An SDC event may corrupt OS structures, which may lead to a system crash. System crashes that are not fail-stop—that is, the effect of the error was not detected and its propagation halted—are usually classified as SDC events.

Finally, a particle strike in a bit can result in both SDC and DUE events. For example, a bit may have an error detection scheme, such as parity, but the parity check could happen a few cycles after the bit is read and used. The hardware can clearly detect the error, but its effect may already propagate to user-visible state. Conservatively, soft errors in this bit can be classified always as SDC. Alternatively, soft errors in this bit can be binned as SDC during the vulnerable interval where its effect can propagate to user-visible state and DUE after parity check is activated. The latter needs careful probabilistic analysis.

1.9.2 SDC and DUE Budgets

Typically, silicon chip vendors have market-specific SDC and DUE budgets that they require their chips to meet. This is similar in some ways to a chip's power budget or performance target. The key point to note is that chip operation is not error free. The soft error budgets for a chip would be set sufficiently low for a target market such that the SDC and DUE from alpha particles and neutrons would be a small fraction of the overall error rate. For example, companies could set an overall target (for both soft and other errors) of 1000 years MTTF or 114 FIT for SDC and 25 years MTTF or about 4500 FIT for DUE for their systems [16]. The SDC and DUE due to alpha particle and neutron strikes are supposed to be only a small fraction of this overall budget.

TABLE 1-1 ■ SDC and DUE Tolerance in Different Application Segments		
	Data Integrity Requirement	**Availability Requirement**
Mission-critical applications	Extremely low SDC	Extremely low DUE
Web-server applications	Moderate SDC tolerated	Low DUE
Back-end databases	Very low SDC	Moderate DUE tolerated
Desktop applications	Higher SDC tolerated	Higher DUE tolerated

Table 1-1 shows examples of SDC and DUE tolerance in sample application servers. For example, databases often have error recovery mechanisms (via their logs) and can often tolerate and recover from detected errors (see Log-Based Backward Error Recovery in Database Systems, p. 317, Chapter 8). But they are often not equipped to recover from an SDC event due to a particle strike. In contrast, in the desktop market, software bugs and device driver crashes often account for a majority of the errors. Hence, processors and chipsets in such systems can tolerate more errors due to particle strikes and may not need as aggressive a protection mechanism as those used in mission-critical systems, such as airplanes. Mission-critical systems, on the other hand, must have extremely low SDC and DUE because people's lives may be at stake.

■ **E X A M P L E**

A system is to be composed of a number of silicon chips, each with an SDC MTTF of 1000 years and DUE MTTF of 10 years (both from soft errors only). The system MTTF budgets are 100 years for SDC and 5 years for DUE. What is the maximum number of chips that can fit into the overall soft error budget?

S O L U T I O N 10 chips can fit under the SDC budget (= 1000/100) and two chips under the DUE budget (= 10/5). Hence, the maximum number of chips that can be accommodated is two chips.

■ **E X A M P L E**

If the effective FIT rate of a latch is 0.1 milliFIT, then how many latches can be accommodated in a microprocessor with a latch SDC budget of 10 FIT?

S O L U T I O N Total number of latches that can be accommodated is 100 000 (= 10/0.0001). The Fujitsu SPARC64 V processor (announced in 2003) had 200 000 latches [2]. Modern microprocessors can have as many as 10 times greater number of latches as that in the Fujitsu SPARC64 V. Consequently, it becomes critical to protect these latches to allow the processor to meet its SDC budget.

1.10 Soft Error Scaling Trends

The study of computer design and silicon chips necessitates prediction of future scaling trends. This section discusses current predictions of soft error scaling trends in SRAM, latches, and DRAM cells. As with any prediction, these could be proven incorrect in the future.

1.10.1 SRAM and Latch Scaling Trends

As explained in the Critical Charge (Qcrit) subsection in this chapter, there are two opposite effects that determine the SER of circuits, such as SRAM cells or latches. An SRAM cell typically consists of six transistors that make up two cross-coupled inverters that store the memory value and a couple of transistors that connect the input and output to the bitlines. As transistor and hence SRAM cell size continues to shrink across process technologies, the SER should reduce because a neutron has a less cross-sectional area to strike and cause a malfunction. However, as the cell size decreases, it holds less charge causing the Qcrit to decrease. This makes it easier for the cell to be upset due to a neutron strike. Figure 1.17 shows that these opposing effects mostly cancel each other out, and overall there is a slight decrease in SRAM SER. This suggests that SRAM cells approach the "saturation" region in which any strike would potentially cause a soft error (because of a sufficiently low Qcrit).

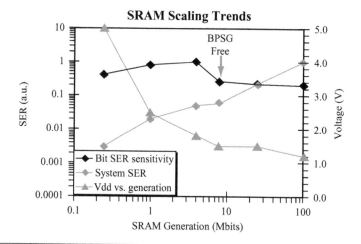

FIGURE 1.17 SRAM scaling trend. a.u. = arbitrary units. BPSG refers to boron contamination that could increase neutron-related errors. The graph shows that the FIT/bit for SRAM cells in recent times (toward the right end of the graph) decreases slowly over time. Vdd is the supply voltage of the chip. Reprinted with permission from Baumann [3]. Copyright © 2002 IEEE.

Overall, the FIT/bit of SRAM storage appears be in the range of 1–10 milliFIT [17,25]. The relatively large variation in the FIT rate arises primarily due to the variation in size, supply voltage, collection efficiency, and other process-related parameters of an SRAM cell. In contrast, the SER for latches appears to be roughly constant (or to vary within a narrow band) over the past few generations since they have not approached the saturation region yet [8].

The SER of other logic devices are, however, predicted to increase over technology generations. SER of logic devices, such as NAND and NOR gates, are typically lower than those of storage elements, such as SRAM cells or latches, because the effects of most alpha particle or neutron strikes are masked. For example, the glitch created at a gate by a strike may not propagate to a forward latch to cause an error. This is known as electrical masking. As the number of logic levels shrinks and the frequency of chips increases, this masking effect is reduced causing more soft errors. Chapter 2 discusses soft error trends in logic devices in greater detail.

Overall, to the first approximation today, the SER of a silicon chip is a function of the number of unprotected latches and SRAM cells in the chip. Because the total number of bits in a microprocessor—both protected and unprotected— roughly doubles every technology generation, one can expect the SER to roughly double, unless significant soft error protection techniques are designed into the microprocessor.

1.10.2 DRAM Scaling Trends

Figure 1.18 shows the scaling trends in DRAMs. DRAM cells are typically made of four transistors (as opposed to six-transistor SRAMs). Large main memory systems are made up of DRAM chips. Typically, a DRAM device consists of 256–1024 megabits in current technology. A number of these devices are, in turn, put in a dual-interlocked memory module (DIMM), which connects to the processor motherboard. DIMMs of 1 gigabyte are not uncommon today in desktops and laptops. Total DRAM capacity in a server machine can be as high as 512 gigabytes.

DRAM and SRAM scaling trends differ significantly. Vendors have successfully reduced the FIT/bit of DRAMs exponentially as the number of bits per DRAM device has increased. The FIT per DRAM device, however, appears to have stayed roughly constant over generations. DRAM vendors have achieved this by digging deeper trenches, adding more stacks, using thinner and alternate dielectrics, and putting bigger capacitors on the DRAM cells. This has allowed the collected charge to decrease without decreasing Qcrit.

Although the DRAM FIT per device has remained constant, large servers with significant amounts of memory may still have significant error rates. Further, because the amount of memory used can be very large, DRAM can also encounter multiple bit errors. Appropriate fault detection and error correction techniques must be used to tackle these errors.

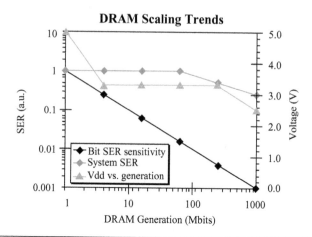

FIGURE 1.18   DRAM scaling trends. a.u. = arbitrary unit. This graph shows the DRAM FIT/bit decreases exponentially with every DRAM device generation, but the FIT/device has remained roughly constant in recent times. Reprinted with permission from Baumann [3]. Copyright © 2002 IEEE.

1.11   Summary

Radiation-induced transient faults are caused by alpha particles found in chip packages and atmospheric neutrons. A soft error is a manifestation of a transient fault. Soft errors due to alpha particles in silicon chips were first reported by Intel Corporation in 1979, and soft errors due to neutrons were first reported by IBM Corporation in 1984. Since then computer system problems caused by alpha particles and neutrons have been reported intermittently.

MTTF is a common metric used to express the SER from alpha particles and neutrons. The MTTF of a system composed of multiple components is the inverse of the sum of the inverse of the component MTTFs. The inverse of MTTF can also be expressed as FIT rates. One FIT represents an error in a billion hours. FIT rates are additive under the exponential failure law, which is commonly used in soft error analysis.

A soft error can be classified either as SDC or DUE. An SDC event arises when corrupted data eventually become visible to the user. A DUE event arises when the system detects a fault. The DUE typically assumes a fail-stop behavior in which an SDC event can never occur. Thus, the SER can be expressed as either the SDC MTTF or DUE MTTF, as the case may be.

Radiation-induced transient faults is one of many silicon reliability problems faced by the chip industry today. Other types of faults include permanent faults caused by EM and gate oxide breakdown and transient faults caused by transistor variability, thermal cycling, and erratic fluctuations of minimum voltage at which a circuit is functional. Radiation-induced transient faults are, however, the only ones that are typically dealt with directly in the field with fault detection and error correction circuitry.

Although alpha particles and neutrons are both classified as radiation, they arise from different sources. Alpha particles arise from chip packaging material. In contrast, neutrons emanate when primary, secondary, and tertiary cosmic rays collide with the atmosphere. Neutrons from galactic particles, instead of solar particles, are the dominant source of neutrons on the earth's surface.

An alpha particle causes a malfunction in a circuit's operation when it penetrates a crystal and loses sufficient energy in a transistor's substrate to create electron–hole pairs that are swept into the diffusion regions of the transistor. The amount of energy lost or released by an alpha particle per unit track length in a silicon crystal is known as the stopping power. Stopping power varies with kinetic energy of an alpha particle. The greater the kinetic energy of an alpha particle, the less is its stopping power and hence the amount of energy released in a silicon crystal.

Neutrons do not interact with the silicon crystal to create electron–hole pairs. Instead, neutrons can undergo inelastic collisions with the silicon crystal, which can produce alpha particles and carbon (^{12}C) nuclei. A carbon nucleus has a low energy but a high stopping power in a silicon crystal. This carbon nucleus can generate sufficient electron–hole pairs to cause a transistor malfunction.

The charge generated by alpha particles and neutrons must be higher than the minimum charge necessary to cause circuit malfunction. This minimum charge is known as the critical charge and expressed as Qcrit. As transistors shrink in size with every process technology generation, they hold less charge, thereby causing a lower Qcrit and making them more vulnerable to a particle strike. However, with every technology generation, the transistors also shrink in size making them harder to strike. For latches, these effects often cancel out, roughly giving a constant FIT/bit over technology generations. For SRAM cells, the transistor has shrunk to an extent that most particle strikes on such a cell will cause a bit flip. Hence, the SER of an SRAM cell continues to decrease with every process technology. DRAMs have, however, managed to aggressively decrease the FIT/bit using various optimizations not typically available for SRAM cells or latches.

1.12 Historical Anecdote

The first half of the 20th century saw a flurry of research into radiation physics. In the 1890s Wilhelm Conrad Roentgen discovered X-rays, for which he received the first Nobel Prize for physics in 1902. Scientists discovered that X-rays and other radioactive materials ionize gases, which enabled these gases to conduct electricity. This effect was used to measure radiation using electroscopes. Scientists soon discovered that the charge in these electroscopes leaked away over time, no matter how they designed the electroscopes or what shielding they used around them. In 1912, an Austrian physicist named Victor Hess took an electroscope to a higher altitude in balloon flights and observed that the intensity of the ionizing radiation initially decreased with altitude, but by about 5000 feet, the radiation was more intense than that at sea level. Victor Hess surmised that this was due to cosmic radiation coming from beyond the earth's atmosphere. Since then, the change in cosmic radiation flux with altitude has typically been used as the signature to identify cosmic rays. The term "cosmic rays" was, however, introduced later in 1926 by an American physicist named Robert Millikan.

References

[1] M. Agostinelli, J. Hicks, J. Xu, B. Woolery, K. Mistry, K. Zhang, S. Jacobs, J. Jopling, W. Yang, B. Lee, T. Raz, M. Mehalel, P. Kolar, Y. Wang, J. Sandford, D. Pivin, C. Peterson, M. DiBattista, S. Pae, M. Jones, S. Johnson, and G. Subramanian, "Erratic Fluctuations of SRAM Cache Vmin at the 90 nm Process Technology Node," in *IEEE International Electron Devices Meeting (IEDM)*, pp. 655–658, December 2005.

[2] H. Ando, Y. Yoshida, A. Inoue, I. Sugiyama, T. Asakawa, K. Morita, T. Muta, T. Motokurumada, S. Okada, H. Yamashita, Y. Satsukawa, A. Konmoto, R. Yamashita, and H. Sugiyama, "A 13 GHz Fifth Generation SPARC64 Microprocessor," in *IEEE Journal of Solid State Circuits*, Volume 38, Issue 11, pp. 1896–1905, November 2003.

[3] R. Baumann, "Tutorial on Soft Errors," in *International Reliability Physics Symposium (IRPS) Tutorial Notes*, IEEE, Dallas, Texas, USA, April 2002.

[4] R. Baumann, T. Hossain, E. Smith, S. Murata, and H. Kitagawa, "Boron as a Primary Source of Radiation in High Density DRAMs," in *IEEE Symposium on VLSI*, pp. 81–82, June 1995.

[5] S. Borkar, "Designing Reliable Systems from Unreliable Components: The Challenges of Transistor Variability and Degradation," *IEEE Micro*, Volume 25, Issue 6, pp. 10–16, November/December 2005.

[6] D. Bossen, "CMOS Soft Errors and Server Design," in *International Reliability Physics Symposium (IRPS) Tutorial Notes*, IEEE, Dallas, Texas, USA, April 2002.

[7] M. W. Friedlander, *A Thin Cosmic Rain: Particles from Outer Space*, Harvard University Press, November 2002.

[8] S. Hareland, J. Maiz, M. Alavi, K. Mistry, S. Walstra, and C. Dai, "Impact of CMOS Process Scaling and SOI on the Soft Error Rates of Logic Processes," in *Symposium on VLSI Technology Digest of Technical Papers*, pp. 73–74, June 2001.

[9] M. S. Gordon, et al., "Measurement of the Flux and Energy Spectrum of Cosmic-Ray Induced Neutrons on the Ground," *IEEE Transactions on Nuclear Science*, Vol. 51, No. 6, Part 2, pp. 3427–3434, December 2004.

[10] B. R. Havekort, et al., *Performability Modelling: Techniques and Tools*, John Wiley and Sons, 2001.

[11] P. Hazucha and C. Svensson, "Impact of CMOS Technological Scaling on the Atmospheric Neutron Soft Error Rate," *IEEE Transactions on Nuclear Science*, Vol. 47, No. 6, pp. 2586–2594, December 2000.

[12] T. Karnik, P. Hazucha, and J. Patel, "Characterization of Soft Errors Caused by Single Event Upsets in CMOS Processes," *IEEE Transactions on Dependable and Secure Computing*, Vol. 1, No. 2, pp. 128–143, April-June 2004.

[13] JEDEC Standard, "Measurement and Reporting of Alpha Particles and Terrestrial Cosmic Ray-Induced Soft Errors in Semiconductor Devices," *JESD89*, August 2001.

[14] J. Maiz, S. Hareland, K. Zhang, and P. Armstrong, "Characterization of Multi-Bit Soft Error Events in Advanced SRAMs," *Digest of International Electronic Device Meeting (IEDM)*, pp. 21.4.1–21.4.4, December 2003.

[15] T. C. May and M. H. Woods, "Alpha-Particle-Induced Soft Errors in Dynamic Memories," *IEEE Transactions on Electronic Devices*, Vol. 26, Issue 1, pp. 2–9, January 1979.

[16] S. E. Michalak, K. W. Harris, N. W. Hengartner, B. E. Takala, and S. A. Wender, "Predicting the Number of Fatal Soft Errors in Los Alamos National Laboratory's ASC Q Supercomputer," *IEEE Transactions on Device and Materials Reliability*, Vol. 5, No. 3, pp. 329–335, September 2005.

[17] E. Normand, "Single Event Upset at Ground Level," *IEEE Transactions on Nuclear Science*, Vol. 43, No. 6, pp. 2742–2750, December 1996.

[18] D. K. Pradhan, *Fault-Tolerant Computer System Design*, Prentice-Hall, 2003.

[19] G. Reis, J. Chang, N. Vachharajani, R. Rangan, D. August, and S. S. Mukherjee, "Design and Evaluation of Hybrid Fault-Detection Systems," in *International Symposium on Computer Architecture (ISCA)*, pp. 148–159, Madison, Wisconsin, USA, June 2005.

[20] N. Seifert, et al., "Radiation-Induced Soft Error Rates of Advanced CMOS Bulk Devices," in *44th Annual International Reliability Physics Symposium (IRPS)*, pp. 217–225, 2006.

[21] G. R. Srinivasan, "Modeling the Cosmic-Ray-Induced Soft-Error Rate in Integrated Circuits: An Overview," *IBM Journal of Research and Development*, Vol. 40, No. 1, pp. 77–89, January 1996.

[22] J. H. Strathis, "Reliability Limits for the Gate Insulator in CMOS Technology," *IBM Journal of Research and Development*, Vol. 46, No. 2/3, pp. 265–286, March/May 2002.

[23] J. Segura and C. F. Hawkins, *CMOS Electronics: How It Works, How It Fails*, Wiley-IEEE Press, 2004.

[24] H. H. K. Tang, "Nuclear Physics of Cosmic Ray Interaction with Semiconductor Materials: Particle-Induced Soft Errors from a Physicist's Perspective," *IBM Journal of Research and Development*, Vol. 40, No. 1, pp. 91–108, January 1996.

[25] Y. Tosaka, S. Satoh, K. Suzuki, T. Suguii, H. Ehara, G. A. Woffinden, and S. A. Wender, "Impact of Cosmic Ray Neutron Induced Soft Errors, on Advanced Submicron CMOS Circuits," in *VLSI Symposium on VLSI Technology Digest of Technical Papers*, pp. 148–149, June 1996.

[26] C. Weaver, J. Emer, S. S. Mukherjee, and S. K. Reinhardt, "Techniques to Reduce the Soft Error Rate of a High-Performance Microprocessor," in *31st Annual International Symposium on Computer Architecture*, pp. 264–275, June 2004.

[27] A. P. Wood, "Software Reliability from the Customer View," *IEEE Computer*, Vol. 36, No. 8, pp. 37–42, August 2003.

[28] J. F. Ziegler, "Terrestrial Cosmic Rays," *IBM Journal of Research and Development*, Vol. 40, No. 1, pp. 19–39, January 1996.

[29] J. F. Ziegler and W. A. Lanford, "The Effect of Cosmic Rays on Computer Memories," *Science*, Vol. 206, No. 776, 1979.

[30] J. F. Zielger and H. Puchner, *SER—History, Trends and Challenges*, Cypress Semiconductor Corporation, 2004.

Device- and Circuit-Level Modeling, Measurement, and Mitigation

2.1 Overview

Analysis of SERs in silicon chips requires an understanding of the impact of alpha particle and neutron strikes on transistor devices, circuit elements, and architectural structures. Chapter 1 covered the basic physics of the interactions of alpha particles and neutrons with transistor devices. Subsequent chapters cover the impact of these errors at the architectural level. This chapter examines how transistors and circuit elements react and respond to alpha- and neutron-induced transient faults. More specifically, this chapter describes current practices to model, measure, and mitigate soft errors in transistor devices and circuit elements.

Computing the SER of a microprocessor or other chipsets requires analyses of two broad areas: the intrinsic FIT rate of the circuits and the devices comprising the chip and the corresponding vulnerability factors by which the intrinsic FIT rate can be derated. Computing the intrinsic FIT rate of a circuit element is a two-step process: first one must compute the Qcrit or critical charge that the charge released by an alpha particle or neutron strike must overcome to cause a circuit malfunction. Thereafter, the Qcrit must be mapped to a corresponding FIT rate for the circuit element. The general procedure to compute the FIT rate in such a fashion, as described in this chapter, applies to memory elements, latches, and logic gates.

Once the intrinsic FIT rate is computed, it needs be derated by a variety of vulnerability factors. For example, if the intrinsic FIT rate of an adder is 10 milliFIT, but the adder itself is not vulnerable 70% of the time, then the intrinsic FIT rate needs to be multiplied by 0.3 to compute a derated FIT rate of 3 milliFIT. Subsequent chapters explain how to model this effect at the architectural level. Typically, however, at the circuit level, unclocked circuit elements, such as SRAM cells, are typically vulnerable 100% of the time, whereas clocked elements, such as latches, static logic, and dynamic logic, may not be vulnerable 100% of the time. This chapter describes how to model a variety of such masking effects and vulnerability factors in clocked circuit elements and the impact of transient faults on clock circuits and how they create soft errors.

The soft error models must be calibrated and validated with measurements. Because soft errors typically occur once in several years in a single chip, the occurrence of errors needs to be accelerated to measure them within a short period of time. This can be accomplished either by collecting error data from numerous chips and computers or by increasing the flux of the generated alpha particles and neutrons. This chapter describes how such measurements are done today.

Finally, this chapter describes some state-of-the-art device- and circuit-level soft error mitigation techniques. This is a rich area of research. Both industry and academia are actively pursuing such mitigation techniques.

2.2   Modeling Circuit-Level SERs

Modeling the SER of a chip requires a method to compute the intrinsic SER of a circuit. Then, the intrinsic error rate is derated by various vulnerability factors. These vulnerability factors arise from a certain resilience of circuits to soft errors as well as from various masking effects inherent in the circuit's operation. To explain how alpha particles and neutrons interact with circuit elements, this section first describes the concept of Qcrit (or critical charge) and how it can be mapped to the FIT rate of a circuit. Then, this section describes the timing vulnerability factor (TVF) of latches and masking effects in static logic and dynamic logic. Finally, this section describes how clock circuits can be impacted by alpha or neutron strikes.

For research in architecture design for soft errors, it may be hard to model these effects in detail. However, Normand found the FIT/bit for memory cells to range between 1 and 10 milliFIT/bit [20]. Although FIT rate of individual circuits can vary widely, this book assumes intrinsic SERs in this range, unless specified otherwise.

2.2.1 Impact of Alpha Particle or Neutron on Circuit Elements

An alpha particle or a neutron strike typically manifests itself as a transient disturbance that would usually last less than 100 picoseconds. If this charge disturbance is smaller than the noise margin, the circuit will continue to operate correctly. Otherwise, the disturbed voltage may invert the logic state.

Let us examine an SRAM cell to understand this phenomenon better (see Rabaey et al. [21], for more detail on SRAM cell design). Figure 2.1 shows an SRAM cell made of a pair of cross-coupled inverters. When the wordline is low, the cell holds data in the cross-coupled inverters and the bitlines are decoupled. If a particle strike causes one of the sensitive nodes to transition, then the disturbance may propagate through the inverter and cause a transient disturbance on the second sensitive node. This will cause the second node to propagate the incorrect value, thereby causing both nodes to flip. This results in flipping the state of the bit held in the SRAM cell. Radiation-hardened cell design—described later in this chapter—is one way to correct such bit flips using a regenerative circuit.

An SRAM cell can also encounter a soft error when the wordline is high and the data are being read out through the bitlines. The voltage differential, which is used to sense if the cell holds a value "0" or a "1," can be disturbed causing a corrupted value to be read out.

Other circuit elements, such as DRAM cells, register file cells, latches, static logic gates, and dynamic logic gates, are affected in similar ways by particle strikes. The size of these cells, number of ports, nature of their operation, etc. affect the degree to which a particle strike can introduce a disturbance in each circuit's operation. The next section explains how to reason about the rate at which a particle strike will introduce a sufficiently large disturbance to cause a circuit element to malfunction.

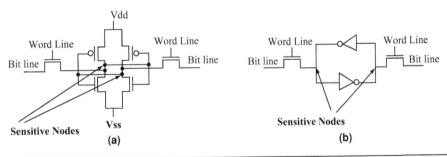

FIGURE 2.1 Nodes in an SRAM cell most sensitive to an alpha or a neutron strike. (a) A transistor-level diagram. (b) The same figure at a logic level, in which the two cross-coupled inverters represent the memory element. The bitlines are complements of each other, whereas the wordlines are the same for both.

2.2.2 Critical Charge (Qcrit)

Computing and modeling the intrinsic SER experienced by a circuit element, such as an SRAM cell or a latch, due to the disturbance introduced by a particle strike involves a two-step process. First, the critical charge (Qcrit) of the circuit element must be computed. Then, the Qcrit must be mapped to the SER expressed in FIT.

Critical charge or Qcrit is the minimum charge that must be deposited by a particle strike to cause a circuit to malfunction. Qcrit is usually computed using integrated circuit simulators, such as SPICE[1], by injecting current pulses into the sensitive nodes of a circuit. These pulses represent the current generated from electron–hole pairs created by an alpha particle or a neutron strike. The smallest charge corresponding to an injected current pulse that inverts the state of a circuit element is the Qcrit of the circuit. Because charge = capacitance × voltage[2], Qcrit depends on the supply voltage (Vcc). Hazucha et al. [11] found that raising the Vcc from 2.2 to 3.3 V for an SRAM cell in a 0.6-μm process raises the Qcrit from 51.5 to 91.4 fC. It is also generally accepted that Qcrit is very weakly dependent on temperature.

This section discusses two broad methodologies to map Qcrit to an FIT rate. The first one involves empirical measurements that help map Qcrit to FIT. The second method uses simulation for the same.

Semiempirical Mapping of Qcrit to FIT

Once Qcrit is computed for a specific circuit element, it needs to be mapped into a SER expressed in FIT. This mapping can be derived by combining physics-based models and experimental data. This section describes three such methods.

Hazucha and Svensson Model   One can start from an equation, such as the one proposed by Hazucha and Svensson [10]:

$$\text{Circuit SER} = \text{Constant} \times \text{Flux} \times \text{Area} \times e^{-\frac{Q_{crit}}{Q_{coll}}}$$

Flux is the alpha or neutron flux experienced by the circuit, Area is the effective diffusion area, and Qcoll is the collection efficiency (ratio of charge collected and charge generated). The parameters of the equation (e.g., Constant, Qcoll) can be derived empirically using accelerated tests. For alpha particles, such accelerated

---

[1]SPICE stands for Simulation Program with Integrated Circuit Emphasis. It was developed by a group of researchers in the mid-70s at the University of California, Berkeley and has been continuously enhanced since then for improved speed, accuracy, and usability in both digital and analog circuit designs. SPICE and its various offshoots are by far the most popular circuit simulators available.

[2]For circuits with feedback loops, the charge held may have a different dependence on voltage and current.

tests can be performed using a radioactive thorium foil (see later sections of this chapter). For neutrons, the accelerated neutron tests can be performed in particle accelerators, such as the one available in the Weapons Neutron Research (WNR) Facility at the Los Alamos Neutron Science Center (LANSCE). The maximum WNR flux is 10^8 times greater than that experienced at sea level in the atmosphere. Thus, soft errors can be captured easily within seconds to hours by exposing SRAM arrays or test chips under the WNR neutron source. Besides the advantage of vastly increased flux, the WNR neutron beam also offers an energy distribution that closely matches that of atmospheric neutrons, which makes it attractive for chip manufacturers. The energy distribution of the WNR facility is discussed later in this chapter.

Such semiempirical mapping of Qcrit to FIT is a popular and fairly precise method to compute the FIT rate of CMOS circuits. However, because the parameters of the equation depend on a specific process generation, the equation must be calibrated for each new technology generation.

Burst Generation Rate Method   The burst generation rate (BGR) method proposed by Ziegler and Lanford [29] is based on two key parameters: the *sensitive volume* (SV) and neutron-induced *recoil energy* (E-recoil). In this method, an upset is said to occur if the burst of charge generated by neutron–silicon interactions within the SV of a device is greater than Qcrit. E-recoil is expressed as:

$$\text{E-recoil} = \text{Qcrit} \times 22.5,$$

where E-recoil and Qcrit are expressed in MeV and pC, respectively. Then, the upset rate is computed as

$$\text{Upset rate} = \text{Qcoll} \times \text{SV} \times \int_{E-\text{neutron}} \left(\text{BGR(E-neutron, E-recoil)}\frac{dN}{dE} \right) dE,$$

where dN/dE is the differential neutron flux spectrum (expressed in neutrons/ cm^2-hour-MeV), E-neutron is the neutron energy (expressed in MeV), BGR(E-neutron, E-recoil) is the energy deposited in silicon by neutron interactions (expressed in $cm^2/\mu m^3$), and Qcoll is the collection efficiency (ratio of charge collected and charge generated). Empirical heavy ion testing—often done in particle accelerators—is used to obtain and tabulate the BGR values as a function of E-neutron and E-recoil. The integration itself is performed numerically using the experimental BGR data.

▮ E X A M P L E

Compute the approximate FIT rate of a 1-megabyte SRAM device using the BGR method. Assume that the SV per bit is $1\mu m^3$, collection efficiency is 75%, and a total BGR is 10^{-13} $cm^2/\mu m^3$.

SOLUTION Figure 1.10 (p. 25) shows that the total neutron flux is 14 neutrons/cm^2-hour (integrated across all the energy ranges). Simplifying the BGR equation shows that the upset rate per bit = $0.75 \times 1 \times 10^{-13} \times 14 \times 10^9$ FIT = 1.1 milliFIT/bit. Then, the approximate FIT rate of a 1-megabyte SRAM device = $8 \times 2^{20} \times 0.0011$ FIT $\sim= 9$ kiloFIT. It should be noted that this is an approximation. To compute the exact FIT rate, the product of BGR and differential flux for each energy level must be integrated.

Neutron Cross-Section Method   To compute the device upset rate using the BGR method, one must compute a device's SV, which is often difficult to ascertain. Instead, the neutron cross-section (NCS) method proposed by Taber and Normand [26] tries to avoid using the SV parameter (as well as Qcrit), by directly correlating the neutron environment parameters, such as flux and energy, with the device upset rate. The NCS method expresses the upset rate as

$$\text{Upset rate} = \int_{E\text{-neutron}} \left(\sigma \frac{dN}{dE} \right) dE$$

This equation replaces Qcoll, SV, and BGR, used in the BGR method, with a single variable σ denoting the neutron cross section. The neutron cross section is defined as the probability that a neutron with energy E-neutron will interact and produce an upset in a semiconductor device. It is expressed in cm^2/device or cm^2/bit. These probabilities are generated for specific device types using accelerated neutron and/or proton testing.

Using Simulation Models to Map Qcrit to FIT

To avoid such a detailed calibration, Murley and Srinivasan [19] had proposed modeling the charge collection phenomenon from first principles. In this methodology, alpha and neutron strikes on devices are simulated from first principles. In cases where simulations result in a collected charge greater than Qcrit, the circuit is assumed to malfunction. This gives the probability of an upset, given a certain alpha or neutron flux. This can then be easily converted into a FIT rate. This methodology, however, requires a detailed knowledge of the process technology and how that can interact with alpha particles and neutrons.

Two key concepts are typically used to model the interaction of alpha particles and neutrons with transistor devices and circuit elements. First, as described in Chapter 1, an alpha or a neutron strike directly or indirectly results in ionization that creates electron–hole pairs. If the charge collected from the electron–hole pairs is greater than Qcrit, then the circuit can potentially malfunction. The particle strike simulations must take into account that particles can come at any incident angle, strike any node within a circuit, and result in current pulses of various

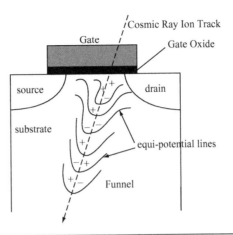

FIGURE 2.2 Field funneling effect in bulk CMOS transistor.

shapes. These can result in very long simulation times, so typically such simulations use various approximations and sampling techniques, such as the Monte Carlo method.[3]

Second, an alpha or a neutron strike can cause a transient distortion of the electric field in the depletion region of a transistor. This phenomenon is known as "field funneling" because the equipotential lines are stretched in the shape of a funnel along the radiation track (Figure 2.2). Field funneling increases the total amount of charge collected by the radiated junction resulting in a higher SER. Field funneling also produces a sharp peak in the disturbed current. This effect is not as critical for DRAM cells, where the cells are periodically refreshed and the charge collection times are relatively small compared to the refresh cycle (in nanoseconds). But in SRAM cells, the stabilization is produced within a few tens of picoseconds by the cross-coupled inverter circuits. Hence, a sharper pulse with a few picosecond width will deposit more charge in a shorter time than a slower rising pulse. Hence, SRAM SER simulations must consider the shape of the pulse generated by an alpha particle or a neutron strike.

Once the intrinsic FIT rate of a circuit is computed, it must be derated by the appropriate vulnerability factors to compute the circuit-level SER. The next few sections will describe how to compute such vulnerability factors for various types of circuits.

---

[3]Monte Carlo simulations are usually stochastic algorithms that rely on random numbers to effectively simulate part of a large experimental space.

2.2.3 Timing Vulnerability Factor

Timing Vulnerability Factor (TVF) is the fraction of time a circuit element is vulnerable to upsets. The circuit SER computed in the last section must be multiplied by the circuit's TVF to obtain the circuit's SER. An SRAM cell in a processor cache usually has a TVF of 100% (or 1, if expressed as a fraction) because any strike during a clock cycle can change the value stored in the SRAM cell. However, flip-flops and latches are clocked elements and may have a TVF less than 100%. The rest of the section discusses how to reason about TVF of edge-triggered flip-flops and edge-triggered latches. To better explain the TVF of flip-flop or latch, this section describes the operation of a master–slave flip-flop first.

Figure 2.3 shows a flip-flop and its corresponding timing diagram. When the clock transitions from low to high, the flip-flop latches in the data at input d. During the high phase of the clock, the flip-flop is in the hold mode, maintaining the value of d (just before the rising clock edge) at the output q. When the clock transitions to the low phase, the output q is typically driven to the next logic element, latch, or flip-flop. The storage nodes of the flip-flop are typically vulnerable to soft errors when the flip-flop is sampling and holding data during this high phase of the clock. When the clock phase is low, the flip-flop is driving data to the next stage and is able to recover from a particle strike. Consequently, in this case flip-flop's TVF is roughly 50% (half of the clock phase). For a master–slave flip-flop operating in sequence, the same concepts apply, except that the master is vulnerable in one clock phase while the slave is vulnerable in the alternate clock phase.

Reasoning about TVF of flip-flops and latches in modern microprocessors is slightly more involved [24]. Unlike lower frequency designs (e.g., below 1 GHz), higher frequency latches start driving data during the high clock phase. Therefore, the alpha or neutron strike causing the upset must occur sufficiently early in the high clock phase for the signal to propagate to the next flip-flop or latch through the forward logic in the path to the next flip-flop or latch. The longer the forward path, the greater the amount of time it takes for the error to propagate to the next flip-flop or latch. Hence, shorter is its window of vulnerability and smaller is the TVF (Figure 2.4).

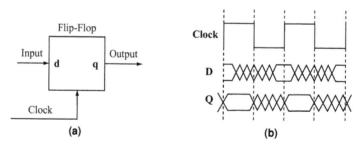

(a)   (b)

FIGURE 2.3   (a) Edge-triggered flip-flop. (b) Timing diagram of the flip-flop.

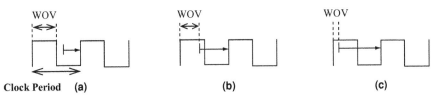

FIGURE 2.4 Timing diagram of a flip-flop. The arrow represents the time it takes to propagate through the forward logic to reach the next forward latch. WOV = window of vulnerability. TVF = WOV/clock period. (a) Only the high clock phase is vulnerable. (b) TVF < 50% because the path length to cover in the forward path is greater than 50% of the clock period. (c) TVF < 50%, but significantly smaller than that in (b) because the path length to be covered is significantly longer.

EXAMPLE

If the frequency of a processor is 2 Ghz, with an average propagation delay of 400 picoseconds, what is the TVF of an average flip-flop in the design?

SOLUTION The clock cycle of a 2-GHz design is 500 picoseconds. Hence, the TVF = (500 − 400)/500 = 20%.

EXAMPLE

Clock gating schemes are used aggressively to reduce power consumed by a processor. Assume a design based on master–slave flip-flops in which the master flip-flop holds the data and the slave is shut off during clock gating. Also, assume that each flip-flop has a TVF of 20%. How much will the TVF increase if the processor has all its flip-flops clock gated for 40% of the time?

SOLUTION Sixty percent of the time the TVF of each flip-flop is 20%. Forty percent of the time the master–slave flip-flop is clock gated during which the master's TVF is 100% and the slave's TVF is 0%. The average TVF during the 40% duration is 50%. Therefore, the average TVF of a flip-flop in this design is (0.6 × 20% + 0.4 × 50%) = 32%. Thus, the TVF increases from 20% to 32%, thereby raising the SER by 60%.

In reality, TVF depends on a number of different components besides propagation delay through the forward logic path. TVF also depends on the setup time of the flip-flop or latch, the clock rise, and fall time, as well as on the clock skew.

Figure 2.5 shows the variation of TVF as a function of propagation delay for a specific set of parameters.

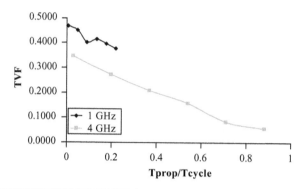

FIGURE 2.5   TVF as a function of Tprop (propagation delay) and clock frequency. TVF is lower for higher frequencies. Reprinted with permission from Seifert and Tam [24]. Copyright © 2004 IEEE.

Figure 2.6 shows another interesting point about TVF. For a given design in a high-frequency processor, TVF decreases with an increasing clock frequency. This is because propagation delay in current microprocessors does not decrease proportionately with an increase in clock frequency. Hence, the window of vulnerability decreases, thereby decreasing the TVF.

2.2.4 Masking Effects in Combinatorial Logic Gates

Logic gates are building blocks of modern silicon chips, including complex microprocessors. Figure 2.1 shows an example of an SRAM cell built from inverters, which are perhaps the simplest of logic gates. Figure 2.7 shows an adder circuit created from a combination of XOR, AND, and OR gates. See Rabaey et al. [21] for an explanation of how these gates are created from basic CMOS transistors.

A malfunction due to a particle strike in one of these logic gates must reach and be captured in the forward memory element (LATCH_S or LATCH_C_{out} in Figure 2.7) for the malfunction to cause an error.[4] Otherwise, the effects are masked and do not cause a malfunction in the full circuit's operation. Thus, evaluating the SER of a logic gate consists of evaluating the Qcrit of each gate, mapping the Qcrit to the appropriate FIT rate using the method described earlier in this chapter, and evaluating whether the fault introduced in the gate's operation will be masked or actually reach the forward latch. Alternatively, using circuit simulations, it is also possible to evaluate the average Qcrit of a whole circuit itself, such as the

---

[4]Whether the error latched in actually causes a user-visible error depends on the architectural design and operation of the chip itself. Chapter 3 discusses this effect in detail.

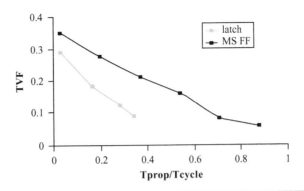

FIGURE 2.6 Variation of TVF with Tprop (propagation delay). Tcycle = clock period. MS FF = master–slave flip-flop. Reprinted with permission from Seifert and Tam [24]. Copyright © 2004 IEEE.

adder circuit shown in Figure 2.7. Then, the average Qcrit should incorporate all the masking effects.

In today's microprocessors, greater than 90% of the radiation-induced faults in logic gates could potentially be masked. Nevertheless, faults in logic gates cannot be ignored because of three reasons. First, a modern billion-transistor microprocessor is composed of tens to hundreds of millions of logic gates. Even a small contribution to the FIT rate from logic can add up creating a significant overall FIT rate for the microprocessor itself. Second, as discussed later in this section, the effect of this masking decreases with new technology generations. This may cause the overall contribution to the FIT rate from logic gates to come close to and possibly even exceed that caused by latches. Third, it is significantly more difficult to architecturally protect these logic gates compared to SRAM cells used in processor caches. Simple ECCs, such as parity, are difficult to implement for logic blocks. Circuit-level techniques, such as radiation hardening, that can protect logic blocks are discussed later in this chapter.

There are three kinds of masking commonly observed in logic blocks [1,16,25]:

- *Logical masking.* A particle strike can be logically masked if it affects a portion of the circuit that does not logically affect the final outcome of the circuit. For example, if the first input of the OR gate in Figure 2.7 is one, then the second input is a "don't care" because the final result will always be one. Hence, a strike at the second input node of the OR gate, when the first input node is one, is logically masked.

- *Electrical masking.* A particle strike can be electrically masked if the pulse created by the strike attenuates before it reaches the forward latch. For example, in Figure 2.7, if the particle strikes c_{in} at the input of the first AND gate, but this pulse attenuates by the time it reaches the OR gate, then the error introduced by the pulse will never reach the forward latch LATCH_C_{out}.

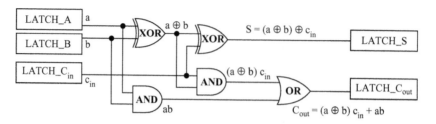

FIGURE 2.7 Logic diagram for a full-adder. A and B are the inputs, C_{in} is the carry-in from a previous full-adder, S is the sum, and C_{out} is the carry-out to the next adder. $\oplus$ denotes the XOR operation.

■ *Latch-window masking.* A particle strike can also be masked if the resulting pulse does not reach the forward latch at the clock transition (or latching window) where the latch captures its input value. For example, in Figure 2.7, if a particle creates an erroneous pulse at the input of the OR gate, but the propagated pulse does not reach the forward latch LATCH_C_{out} at the point where the latch reads the value of the input, then the error will be masked.

To accurately compute the SER of logic blocks, it is essential to model each of these masking effects.

Modeling the Effects of Masking in Logic

When a particle strikes a sensitive node of a circuit, it produces a current pulse with a rapid rise time but a more gradual fall time. Hence, the first step in modeling the masking effects is to model this current pulse $I(t)$ as a time-dependent current source [6, 28]:

$$I(t) = \frac{2}{p} \times \frac{Q}{T} \times \sqrt{\frac{t}{T}} \times e^{-\frac{t}{T}}$$

where Q is the amount of charge collected from a particle strike and the time constant T is a function of the CMOS process. A smaller T results in a shorter, but more intense, pulse compared to the pulse produced by a larger T. The square root function captures the rapid rise in the current pulse, whereas the negative exponential term captures the gradual fall of the pulse. Typically, both T and Q decrease with each successive technology generation.

This current pulse can now be used to drive circuit simulators, such as SPICE, to gauge the impact of a particle strike on a logic gate.

Logical Masking   Conceptually, computing the effect of logical masking is relatively straightforward. It involves injecting erroneous current pulses into different parts of a logic block and simulating its operation for various inputs or benchmarks.

A random sample of nodes and pulses is typically selected to avoid simulating the logic block under every different configuration of inputs and error pulses. Alternatively, logical masking can also be modeled in a logic-level simulator by flipping inputs from zero to one or vice versa. The latter method is much faster because it does not involve detailed simulation of a current pulse and its effect on the logic.

Electrical Masking   Computing the effects of electrical and latch-window masking is a little more involved. As the current pulse traverses through the cascade of gates, its strength continues to attenuate. More specifically, the rise and fall times of the pulse increase and its amplitude decreases. The increase in rise and fall times of the pulse results from circuit delays caused by the switching delay of the transistors. The decrease in amplitude may occur if and when a gate turns off before the output pulse reaches its full amplitude. This can happen if an input transition occurs before the gate has completely switched from its previous transition. This causes the gate to switch in the opposite direction before reaching the peak amplitude of the input pulse, thereby degrading the output pulse. This effect cascades from one gate to the next, thereby slowly attenuating the signal. If the signal completely attenuates before reaching the forward latch, then the forward latch does not record an erroneous value, and the error is said to be electrically masked. Shivakumar et al. [25] used the rise and fall time model of Horowitz [12] and the logical delay degradation model of Bellido-Diaz et al. [2] to compute the impact of electrical masking through a logic block.

Latch-Window Masking   An edge-triggered latch is only vulnerable to latching in a propagated error during a small latching window around its closing clock edge (Figure 2.8). This latching window is effectively the sum of the setup time and hold time of the latch. The setup time is the minimum amount of time before the clock edge for which data to be latched in must be valid. The hold time is the minimum amount of time after the clock edge that the data must be valid for the latch to correctly read it in. Pulses that completely overlap the latching window will always cause an error in the latch. Pulses that are not overlapped with the latching window will always be masked. Pulses that partially overlap with the latching window may or may not be masked. Shivakumar et al. [25] believe that errors caused by partially overlapped pulses are a secondary effect.

Assume c = clock cycle, d = pulse width, and w = width of latch window. If soft errors due to partially overlapped pulses are ignored, then the probability of a soft error can be expressed as

- If $d < w$, Probability(soft error) = 0 because the pulse cannot span the entire latch window.

- If $w \leq d \leq c + w$, Probability(soft error) = $(d - w)/c$ because the pulse must arrive in the interval $(d - w)$ just prior to the latching window.

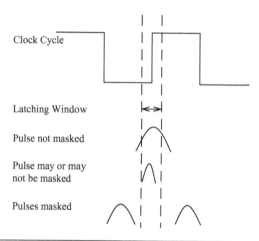

FIGURE 2.8 Latch-window masking.

- If $d > c + w$, Probability(soft error) = 1, the pulse is guaranteed to overlap with at least one latching window. It should be noted if $c < d < c + w$, then d can overlap with two consecutive latching windows and still not cause a soft error.

It should be noted that the latch-window masking reduces the fault rate of logic gates. In contrast, a strike to a latch may get masked if the latch drives data to its output. This reduces the TVF of the latch. This latter masking effect reduces the fault rate of the latch and not the fault rate of the logic gates feeding it.

Putting All These Together   To appropriately model the SER of combinatorial logic gates, all the three masking effects must be taken into account. A fully exhaustive model would simulate charge collections of all different magnitudes and at different nodes of the logic circuits (e.g., as in the TIme DEpendent Ser Tool called TIDEST [23]) and then study the masking effects for each of these cases. A fully exhaustive simulation model can be very precise but can also lead to extremely long simulation times even for small circuits. Hence, sampling methods, such as Monte Carlo simulations, are typically used to reduce the simulation space.

Zhang and Shanbhag [28] proposed an alternate approximation to reduce the simulation times required to compute the masking effects. In this method, logical masking effects were computed using fault injection into a logic-level simulator, which is significantly faster than a circuit simulator. Then, the electrical and latch-window masking effects were computed using a circuit simulator. For each circuit encountered in a chip, they first extracted the path that the resulting error from a particle strike would propagate through. They mapped this path to an equivalent chain of inverters. The electrical masking and latch-window masking effects were computed ahead of time for representative inverter chains. Hence, the effects of electrical and latch-window masking in these circuits become simply a table lookup. The authors found that this approximation introduced less than 5% error in the SER prediction compared to Monte Carlo-based simulation approaches. Overall, these three techniques—using logic-level simulation for logic masking, extracting the path the error propagates through, and mapping the path to an equivalent inverter chain—speed up the masking simulations by orders of magnitude over using brute-force circuit simulation. Other researchers (e.g., Gill et al. [7]) are exploring other options to further reduce this simulation time.

Impact of Technology Scaling   As feature size decreases, the relative contribution of logic soft errors may continue to increase. This is because of three reasons. First, logic gates are typically wider devices than memory circuits, such as SRAM cells. But technology scaling more rapidly decreases the size and Qcrit of logic gates than that of SRAM cells.

Second, the effect of electrical masking will decrease with technology scaling. This is because fewer error pulses will attenuate as the frequency of these gates continues to increase.

Third, a higher degree of pipelining, if used by high-end microprocessors and chipsets, will decrease the clock cycle without significantly changing the setup time and hold time of latches. Recently, microprocessors have moved toward shallower pipelines to avoid excessive power dissipation and design complexity. Nevertheless, after this sharp change toward shallower pipelines, the number of pipeline stages in a processor continues to increase again. This will decrease the amount of latch-window masking experienced by a circuit. Overall, Shivakumar et al. [25] predicted that the SER from logic gates rises exponentially. But the jury is still out on this issue.

Masking Effects in Dynamic Logic

Evaluating the masking effects in a dynamic logic gate is a little more involved than computing the same for a static logic gate. Figure 2.9 shows a dynamic logic gate evaluating the NAND function. The operation of a dynamic gate is typically divided into two major phases: *precharge* and *evaluation*. The mode of operation is determined by the clock signal. When CLK = 0, the precharge transistor precharges the output node OUT to Vdd and the pull-down path is off. When CLK = 1, the precharge transistor turns off and the evaluation transistor is on. The output discharges conditionally based on the input values and the pull-down topology. The dynamic gate in Figure 2.9, for example, will discharge if all the three inputs—IN1, IN2, and IN3—are one.

Dynamic gates are mostly vulnerable to particle strikes during the evaluation phase. The precharge phase drives the output strongly and hence makes the gate resilient to soft errors during that period. Thus, the TVF of the dynamic gate is roughly 50%. Further, even during the evaluation phase, the pull-down topology

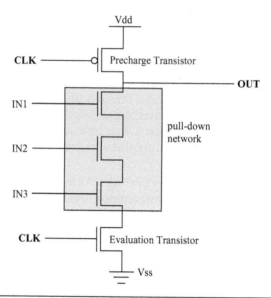

FIGURE 2.9   A dynamic logic gate evaluating the NAND function of the three inputs (IN1, IN2, and IN3).

and the inputs determine whether a soft error is logically masked. In Figure 2.9, a soft error can affect the output only under two conditions: each input is one or one of three inputs is zero. In the first case in which each input is one, fault in any input will change the outcome. For the second case, only strikes to the input that is zero will change the outcome. There are eight possible input combinations. Assuming a uniform distribution of input values, the SER can be derated by $(1 + (1/3) \times 3)/8 = 25\%$ (i.e., one out of four input combinations is vulnerable). Thus, the circuit-level vulnerability factor for the dynamic gate is $50\% \times 25\% = 12.5\%$.

■ E X A M P L E

Compute the circuit-level vulnerability factor for the dynamic gate computing the NOR function as shown in Figure 2.10. Assume a uniform distribution of input values.

SOLUTION The dynamic gate in Figure 2.10 will discharge if either IN1 or IN2 is one. A soft error can change the outcome only under two conditions: when both inputs are zero or one of the two inputs is one. Hence, the SER can be derated by $(1+(1/2)\times2)/4=50\%$. Multiplying this with a TVF of 50% results in a circuit-level vulnerability factor of $50\% \times 50\% = 25\%$.

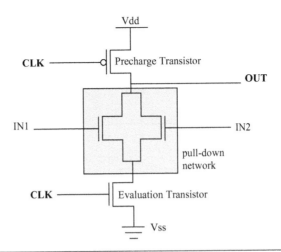

FIGURE 2.10   A dynamic logic gate evaluating the NOR function of two inputs (IN1 and IN2).

Dynamic logic gates sometimes have storage nodes at their outputs to hold the data generated by the dynamic logic. The storage node is referred to as a keeper, whose SER can be computed the same way as a storage node.

2.2.5 Vulnerability of Clock Circuits

Global and local clock networks used in CMOS microprocessors and other chips are also vulnerable to alpha particle and neutron strikes. But errors generated by these clock networks are slightly different from those generated by logic or latches. The two dominant error modes from particle strikes in clock networks are

- **Jitter.** Particle strikes can cause the clock edge to move randomly in time creating jitter in the clock edge (Figure 2.11a). Such jitter can also be caused by noise and process variations. More specifically for radiation, the jitter occurs when a particle strikes a clock node close to when a clock will be asserted. This can result in a violation of the setup time, causing the data to be latched incorrectly.

- **Race.** If a particle strike generates sufficient charge to create a new clock pulse (Figure 2.11b), then this may accidentally trigger unwanted activity in the clocked circuit, such as premature latching of the input data. In a pipelined microprocessor, this can result in data races through successive stages of the clocked latches, eventually causing an erroneous operation.

Whether a particle strike results in a clock jitter or a data race depends on the time between the arrival of data at the input of a clocked latch and the beginning of the setup time prior to the clock edge. This time is normally referred to as the

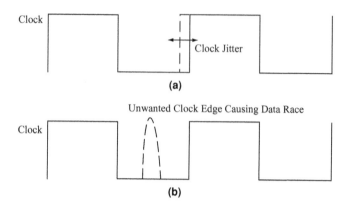

FIGURE 2.11 The impact of particle strikes on the clock network. The strike can result in either a jitter (a) or a data race (b).

"data margin." Figure 2.12 shows this effect. When the margin is low (implying that the data arrive close to the clock edge), then an error can result from the jitter introduced by a particle strike on the clock node (as shown in Figure 2.11a). In contrast, if the data margin is high (implying that the data arrive much earlier than the clock edge), then an error can result from the particle strike if an additional clock pulse is created by the particle strike (as shown in Figure 2.11b).

Modeling the impact of jitter and race is somewhat involved for a simple circuit and gets significantly more computationally expensive for a whole chip. For details on how to model SERs in clock networks, the readers are referred to Seifert et al. [23]. They proposed a faster simulation technique that uses a combination of analytical modeling and simulation using a critical pulse width that is sufficient to upset a latch.

They also demonstrated that the soft error contribution from clock networks is sufficiently low—of the order of <10% of that contributed by flip-flops and latches. This is because the window of vulnerability for a particle to strike and cause a jitter or a race is quite small. Nevertheless, in certain design styles, such as in certain pulsed latch designs in which pulse generators feed few pulsed latches, the clock network can result in as high an SER as 50% of that contributed by the latches in the design.

2.3   Measurement

The previous section discussed how to model and compute the SER of various circuit elements. This section examines techniques to measure the SER of circuit elements and chips. The measurements of SERs are typically used in two ways. First, selected predictions and parameters from the models are calibrated using measurements to increase our confidence in these models. Second, data from measurements

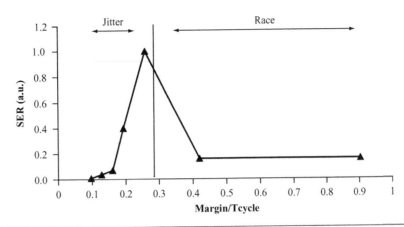

FIGURE 2.12 SER and Qcrit of clock node as a function of data margin for a flip-flop-based design. Tcycle = clock cycle time. a.u. = arbitrary unit. Reprinted with permission from Seifert et al. [23]. Copyright © 2005 IEEE.

are used to create the models, as discussed in the Critical Charge (Qcrit) subsection in this chapter.

Measurements of SERs, however, pose one challenging problem. If the typical FIT rate of a circuit element is in the range of around 1 milliFIT, then one encounters only one error every 10^{12} hours or 114 155 years of operation. Further, to achieve a statistical significance, several errors must be detected before the FIT rate of the circuit element can be predicted. Obviously, it is not possible to measure the FIT rate of an individual circuit element within a reasonable amount of time. Guidelines for such measurements are given in the JEDEC standard [13].

To measure the SER of circuits and chips in a reasonable amount of time and with some degree of statistical significance, practitioners typically use the following methods (or a combination of these):

- *Measure the SER over a large number of elements.* To compute the FIT per bit (often denoted as FIT/bit) of an SRAM cell in a processor's cache, one can measure the SER of the entire cache and then divide the FIT rate by the number of bits in the cache. Similarly, to compute the FIT/latch on a test chip, one can measure the FIT rate over many latches and then divide by the number of latches to obtain the FIT/latch. These numbers can even be aggregated over a large number of computers (e.g., deployed in the field).

- *Increase the flux of alpha particles or neutrons striking the device.* Increasing the flux of alpha particles or neutrons typically causes circuit elements to fail faster. For example, in Los Alamos National Laboratory, the neutron beam produces a maximum neutron flux that is 10^8 times greater than that experienced on the earth's surface. Dividing the observed failure rate by 10^8 gives the approximate atmospheric FIT rate experienced by these elements.

Measurement of radiation-induced SERs must consider three aspects:

- Do the data come from the field or accelerated tests?

- Is the error rate due to alpha particles or neutrons?

- Is the FIT rate of individual circuit elements or that of a full chip being measured?

The rest of this section provides answers to these questions.

2.3.1 Field Data Collection

Field data on soft errors can be collected from error logs of computer systems. Computer systems will typically log information—either in hardware or in software—about both corrected and detected but uncorrected errors experienced by the computer system. Significant information about soft errors can be gleaned from these error logs. Although these error logs capture information about a variety of errors, soft errors can be recognized from their "signature," such as random occurrence in time across machines and an enhanced rate of errors in machines at higher elevations.

Eugene Normand is probably the first to report an analysis of SERs experienced by a number of computer systems in the field and at ground level [20]. He was looking for evidence of neutron-induced errors in such computer systems. In Fermilab computers, Normand found the FIT/bit of the DRAM to be around 0.7 milliFIT, which is 5–10 times greater than that expected by the manufacturers and 500 times greater than that expected from alpha particles. With further analysis, Normand inferred that this error rate must have been due to atmospheric neutrons. In five Nite Hawk computers, Normand found the error rate of DRAM to be around 2.3 milliFIT/bit. He also looked at a CRAY YMP-8's SRAM memory system log and found the error rate to be about 1.3 milliFIT/bit. It should be noted that FIT/bit of DRAM chips have come down significantly since then (see DRAM Scaling Trends, p. 37, Chapter 1).

2.3.2 Accelerated Alpha Particle Tests

Accelerated measurements of soft errors due to alpha particles can be obtained by exposing silicon chips to radioactive isotopes, such as americium-241, uranium-238, or thorium-232 that emit alpha particles. These are commercially available as thin foils or disks. The radioactive nature of the materials makes it imperative that they be handled by trained personnel and stored appropriately to avoid health hazards. Figure 2.13 shows an experimental setup in which a CMOS chip is exposed to a thorium-232 foil.

The radioactive foils emitting the alpha particles must be in close proximity to the CMOS chip. This is because alpha particles are absorbed within 5–8 cm in air. Typically, the top of the circuit die is directly exposed to the foil. Packages

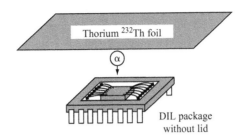

FIGURE 2.13 Measurement setup for soft errors due to alpha particles. DIL package = dual in-line package. Reprinted with permission from Karnik et al. [14]. Copyright © 2005 IEEE.

suitable for such testing are dual-in-line packages—shown in Figure 2.13—and similar wirebond-type packages with pads around the perimeter of the die. This setup requires removing the lid of the package.

Flip-chip packages, however, make it hard to expose silicon chips to these radioactive foils. In flip-chip packages the chips are contacted using solder-ball technology such that the chips face downward toward the module substrate. For these packages, Ziegler and Puchner [30] suggest the use of radioactive fluids that can be introduced at the edge of a chip. Capillary action is often adequate to make the fluid flow under the chip, thereby bringing the silicon chip in close proximity to the alpha-emitting radioactive material.

The foil or fluid used in the experiment must be calibrated by a counter to gauge its alpha emission rate. Then, after exposing the die/chip to the foil or fluid for a predetermined period of time, the total number of errors is counted. For memories, this involves reading out the final data pattern and comparing it to the data pattern originally written into the memory. From the alpha activity of the foil, exposure time, and error count, the SER due to alpha particles under natural conditions can be computed.

The exact alpha particle flux arising from natural conditions varies with the process and packaging technology. Typically, each semiconductor vendor will know the maximum alpha particle rate experienced by the chips.

2.3.3 Accelerated Neutron Tests

Accelerated measurement of soft errors due to neutron particles can be performed by exposing the silicon chips to an energetic neutron or proton beam. The JEDEC standard [13] contains a list of facilities where neutron soft error experiments can be performed. Most of these facilities charge about US$300–US$1000 per hour of beam time. There are three possibilities:

- *White neutron beam.* The preferred method of measuring neutron-induced SER is to use a neutron beam with an energy spectrum similar to that of

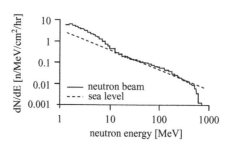

FIGURE 2.14 Neutron energy spectrum at sea level versus the one available from the "white" neutron beam at the WNR Facility in Los Alamos National Laboratory. Reprinted with permission from Karnik et al. [14]. Copyright © 2005 IEEE.

atmospheric neutrons at sea level (Figure 2.14). The beam is referred to as a *white* neutron beam because it contains neutrons with different energy levels. In the WNR Facility at the Los Alamos National Laboratory in the United States, this white beam is produced by directing the 800-MeV protons on a tungsten target. When the protons hit the tungsten target, they shatter the tungsten nucleon (much like when a glass shatters into many pieces) and give rise to neutrons with different energy levels. This mechanism is normally referred to as a *spallation* reaction. At WNR, the spallation reaction produces a total neutron flux (of various energies) of up to 10^8 times higher than the flux at sea level. Hence, the error rate measured at WNR must be reduced by 10^8 times to obtain the atmospheric SER.[5] This beam is widely used by semiconductor companies and memory manufacturers to measure the neutron-induced SER of their parts.

■ *Monoenergetic neutron beam.* The second approach is to use monoenergetic neutrons to obtain the SER for specific energy levels. These numbers can then be combined—a process called deconvolution—to approximate the SER under a white beam. It appears that in recent years, the dependence of the SER on neutron energy levels has come down dramatically, making it easier to carry out the deconvolution. Monoenergetic neutrons can be produced in a particle accelerator, in a nuclear reactor, or via a deuterium (^{2}H)– tritium (^{3}H) reaction. Examples of monoenergetic neutron sources in the United States are the 160-MeV beam at Indiana University Cyclotron Facility, 65-MeV beam in Crocker Nuclear Lab in University of California at Davis, and 14-MeV beam in BOEING.

---

[5]There is some controversy over whether the SER scales linearly with the neutron flux, particularly at the highest flux levels.

■ *Proton beam.* The third approach is to use a proton beam, which is usually more readily available than a neutron beam. It is much easier to produce a mono-energetic and high-intensity proton beam. In contrast, neutrons typically arise as secondary particles following a collision of accelerated charged particles with a target. The use of a proton beam is justified because for energies greater than 50 MeV, a proton beam produces results similar to those produced by neutrons. High-energy protons interact with the nucleus of an atom in the same way as neutrons would (via strong interaction). Low-energy protons behave differently and have a much higher probability of interacting with electrons (via Coulomb interaction). Proton beams may, however, produce less accurate results in the future because the median neutron energy that affects silicon appears to be decreasing with technology scaling [30]. Another disadvantage of the proton beam is that the damage from radiation to the tested parts is typically higher than that due to neutrons. This is a concern if the supply of chips is limited (e.g., test chips or prototypes). The Northeast Proton Therapy Center in Boston, USA, has a 150-MeV proton beam.

Figure 2.15 shows a typical experimental setup with the neutron beam available in WNR. The neutron beam enters from the left, passes through a uranium fission detector chamber, and continues toward the silicon chips mounted on the circuit boards. The boards are carefully aligned with a laser beam to ensure that the chips under test—often referred to as DUT or device under test—are covered by the incident beam.

The exact neutron flux reaching the DUT is tracked by the WNR detectors. Since neutrons penetrate easily through circuit boards, often multiple boards will be set up back to back to get greater bandwidth out of these tests. The neutron flux reduction from each circuit board can be easily calibrated with the use of the WNR detectors. Karnik et al. [14] measured about a 4% reduction in neutron flux for

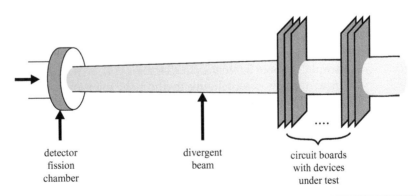

detector   divergent   circuit boards
fission   beam   with devices
chamber   under test

FIGURE 2.15   Experimental setup for a neutron beam experiment. Reprinted with permission from Karnik et al. [14]. Copyright © 2004 IEEE.

every circuit board in the path of the beam. Unlike the WNR beam, some beams are isotropic (i.e., dissipates in all directions). In such cases, specially designed collimators may be necessary to direct the beam to the appropriate DUT. The computer systems that are used to collect the data from the DUT are also not kept in the path of the beam and are often several tens of feet away from the DUT and connected with long cables.

The metallic portions of the DUT and circuit boards may become radioactive after exposure to a neutron or a proton beam. Hence, these DUT and circuit boards must be kept in isolation till these boards are no longer radioactive. This may take a few weeks to several months.

Typically, DUT can be of three types: memory (SRAM or DRAM) chips, specially manufactured test chips, and complex silicon chips, such as microprocessors. SRAM or DRAM chips are usually commercially available but may require expensive tester equipment to write test patterns and to read them after beam exposure to check how many of these failed.

Test chips are often used to measure the intrinsic SER of specific cells, such as latches, manufactured in a specific technology. For example, Hazucha et al. [8] created test circuits that have a number of latches connected as a shift register. These were mounted on specially designed circuit boards that can be controlled from a laptop. Test patterns are shifted in to these latches, which are exposed to the beam. Subsequently, the test patterns are read to determine which latches experienced bit flips. Yamamoto et al. [27] described another soft error test chip that arranges latches like SRAM cells.

Complex silicon chips, such as microprocessors, can also be exposed to such a beam, as long as the incident beam covers the portions of the microprocessor of interest. Two types of measurements are typically performed with such a beam. Intrinsic SER of large arrays, such as caches, can be measured by loading the cache with a test pattern and computing how many of the test bits flip under the beam. Microprocessors often have debug ports that allow one to load test patterns into such large arrays.

Alternatively, the MTTF of microprocessors could be measured using a neutron beam, as long as the beam covers the entire microprocessor. The WNR beam, for example, is about 3.5 inches in diameter [14], which is sufficiently large to cover a typical microprocessor. In this method, the microprocessor under test is exposed to a neutron beam until a system crash or a data corruption is observed. Such experiments can be expensive because microprocessors may take seconds to minutes before a failure may be observed, and this procedure must be repeated hundreds of times to achieve a statistical significance. Constantinescu [5] described one such experiment with an Itanium® processor running Microsoft Windows NT 4.0 and a Linpack benchmark. The aim of the modeling techniques described in this book is to predict this measured MTTF of a full-blown silicon chip, such as a microprocessor.

Modeling and measurement of device and circuit-level soft errors allow one to identify candidate hardware structures that may need protection. The next section discusses such protection mechanisms available at the device and circuit levels.

2.4 Mitigation Techniques

Designers try to accurately model and measure the SER of a chip, so that appropriate design techniques can be used to mitigate it. The rest of the book talks about how to model mitigate SER at the logic and architectural levels. This section provides several examples of soft error mitigation techniques used in CMOS devices and circuits.

2.4.1 Device Enhancements

Two of the most effective device-enhancement schemes to reduce the SER are the triple-well and silicon-on-insulator (SOI) technologies. Ziegler and Puchner [30] described a number of other techniques, such as buried-layer implants and the use of epitaxial substrate wafers, to reduce the SER of silicon devices. The impact of the latter techniques can vary depending on the specific technology and device characteristics.

Triple-well technology is nowadays commonly used in deep submicron CMOS technology to improve transistor performance. As Figure 2.16 shows, a triple-well device provides a complete electrical isolation for nMOS devices in a p-type substrate. This reduces substrate noise currents, thereby improving the performance of the nMOS device. Triple-well processes help reduce the SER because the deep n-well sweeps away some of the electrons generated from an alpha particle or a neutron strike. This helps reduce the number of electrons collected by the drain of the nMOS device, thereby reducing the probability of an upset.

An SOI process also improves transistor performance in the deep submicron technology. The SOI process introduces a buried oxide between the source (and drain) and the substrate (Figure 2.17). This eliminates the junction capacitance between the source (or drain) and substrate, thereby speeding up the device. At the same time, SOI significantly reduces the SV, which reduces the amount of

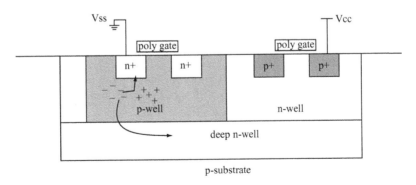

FIGURE 2.16 Triple well. Many of the electrons generated from an alpha particle or a neutron strike can be swept away by the deep n-well.

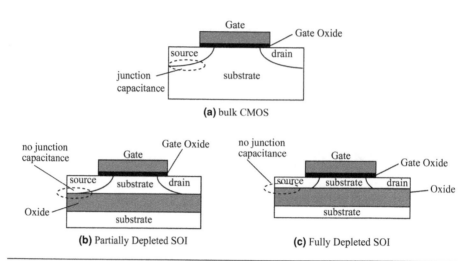

FIGURE 2.17   (a) Junction capacitance slows down the switching speed of a bulk CMOS transistor. The absence of junction capacitance speeds up both a partially depleted SOI transistor (b) and fully depleted SOI transistor (c).

charge collected from a particle strike, thereby reducing the vulnerability to soft errors.

There are two types of SOI processes—partially depleted and fully depleted. Partially depleted SOI is already being used by IBM in its manufacturing processes. IBM's experiments [4] have shown that their partially depleted SRAM cells can have 5 times lower SER than IBM's comparable bulk CMOS SRAM cells. But there are no current data on the effectiveness of SOI in reducing SER in latches and logic. The reduction in SER from partially depleted SOI (compared to bulk) could be offset by a phenomenon called bipolar junction effect, in which an additional current could be activated from the drain to the source due to the presence of the floating body.

As Figure 2.17c shows, fully depleted SOI has the lowest sensitive region compared to bulk or partially depleted SOI and consequently could achieve the greatest reduction in SER. It also does not have the bipolar junction effect because the substrate above the buried oxide is fully depleted and therefore excess charge generated in the substrate would be swept away to the drain. Nevertheless, manufacturing fully depleted SOI devices in large volumes is still a significant challenge faced by the industry.

2.4.2 Circuit Enhancements

Designers commonly use two tricks to reduce the soft error vulnerability of circuit elements: increasing the capacitance of a device, thereby increasing its Qcrit, and using radiation-hardened cells. This section discusses these two techniques.

Increasing Capacitance of Circuit Elements

The most common and perhaps obvious way to reduce the SER of a circuit is to increase the capacitance of its diffusion areas. Increasing the capacitance raises the Qcrit because the total charge at a node is the product of its capacitance and voltage. The capacitance can be raised usually by increasing the size of the device. Alternatively, an explicit capacitor can be added to the diffusion area of the device. Figure 2.18a shows the structure of a capacitor that can be attached to various nodes

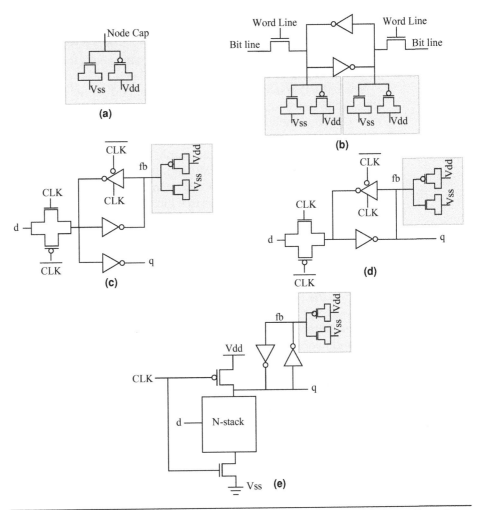

FIGURE 2.18 Reduction of SER by introducing capacitance. (a) Capacitor structure. (b) Addition of capacitance to an SRAM cell. (c) Addition of capacitance to the feedback node of a latch. (d) Addition of capacitance to the feedback node and to the output of a latch. (e) Addition of capacitance to the keeper cell of the dynamic node.

in a circuit to raise the sensitive node's capacitance. For example, Figure 2.18b shows how two capacitors can be added to an SRAM cell to reduce its SER.

Figure 2.18c–e shows how capacitors can be added to three latches commonly used in a high-performance microprocessor design. The input and output of these latches are typically denoted by d and q, respectively. The delay from d to q is referred to as the d-to-q time. The first latch in Figure 2.18c—called *path-exclusive latch*—has a decoupled feedback node (*fb*) and output node (q), whereas the second latch in Figure 2.18d has the feedback node and output node connected. Adding the capacitor to the path-exclusive latch does not affect the delay through the latch (d-to-q) but slows the loop recovery time (d-to-*fb*). Adding the capacitor to the latch in Figure 2.18d slows down both the d-to-q and d-to-*fb* times. Figure 2.18e shows how to add a capacitor to a dynamic cell. A dynamic cell stores its value in a "keeper" circuit, which is a standard cross-coupled latch.

The reduction in the SER and the corresponding reduction in performance and increase in power are directly related to the capacitance added to a node. Karnik et al. [15] have shown that such techniques can reduce the SER of latches three-fold but could slow down the latch. The performance penalty can be hidden by only adding capacitors to latches that are not in a critical path. Using these techniques, the authors estimated an overall chip-level power increase of less than 3%, with an overall SER reduction of 25–32%. Mohanram and Touba [18] and Gill et al. [7] proposed an alternate technique in which they identified the most vulnerable gates in a logic circuit and then added appropriate protection—including capacitors—to reduce the soft error vulnerability. The vulnerability of gates can vary widely because of the logical, electrical, and latch-window masking effects discussed earlier.

Recently, DRAM vendors have been investigating the use of stacked capacitors to reduce the SER of DRAMs. Although this works for DRAMs, it does not work as effectively for SRAMs embedded in microprocessors due to power and area reasons. Often, the area above the SRAMs is also unavailable because of metal lines and hence it may be hard to find a place to put these capacitors on the SRAM cells.

Radiation-Hardened Cells

Radiation hardening is a second circuit technique for reducing the circuit-level SER. Unlike the previous method of increasing the diffusion capacitance that applies to almost any transistor, radiation hardening only applies to storage cells, such as SRAM cells or latches. This section discusses two radiation-hardened cell designs. The underlying principle of the radiation-hardened design is quite simple. It maintains a redundant copy of data, which can not only provide the correct data after a particle strike but also help the corrupted section recover from the upset.[6]

---

[6]Although the term "radiation hardening" could have a broader meaning, radiation-hardened cells typically have this definition in the industry.

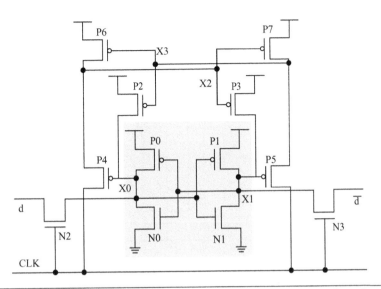

FIGURE 2.19 Radiation-hardened latch design using a pMOS slave latch. The shaded region shows the original latch (without the clock).

Figure 2.19 shows a radiation-hardened latch proposed by Rockett [22]. The main latch consists of the cross-coupled inverters composed of {N0,P0} and {N1,P1}. Two additional transistors—N2 and N3—control the writability of the latch. The redundant cell is a pMOS latch composed of two cross-coupled inverters {P2,P4,P6} and {P3,P5,P7}. The pMOS latch acts as a redundant storage only when the clock (CLK) is low. When the clock is high, data can be read from or written into the main latch.

In this circuit, the nodes X0 through X3 are most sensitive to particle strikes. Because X0 and X1 behave in the same way as do X2 and X3, the different cases are only illustrated with X0 and X2. Four possibilities can arise from a particle strike: (i) X0 is 0 and can transition to 1, (ii) X0 is 1 and can transition to 0, (iii) X2 is 0 and can transition to 1, and (iv) X2 is 1 and can transition to 0. Case (i) cannot occur because a positive upset pulse, which forces X0 to transition from 0 to 1, also shuts off the pMOS transistor P4, while the pMOS transistor P2 restores X0's state. Case (ii) can occur due to a negative upset pulse on the drain of N0 or N2. To prevent this scenario, the cell is designed to make P6 significantly stronger than P4, which avoids reverting the state of the slave latch. Case (iii) is similar to case (i) and is avoided because the corresponding pMOS transistors, P3 and P7, are shut off. Case (iv) cannot generate upsets in pMOS transistors and can only generate positive upset pulses (that is, transitions from 0 to 1 only).

Calin et al. [3] proposed an alternate and perhaps more widely known design called DICE (dual-interlocked CEll). Unlike other radiation-hardened designs, the DICE cell does not rely on device sizing or added capacitance.

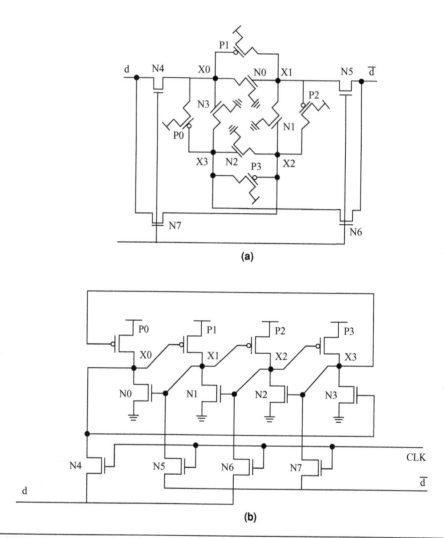

FIGURE 2.20   **DICE (dual-interlocked CEII). (a) Schematic diagram that illustrates the principles of operation of the DICE storage cell. (b) Same as (a), but drawn as the canonical DICE memory cell.**

Figure 2.20 shows a DICE memory cell. It uses twice the number of transistors compared to a traditional SRAM cell. Four nodes—X0, X1, X2, and X3—store the data as two pairs of complementary values: 1010 or 0101. Each X is connected to a pair of N and P transistors, and it also controls the operation of another pair of N and P transistors. For example, X0 is connected to N0 and P0, but it also controls the operation of P1 and N3. These are simultaneously written to and read through the transmission gates. Figure 2.21 shows a radiation-hardened DICE latch. The connections

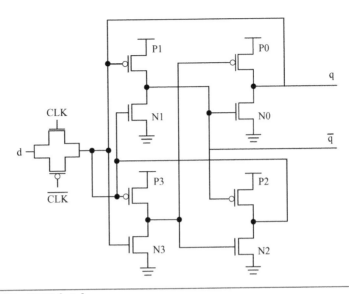

FIGURE 2.21 DICE latch.

and principles of operation of this latch are very similar to that of the DICE memory cell.

In Figure 2.20, the four nodes—X0 through X3—are most sensitive to particle strikes. The soft error immunity of this cell can be illustrated using X0. Two possibilities can arise in the DICE memory cell after a particle strike on X0: (i) X0 is 0 and can transition to 1 and (ii) X0 is 1 and can transition to 0. Case (i) forces X3 to transition to 0. But this shuts off P1 and N2. This isolates X1 and X2, which capacitatively preserve their states. Once the transient pulse generated by the particle strike disappears, X1 and X2 restore the state of X0 and X3 via the transistors N0 and P3. Case (ii) works in the same way. The transition of X0 from 1 to 0 affects X1, but shuts off N3. Corresponding X1 transition from 1 to 0 shuts off P2. X2 and X3 capacitatively preserve the value of the latch and restore the states of X0 and X1 via the transistors N1 and P0.

Estimates vary, but the DICE cell can reduce the SER by 10- to 100-fold [9]. The DICE cell does not reduce the SER of the cell to zero because of two reasons. First, if a single particle strikes two adjacent nodes simultaneously, such as X0 and X1, then the cell cannot recover. Second, although a DICE latch itself can recover from a single strike, a glitch may appear at the output q, which in turn may be captured by subsequent latches on its path.

A DICE cell can be 1.7 to 2 times greater in area than the original cell [3], but Hazucha et al. [9] have shown that with careful design, this overhead can be reduced to less than 1.5-fold. Further, if scan logic is factored in, then the area overhead of a DICE latch could be even lower because the scan logic is not typically

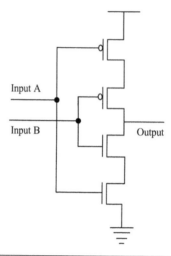

FIGURE 2.22   C-element used for fault detection.

radiation hardened. Interestingly, however, a DICE latch does not affect the d-to-q time, thereby making it attractive for microprocessor pipeline latches that are in critical paths where extra levels of logic introduced by a conventional parity or an ECC could increase the cycle time. Hazucha et al. have also demonstrated that the energy penalty can be reduced to less than 40% by using devices with two different threshold voltages.

Recently, Mitra et al. [17] have proposed another radiation-hardening scheme that detects faults using at-speed scan logic. The basic idea is illustrated in Figure 2.22, which shows a circuit called C-Element. The C-Element acts as an inverter when both inputs A and B are the same. But when the inputs are different, the C-Element does not let either input propagate. Thus, one could detect a transient fault by connecting the redundant storage elements to the inputs A and B. In the presence of scan, the second redundant value could be generated by the scan latch itself. To ensure that the scan latch does not slow down the main latch, it must run at the same speed as the main latch. Hence, this will work as long as the scan latch runs "at speed." The high-end silicon industry, such as the those manufacturing microprocessors, is moving away from at-speed scan because at-speed scan logic ends up consuming too much power and area.

2.5   Summary

To compute the SER of a circuit, one must ascertain the critical charge (Qcrit) of the circuit and the circuit-specific vulnerability. The Qcrit must be mapped to the corresponding SER (typically expressed in FIT) and multiplied by the circuit's vulnerability factor to obtain the circuit-specific SER. Qcrit is the minimum charge that must be deposited by an alpha particle or a neutron strike to cause the circuit to malfunction. Qcrit is usually computed using integrated circuit simulators. Once Qcrit is computed for a specific circuit element, it must be mapped to an SER. There are various models and methods, such as the Hazucha and Svensson model, the BGR method, the NCS method, and Murley and Srinivasan model, which allow one to map Qcrit to its corresponding FIT rate.

Different circuit elements have different levels of vulnerability to soft errors. Latches and flip-flops are only vulnerable to a particle strike when they hold data (as opposed to driving the data). This is commonly expressed as a TVF. Similarly, many strikes in static logic gates are masked before they can reach a forward latch where the error can get latched. The inherent nature of the pull-down network in a dynamic logic circuit can mask many transient faults. Finally, clock circuits are also vulnerable to transient faults, which can either move a clock edge or create a new clock edge, eventually leading to a soft error. These effects must be taken into account to compute the SER for a circuit element.

Soft error models for circuit elements must be validated using measurements. Measurement of soft errors is a challenge because a typical soft error in an individual bit may take hundreds of thousands of years to manifest. Hence, to measure the SER of a device or a circuit in a reasonable amount of time, the error rate is usually measured over a large number of such bits or circuit elements. Alternatively, accelerated tests with a higher alpha or neutron flux can be performed to obtain errors faster. For field data collection, the luxury of increasing the flux does not exist, so the measurements are done typically over a large number of bits or parts. Even for accelerated tests, which use thorium foils for alpha particles or particle accelerators for neutron measurements, the error rates are measured over several bits or parts to ensure the results are statistically significant.

Modeling and measurement expose vulnerable circuits in a design. These vulnerable circuits could be protected by architectural techniques as well as by device and circuit techniques. Architectural techniques to protect against soft errors are described in later chapters. Device-level protection techniques include triple-well and SOI technologies. Triple-well transistors reduce the SER because their deep n-well sweeps away some of the electrons generated from an alpha particle or a neutron strike. SOI reduces the SER by reducing the SV (in the substrate) that is vulnerable to particle strikes.

At the circuit level, there are two techniques to reduce the SER. The first one is to add capacitance to vulnerable nodes in a circuit, thereby increasing its Qcrit. The second method is to create a radiation-hardened cell. A radiation-hardened cell usually uses extra transistors that restore the state of the original circuit in the case of a particle strike.

2.6   Historical Anecdote

Linear particle accelerators, such as the one available in the LANSCE, are used routinely to test the vulnerability of silicon chips to neutron-induced soft errors. One purpose of these accelerators is to accelerate a particle with sufficient energy toward a nucleus, so that it can overcome the electrical repulsion of the nucleus and thereby "crack" it open. For example, in LANSCE when an accelerated proton beam hits a tungsten nucleon, the nucleon breaks open giving rise to neutrons of various energies.

Work on linear particle accelerators started in the early 20th century. In the 1930s in England, John Cockcroft and Ernest Walton started experimenting with accelerating protons down an 8-feet tube by building up a potential of 800 kV. This was sufficient to disintegrate a lithium nucleus into two alpha particles. Greater energy levels were, however, required to shatter other types of nucleons. Around the same time, Robert Van de Graff developed the famous Van de Graff generator that could generate voltage difference as high as 1.5 MV. Maintaining such high voltages proved to be difficult, so researchers started exploring the idea of accelerating particles by using a lower voltage more than once. In Germany, Rolf Wideroe invented exactly such a scheme. By extending Wideroe's scheme, in the United States, Ernest Lawrence and David Sloan managed to accelerate mercury ions to energies of a million volts.

The linear accelerators developed thus far are impractical for particles lighter than mercury ions, such as alpha particles. This is because several meters long vacuum tubes were required for lighter particles. Instead, Ernest Lawrence proposed the use of a circular path in which particles—guided by magnetic and electric fields—would accelerate as they spiral outward from the center to the edge of the circle. Lawrence overcame several practical obstacles to finally create the device called the cyclotron that could create million-volt projectiles out of light particles. Since then, linear particle accelerators and cyclotrons have evolved in sophistication. Today, they can be several miles in length and can accelerate lighter particles to energies of several hundred million volts.

References

[1] M. P. Baze and S. P. Buchner, "Attenuation of Single Event Induced Pulses in CMOS Combinational Logic," *IEEE Transactions on Nuclear Science*, Vol. 44, No. 6, pp. 2217–2223, December 1997.

[2] M. J. Bellido-Diaz, J. Juan-Chico, A. J. Acosta, M. Valencia, and J. L. Heurtas, "Logical Modeling of Delay Degradation Effect in Static CMOS Gates," *IEE Proceedings Circuits, Devices, and Systems*, Vol. 147, No. 2, pp. 107–117, April 2000.

[3] T. Calin, M. Nicolaidis, and R. Velazco, "Upset Hardened Memory Design for Submicron CMOS Technology," *IEEE Transactions on Nuclear Science*, Vol. 43, No. 6, pp. 2874–2878, December 1996.

[4] E. H. Cannon, D. D. Reinhardt, M. S. Gordon, and P. S. Makowenskyj, "SRAM SER in 90, 130 and 180 nm Bulk and SOI Technologies," in *Reliability Physics Symposium Proceedings, 2004. 42nd Annual. 2004 IEEE International*, pp. 300–304, 25–29 April 2004.

[5] C. Constantinescu, "Neutron SER Characterization of Microprocessors," in *International Conference on Dependable Systems and Networks (DSN)*, pp. 754–759, July 2005.

[6] L. B. Freeman, "Critical Charge Calculations for a Bipolar SRMA Array," *IBM Journal of Research and Development*, Vol. 40, No. 1, pp. 119–129, January 1996.

[7] B. S. Gill, C. Papachristou, F. G. Wolff, and N. Seifert, "Node Sensitivity Analysis for Soft Errors in CMOS Logic," in *International Test Conference*, paper 37.2, pp. 1–9, November 2005.

[8] P. Hazucha, T. Karnik, J. Maiz, S. Walstra, B. Bloechel, J. Tschanz, G. Dermer, S. Hareland, P. Armstrong, and S. Borkar, "Neutron Soft Error Rate Measurements in a 90-nm CMOS Process and Scaling Trends in SRAM from 0.25-μm to 90-nm Generation," in *IEDM '03 Technical Digest, IEEE International*, pp. 21.5.1–21.5.4, 8–10 December, 2003.

[9] P. Hazucha, T. Karnik, S. Walstra, B. A. Bloechel, J. W. Tschanz, J. Maiz, K. Soumyanath, G. E. Dermer, S. Narenda, V. De, and S. Borkar, "Measurements and Analysis of SER-Tolerant Latch in a 90-nm Dual-V_T CMOS Process," *IEEE Journal of Solid-State Circuits*, Vol. 39, No. 9, pp. 617–620, September 2004.

[10] P. Hazucha and C. Svensson, "Impact of CMOS Technology Scaling on the Atmospheric Neutron Soft Error Rate," *IEEE Transactions on Nuclear Science*, Vol. 47, No. 6, pp. 2586–2594, December 2000.

[11] P. Hazucha, C. Svensson, and S. A. Wender, "Cosmic-Ray Soft Error Rate Characterization of a Standard 0.6-μm CMOS Process," *IEEE Journal of Solid-State Circuits*, Vol. 35, No. 10, pp. 1422–1429, October 2000.

[12] M. A. Horowitz, *Timing Models for MOS Circuits*, Technical Report SEL83-003, Integrated Circuits Laboratory, Stanford University, 1983.

[13] JEDEC standard JESD89, *Measurement and Reporting of Alpha Particles and Terrestrial Cosmic-Ray-Induced Soft Errors in Semiconductor Devices*, August 2001.

[14] T. Karnik, P. Hazucha, and J. Patel, "Characterization of Soft Errors Caused by Single Event Upsets in CMOS Processes," *IEEE Transactions on Dependable and Secure Computing*, Vol. 1, No. 2, pp. 128–143, April–June 2004.

[15] T. Karnik, S. Vangal, V. Veeramachaneni, P. Hazucha, V. Errguntla, and S. Borkar, "Selective Node Engineering for Chip-Level Soft Error Rate Improvement," in *2002 Symposium on VLSI Circuits Digest of Technical Papers*, pp. 204–205, June 2002.

[16] P. Liden, P. Dahlgren, R. Johansson, and J. Karlsson, "On Latching Probability of Particle Induced Transient in Combinatorial Networks," in *24th Symposium on Fault-Tolerant Computing (FTCS)*, pp. 340–349, June 1994.

[17] S. Mitra, N. Seifert, M. Zhang, Q. Shi, and K. S. Kim, "Robust System Design with Built-In Soft-Error Resilience," Vol. 38, No. 2, pp. 43–52, *IEEE Computer*, February 2005.

[18] K. Mohanram and N. A. Touba, "Cost-Effective Approach for Reducing Soft Error Failure Rate in Logic Circuits," in *International Test Conference*, Sep. 30–Oct. 2, 2003.

[19] P. C. Murley and G. R. Srinivasan, "Soft-Error Monte Carlo Modeling Program, SEMM," *IBM Journal of Research and Development*, Vol. 40, No. 1, pp. 109–118, January 1996.

[20] E. Normand, "Single Event Upset at Ground Level," *IEEE Transactions on Nuclear Science*, Vol. 43, No. 6, pp. 2742–2750, December 1996.

[21] J. M. Rabaey, A. Chandrakasan, and B. Nikolic, *Digital Integrated Circuits*, Prentice Hall, 2003.

[22] L. Rockett, "An SEU Hardened CMOS Data Latch Design," *IEEE Transactions on Nuclear Science*, Vol. NS-35, No. 6, pp. 1682–1687, December 1988.

[23] N. Seifert, P. Shipley, M. D. Pant, V. Ambrose, and B. Gill, "Radiation-Induced Clock Jitter and Race," in *International Reliability Physics Symposium*, pp. 215–222, April 2005.

[24] N. Seifert and N. Tam, "Timing Vulnerability Factors of Sequentials," *IEEE Transactions on Device and Materials Reliability*, Vol. 3, No. 4, pp. 516–522, September 2004.

[25] P. Shivakumar, M. Kistler, S. W. Keckler, D. Burger, and L. Alvisi, "Modeling the Effect of Technology Trends on the Soft Error Rate of Combinatorial Logic," in *International Conference on Dependable Systems and Networks*, pp. 389–398, June 2002.

[26] A. Taber and E. Normand, "Single Event Upset in Avionics," *IEEE Transactions on Nuclear Science*, Vol. 40, No. 2, pp. 120–126, April 1993.

[27] S. Yamamoto, K. Kokuryou, Y. Okada, J. Komori, E. Murakami, K. Kubota, N. Matsuoka, and Y. Nagai, "Neutron-Induced Soft Error in Logic Devices Using Quasi-Monenergetic Neutron Beam," in *42nd Annual International Reliability Physics Symposium*, Phoenix, pp. 305–309, April 2004.

[28] M. Zhang and N. R. Shanbhag, "A Soft Error Rate Analysis (SERA) Methodology," in *International Conference on Computer Aided Design*, pp. 111–118, November 2004.

[29] J. F. Ziegler and W. A. Lanford, "Effect of Cosmic Rays on Computer Memories," *Science*, Vol. 206, No. 4420, pp. 776–788, November 1979.

[30] J. F. Zielger and H. Puchner, *SER—History, Trends and Challenges*, Cypress Semiconductor Corporation, 2004.

Architectural
Vulnerability Analysis

3.1 Overview

Chapter 2 examined device- and circuit-level techniques to model soft errors. This chapter discusses the concept of Architectural Vulnerability Factor (AVF) to model soft errors at the architectural level. AVF is the fraction of faults that show up as user-visible errors. The circuit- and device-level SERs need to be derated by the AVF to obtain the observed SERs of silicon chips. The higher the AVF of a bit, the higher is its vulnerability to cause soft errors. Hence, architects use AVFs to identify candidates to add error protection schemes.

The basic question that underlies the AVF computation is whether a particle strike on a bit matters. If a bit flip does not produce an incorrect output, then the bit flip does not matter and fault is masked. The fraction of bit flips that affect the final outcome of a program is captured by the bit's AVF. AVF across bits in a silicon chip can be as low as a few percent, which will reduce the intrinsic device- and circuit-level SER by 100-fold. AVF of different structures can vary widely, which makes AVF an important metric to identify structures that need protection.

This chapter discusses the basics of AVF analysis, its relationship to SDC and DUE of a chip, Architecturally Correct Execution (ACE) principles, and how ACE principles can be used to compute per-structure AVFs using Little's law as well as performance simulation models. Chapter 4 describes advanced techniques to compute AVF of structures.

3.2 AVF Basics

As discussed in the previous two chapters, an alpha particle or a neutron strike can induce a malfunction in a transistor's operation. This malfunction can manifest itself in a variety of ways, such as change in the output of a gate or a bit flip in a latch or a memory cell. Not all these errors, however, manifest themselves as user-visible errors. Consider the following instruction sequence:

1 R3 = R2 OR R1

2 R1 = R2 + R2.

(1) ORs the values of registers R1 and R2 and puts the result in register R3. If R2 = 1 and R1 = 0, then R3 = 1. Even if a particle strike on register R1's least significant bit changes the value of R1 to 1, R3 will still be 1. (2) immediately overwrites R1 and hence an error in R1 will not manifest itself to the user.

There are a number of instances during the execution of a program where a bit flip in a latch, a memory cell, or a logic gate gets masked and does not result in a user-visible error. Chapter 2 discussed two examples of such masking: TVF and logical masking. TVF captures the fraction of time a clocked storage element, such as a latch, is vulnerable to an upset. Logical masking arises when a logic gate logically masks a bit flip (e.g., if one input of an OR gate is one, then a strike on the other input is logically masked).

In general, the intrinsic FIT rate of a transistor or a group of transistors needs to be derated by a number of vulnerability factors that specify the probability that an internal fault in a device's operation will result in an externally visible error. This chapter describes the concept of the AVF, which is central to the design of architectural solutions to soft errors. AVF has also been called logic derating factor, architectural derating factor, and soft error sensitivity.

AVF expresses the probability that a user-visible error will occur given a bit flip in a storage cell [8]. For example, a bit flip in a branch predictor will never show up as a user-visible error. Hence, the branch predictor's AVF = 0%. In contrast, a bit flip in a committed program counter will almost invariably crash a program. Hence, a program counter's AVF is close to 100%. Computing the AVF of other microarchitectural structures, such as an instruction queue or processor cache, is more involved because the value stored in a bit in such a structure may sometimes be required to be correct for ACE and at other times the value may not matter for correct execution. The AVF of such structures can vary anywhere from 0% to 100%. Consequently, AVFs can significantly change the overall FIT rate of a chip. This chapter discusses how to compute the AVF of such complex structures.

Implicit in the definition of AVF is the concept of "scope" as explained in Chapter 1 (Figure 1.3). A bit flip can get masked in an inner microarchitectural scope (e.g., bit flip in branch predictor bits) or in an inner architectural scope (e.g., bit flip in architectural register that is overwritten before being read) or may not matter to the user, even if it is visible (e.g., bit flip causing a pixel on the screen to change color for a fraction of a second). Discussions on AVF in this book exclude the last case because that is subjective to the user and hard to quantify. One could conceptually define a new vulnerability factor—perhaps called the perceptual vulnerability factor—that takes into account whether such bit flips will matter to the user or not. As may be obvious by now, the term "user visible" here refers to the scope where the error is observed and corresponding AVF is computed.

3.3 Does a Bit Matter?

As may be apparent by now, the SER and AVF are properties of a bit in a chip and not of a program itself. That is, we cannot talk about a program's AVF, but we can talk about a bit's SER or AVF. To compute the AVF of a bit, one can pose the following question: Does the bit matter? More specifically, does the value of the bit in a specific cycle affect the final output of a program? If the bit in a particular cycle has no effect on the correct execution of a program, then the bit in the specified cycle can be flipped by a particle strike without disrupting the program. These concepts are formalized in the next section.

Figure 3.1 shows a flowchart from when a bit gets struck to when one needs to decide whether a bit matters or not. If the faulty bit is not read, then it cannot cause

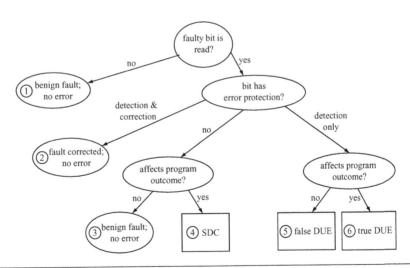

FIGURE 3.1 Classification of the possible outcomes of a faulty bit in a microprocessor (same as Figure 1.16 on p. 33). Reprinted with permission from Weaver et al. [15]. Copyright © 2004 IEEE.

an error and therefore is a benign fault (case 1 in Figure 3.1). However, if the faulty bit is read, then one needs to ask whether the bit has error protection. If the bit has error detection and correction (e.g., like ECC), then the fault is corrected, causing no user-visible error (case 2). If the bit has no error protection—that is, neither error detection nor correction—then one needs to ask whether the bit flip affected the program outcome. If the answer is no, then the bit does not matter and that leads to case 3. However, if the bit flip does affect the program outcome, then it causes what is known as an SDC event (case 4). Now, if the bit only has error detection (e.g., parity bit without the ability to recover from an error), then it prevents data corruption but can still cause the program to crash. Then, irrespective of whether the bit matters or not, the program usually will be halted and crashed as soon as the error is detected. Such error detection events, typically visible to the user, are called DUE. If an error is declared to be a DUE, then it means that it can under no circumstances cause an SDC. Thus, the definition of DUE has the implicit notion of a fail-stop system.

Figure 3.1 shows that DUE events can be further broken down into false and true DUE events. False DUE events (case 5) are those DUE that could have been avoided if there was no error detection mechanism to begin with. For example, certain bits of a wrong-path instruction may not cause an error. In the absence of an error detection mechanism, a flip in such a bit would have gone unnoticed and would not have created any user-visible error. However, because the error detection mechanism detects the error and possibly reports it, the program or the system may be unnecessarily brought down. A bit flip that matters and is detected by the system is a true DUE event (case 6). As may be obvious by now, protecting a bit with an error detection mechanism moves category 3 to 5 and 4 to 6. This is explored in greater detail in the next section.

The discussion in the rest of this chapter assumes a single-bit fault model. The concepts, however, apply directly to spatial multibit faults. Multibit faults manifest themselves as either spatial faults or temporal faults. Spatial faults are those in which a single alpha or neutron strike upsets multiple contiguous bits. Temporal multibit faults are a function of the error detection code and occur when the number of independent strikes to the protected bits overwhelms the coding scheme. For spatial multibit faults, one can treat each bit independently and hence compute its vulnerability. Analyzing the vulnerability for temporal multibit faults is a little more involved. Both the definition and the analysis of temporal multibit faults are discussed in Chapter 5 (see Scrubbing Analysis, p. 190).

3.4   SDC and DUE Equations

This section examines how to compute per-bit, per-structure, and chip-level SDC and DUE FITs and their relationships to the corresponding AVFs.

3.4.1 Bit-Level SDC and DUE FIT Equations

Mathematically, one can express and SDC FIT rate of a storage cell as follows:

$$\text{SDC FIT} = \text{SDC AVF} \times \text{TVF} \times \text{intrinsic FIT}$$

The intrinsic FIT rate of a cell is its device-level SER and includes any extra derating, such as the ones that may be necessary for a dynamic cell (see Masking Effects in Dynamic Logic, p. 57). The SDC AVF is often referred to simply as the AVF.

Similarly, the DUE FIT of a bit can be expressed as

$$\text{DUE FIT} = \text{DUE AVF} \times \text{TVF} \times \text{intrinsic FIT}$$

Thus, as stated in the last section, the device-level error rate (TVF × intrinsic FIT) must be multiplied or derated by the AVF to get the effective FIT rate of the bit.

The above equations explain why AVF forms the foundation of architectural solutions to soft errors. If one protects a bit with an error detection scheme, such as parity, then one can mark its SDC AVF = 0. Consequently, the SDC FIT of the bit is zero. Similarly, if one protects the bit with an error correction scheme, such as ECC, then its DUE AVF = 0, which makes its DUE FIT $\sim = 0$. There are yet other schemes that will reduce the AVF somewhat but not to zero. Also, as will be seen soon, the DUE AVF is a function of the error detection scheme, whereas the SDC AVF is not.

The above equations only express the SDC and DUE FIT of a storage cell, such as a latch or a memory cell, but not that of logic gates. As explained in Chapter 2 (see Masking Effects in Combinatorial Logic Gates, p. 52), most faults in logic gates get masked. One way to capture the effect of faults in gates is to increase the intrinsic FIT rate of the forward latch by the FIT contribution of the logic gates feeding the latch. Thus, the intrinsic FIT may have two contributions—one from faults due to direct particle strikes and the other from logic gate faults that propagated to the forward latch. In both cases, the AVF remains the same.

AVF plays an important role in deciding whether an error protection scheme is necessary. A conservative estimate of AVF (e.g., 100%) will unnecessarily overprotect the chip and devote precious silicon resources that could otherwise be used to improve the performance or add other features. In contrast, not adding sufficient protection will leave the chip with a level of unreliability that the market would not like. Hence, it is critical to accurately compute the AVF of various bits in the processor and chipsets.

■ E X A M P L E

What is the SDC FIT of a bit with an AVF of 15% and FIT/bit of 0.001?

S O L U T I O N The SDC FIT = $0.15 \times 0.001 = 150$ microFIT.

3.4.2 Chip-Level SDC and DUE FIT Equations

To compute the chip-level FIT, one can sum the SDC and DUE FIT rates of all devices in the chip. That is,

$$\text{SDC FIT of chip} = \sum_{i \text{ over all storage cells}} \text{SDC FIT of cell}_i$$

$$\text{DUE FIT of chip} = \sum_{i \text{ over all storage cells}} \text{DUE FIT of cell}_i$$

These summations work if one assumes that the SERs follow the exponential failure law (see Dependability Models, p. 11), which is true in general for soft errors.

■ EXAMPLE

What is the SDC FIT of a chip with 10 million SRAM cells and 1 million latches? Assume that the FIT rate of logic elements is negligible, intrinsic FIT rates of an SRAM cell and a latch are both 1 milliFIT/bit. Assume that the AVF is 30% for SRAM cells and 20% for latches. Also, assume that the TVF of SRAM cells and latches is 100% and 50%, respectively.

SOLUTION The total SDC FIT $= 10\,000\,000 \times 0.3 \times 1 \times 0.001 + 1\,000\,000 \times 0.2 \times 0.5 \times 0.001 = 3100$ FIT. This is equivalent to an SDC event every 37 years.

■ EXAMPLE

A system manufacturer deemed that 37 years of SDC MTTF is too high for a 1000-system cluster being built. Such a 1000-system cluster would have an aggregate SDC MTTF of roughly 2 weeks. The chip manufacturer agreed to protect 90% of the SRAM cells and latches with parity bits. What will be the resulting SDC and DUE FIT rates? Assume that DUE AVFs of parity-protected SRAM cells and latches are 30% and 20%, respectively. Also, assume that the chip manufacturer adds one bit of parity for every 10 storage cells.

SOLUTION The per-system SDC FIT $= (1 - 0.9) \times 3100$ FIT $= 310$ FIT (or about 368 years). The aggregate SDC FIT rate of the 1000-processor cluster would be 4.4 months. The per-system DUE FIT $= 0.9 \times 1.1 \times (10\,000\,000 \times 0.3 \times 1 \times 0.001 + 1\,000\,000 \times 0.2 \times 0.5 \times 0.001) = 3069$ FIT (or about 37 years). The 1.1 multiplier arises from the extra bits of parity one needs to add. The aggregate 1000-processor system would now have a DUE MTTF of roughly 2 weeks. This may be acceptable to the system manufacturer because data integrity was more critical than system uptime.

To compute the chip-level SDC and DUE FIT rates, intrinsic FIT/bit rates and per-bit AVF numbers are required. Computing these numbers on a per-bit basis can, however, be cumbersome given that there are hundreds of millions of cells in a modern microprocessor and chipset. Fortunately, for most structures, such as a cache or an instruction queue, the FIT/bit of the basic storage cell used to create the structure is roughly the same. The per-bit AVF and TVF may vary widely, but one can create average AVFs and TVFs across all bits in a structure. The FIT/structure is computed as the sum of FIT/bit over all its bits. Hence, the per-structure SDC FIT can be approximated as average $AVF_{structure} \times$ average $TVF_{structure} \times$ FIT/structure. Similarly, per-structure DUE FIT $\sim=$ average DUE $AVF_{structure} \times$ average $TVF_{structure} \times$ FIT/structure. Then, one can express the SDC and DUE FIT equations as:

$$\text{SDC FIT of chip} = \sum_{i \text{ over all structures on the chip}} \text{SDC FIT of structure}_i$$

$$\text{DUE FIT of chip} = \sum_{i \text{ over all structures on the chip}} \text{DUE FIT of structure}_i$$

The above notion of computing an average AVF for a structure can also be extended to the entire chip so that one could compute an average per-bit SDC and DUE AVF across the chip. This can easily give the chip-level SDC and DUE FIT rates. But the chip-level AVF may not be as useful in determining the structures that are more vulnerable to soft errors and therefore may need enhanced protection.

The AVF also varies by benchmarks, so one could enumerate the SDC and DUE FIT of a chip for each benchmark. Fortunately, however, AVF is a separable term because the circuit-level FIT (TVF and intrinsic FIT) is mostly independent of the benchmark and AVF. Hence, one can talk about average SDC FIT and average per-structure AVFs across benchmarks. For example, if a hypothetical chip consists of two structures, A and B, and one estimates the SDC FIT and AVFs for benchmarks b1 and b2, then one can express SDC FIT as

$$\text{SDC FIT}_{b1} = \text{SDC AVF}_{A,b1} \times \text{Circuit FIT}_A + \text{SDC AVF}_{B,b1} \times \text{Circuit FIT}_B$$

$$\text{SDC FIT}_{b2} = \text{SDC AVF}_{A,b2} \times \text{Circuit FIT}_A + \text{SDC AVF}_{B,b2} \times \text{Circuit FIT}_B$$

Hence,

$$\text{Avg SDC FIT} = \text{Avg SDC AVF}_A \times \text{Circuit FIT}_A + \text{Avg SDC AVF}_B \times \text{Circuit FIT}_B$$

where Circuit FIT = TVF $\times$ intrinsic FIT, Avg = Average, Avg SDC FIT = $(\text{SDC FIT}_{b1} + \text{SDC FIT}_{b2})/2$, Avg SDC $AVF_A = (\text{SDC AVF}_{A,b1} + \text{SDC AVF}_{A,b2})/2$, and Avg SDC $AVF_B = (\text{SDC AVF}_{B,b1} + \text{SDC AVF}_{B,b2})/2$.

3.4.3 False DUE AVF

As Figure 3.1 shows, false DUE events arise when the occurrence of an error is reported incorrectly (since the error would have been masked in the absence of an error detection mechanism). This can happen, for example, if one incorrectly flags an error in the non-opcode bits of a wrong-path instruction in a microprocessor. Then, mathematically, one can express the per-bit total DUE AVF as

$$\text{DUE AVF} = \text{true DUE AVF} + \text{false DUE AVF}$$

The true DUE AVF can be expressed as

$$\text{true DUE AVF for a bit (with error detection)} = \text{SDC AVF for the same bit}$$
$$\text{(prior to adding error detection)}$$

In other words, adding error detection converts the original SDC AVF to true DUE AVF.

Thus, adding an error detection mechanism has two effects. First, it introduces the false DUE AVF component. Second, it also adds extra bits, which can be vulnerable to strikes. For example, errors in the parity bits are false DUE events.

■ E X A M P L E

One vendor is interested in the total SER and expresses it as the sum of SDC and DUE FIT. The SDC FIT for the chip before adding parity is 3100 and 0 FIT, respectively. After adding parity to 90% of the storage cells, the SDC and DUE FIT are 310 and 3069 FIT, respectively. Compute the total SER FIT before and after adding the parity bits.

S O L U T I O N Before adding parity, the per-system SDC and DUE FIT were 3100 and 0 FIT, respectively. The total SER FIT was 3100 FIT. The introduction of parity resulted in the SDC and DUE FIT rates of 310 and 3069 FIT, respectively. Hence, the total per-system SER FIT with parity is 3379 FIT. Interestingly, adding parity to 90% of the storage cells decreased the SDC FIT by 90% but increased the overall SER by 9%. For applications that care more about data integrity, rather than system uptime, this is often a reasonable compromise.

The true DUE AVF is independent of the specific error detection mechanism. But the false DUE AVF is a function of the error detection mechanism introduced. The false DUE AVF can be negligible for some structures and detection schemes but can be as high as 10 times that of the true DUE AVF in yet other ones. The next section examines how cycle-by-cycle Lockstepping can significantly enhance the false DUE rate.

3.4.4 Case Study: False DUE from Lockstepped Checkers

Lockstepping is a well-established technique that can detect faults in microprocessors and in full systems. Figure 3.2 shows an example of two Lockstepped processors. Each processor's state is initialized in exactly the same way, and both processors run identical copies of the program. In each cycle, signals from Processor0 and Processor1 are compared at the output comparator to check for faults. When the output comparator detects a mismatch, it triggers appropriate actions, such as halting the processors and forcing a reboot or triggering hardware or software recovery actions. To ensure that both Processor0 and Processor1 execute identical paths of the program they are executing, the inputs to both processors must also be appropriately replicated. In this example, the memory and/or disk are not protected with Lockstepping and hence alternate mechanisms, such as error codes, may be necessary to protect them from errors.

Lockstepping by itself is purely a fault detection mechanism. It can reduce the SDC FIT to almost zero for components it is covering—Processor0 and Processor1, for example, in Figure 3.2. However, in the absence of any recovery mechanism,

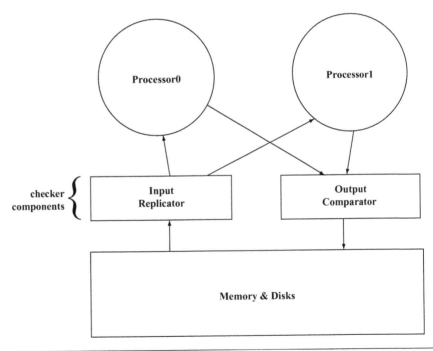

FIGURE 3.2 Lockstepped CPUs. The output comparator and input replicator are components of the Lockstep checker.

Lockstepping can significantly increase the false DUE. This will be illustrated using a branch predictor, which in the absence of Lockstepping has an SDC AVF of zero. For more details about branch predictors, please see Hennessy and Patterson [6].

A fault in a branch predictor usually does not cause incorrect execution. Branch predictors are commonly used by pipelined processors to predict the direction and target of a branch instruction before the branch is executed (usually much later) in the pipeline. The processor continues fetching instructions based on this prediction. When the branch eventually executes and produces the correct branch direction and target, the processor ensures that the original prediction was indeed correct. If the prediction was incorrect, then the processor throws away the executed (but uncommitted) state and restarts execution from the target of the branch. Hence, a fault in the branch predictor can only lead to a misdirection that may affect the performance only slightly (for most microarchitectures) but will not affect the correct execution of the processor. Hence, a branch predictor's SDC AVF is zero.

In a Lockstepped system, a fault in a branch predictor may cause the two Lockstepped processors to produce the same output in different cycles. Assume that Processor0 encounters a branch that is mispredicted due to a particle strike in its branch predictor, whereas the same branch in Processor1 is predicted correctly. This causes a timing mismatch in the two processors, causing them to commit the branch instructions in different cycles. In many cases, this may even throw the two processors to go down two different, but correct, paths of the same program they are running. Executions on both processors are correct, but the Lockstepped checker signals an error because of the mismatch. This is a false DUE event.

■ E X A M P L E

What is the false DUE rate arising from the branch predictors in a pair of Lockstepped processors? Assume that the branch predictor has 100 000 bits, an average false DUE AVF of 10%, and 1 millFIT/bit as the intrinsic FIT/bit.

SOLUTION The total false DUE FIT contribution of the two branch predictors = $2 \times 100\,000 \times 0.1 \times 0.001 = 20$ FIT.

■ E X A M P L E

Assume that each of the Lockstepped processors has an ECC-protected writeback data cache of 8 megabytes. ECC corrects the errors in the data cache. Assume 8 bits of ECC per 64 bits of data (or, 1 bit of code per 8 bits of data). Ignore the data cache tags for this example. Also, assume if an error is detected by the ECC code, then the processor takes an extra cycle to correct it (often called out-of-band correction). If this mode cannot be changed or turned off,[1] what is the false DUE rate from the data cache in a Lockstepped system? Assume that the data cache had an original SDC DUE AVF of 10% and FIT/bit of 1 milliFIT.

SOLUTION Every time an error is detected and corrected by the ECC code, the processors can go out of Lockstep mode. The potential false DUE rate $= 2 \times 2^{23} \times (8 + 1) \times 0.1 \times 0.001 \sim = 16\,100$ FIT (~ 7.6 years of MTTF) just from the 8 megabyte cache.

As the two examples above show, false DUE AVF contributions from structures that would otherwise not cause an SDC or a DUE in the absence of Lockstepping may actually end up contributing significantly to the false DUE FIT rate of Lockstepped processors. Chapter 7 discusses recovery mechanisms that help reduce the false DUE AVF.

3.4.5 Process-Kill versus System-Kill DUE AVF

Until now, it was assumed that on a DUE event, the entire system goes down. Fortunately, this is not always the case. If the OS can determine that the hardware error is isolated to a specific process, then it only needs to kill the user process experiencing the error, but the system can continue operating. For example, if the hardware detects a parity error on an architectural register file and reports it back to the OS, then the OS can kill the current user process and continue normal operation. Such DUE events are called process-kill DUE events. Of course, if the current process experiencing the error happens to be a kernel process, then the OS may not have any choice but to crash the machine. The latter is called system-kill DUE events.

EXAMPLE

A system manufacturer determines that OS hooks can be included to avoid crashing of the system in certain instances when an error is detected. A total DUE FIT of 100 FIT was anticipated. It also determined that in about 40% of the cases, it can kill the process experiencing the error. OS kernel processes were expected to be running 20% of the time. What is the process- and system-kill DUE FIT rate of the system?

SOLUTION Total process-kill DUE FIT $= 0.8 \times 0.4 \times 100 = 32$ FIT. Total system-kill DUE FIT $= 100 - 32 = 68$ FIT.

---

[1]See Chapter 5.

3.5   ACE Principles

As illustrated in the previous sections of this chapter, AVF answers the question whether a bit matters to the final outcome of a program. The concept of ACE formalizes this notion. Let us assume that a program runs for 10 billion cycles through a microprocessor chip. Out of these 10 billion cycles, let us assume that a particular bit in the chip is only required to be correct 1 billion of those cycles. In the other 9 billion cycles, it does not matter what the value of that bit is for the program to execute correctly. Then, the AVF of the bit is $1/10 = 10\%$. The bit is an ACE bit—required for ACE—for 1 billion cycles. For the rest of the cycles, the value of the bit is unnecessary for ACE and therefore termed un-ACE.

3.5.1   Types of ACE and Un-ACE Bits

Broadly, there are two types of ACE (or un-ACE) bits in a machine: microarchitectural and architectural. Microarchitectural ACE bits are those that are not visible to a programmer but can still affect the final outcome of a program. Some microarchitectural bits, such as the branch predictor and replacement policy bits, are often inherently un-ACE. Other microarchitectural bits, such as those in the pointer for an instruction queue, may be ACE part of the time when it ends up pointing to incorrect entries in the queue.

Architectural ACE bits are those that are visible to a programmer. A faulty value in one of these bits would directly affect the user-visible state. To understand how to identify architectural ACE or un-ACE bits, let us examine the dynamic dataflow graph of a program. A program's dynamic dataflow graph shows how data are consumed and generated according to the instructions of a program as the instructions execute. The graph begins with one or more program inputs and ends with the production of one or more program outputs, possibly observable by a user (see Figure 3.3). For the program to produce the correct output, the path from the input to the output must be executed correctly. This is the ACE path. Instructions on the ACE path are denoted using ACE cards in Figure 3.3. Dynamically executed instructions that do not contribute to the final outcome of the program are not required to produce the correct output.

The path with the "king" card in Figure 3.3 is an example of an architectural un-ACE path in the program because this path does not affect whether the program would have executed correctly. As will be seen later, certain bits (e.g., opcode bits) in an un-ACE instruction could still be ACE because a strike on such bits may cause the program to take an incorrect path. However, for simplicity of expression (but not for AVF calculation), the instruction is called un-ACE.

The notion of architectural ACE-ness is transitive. If an ACE store instruction stores a byte into memory, then at least one bit of that byte is ACE. For simplicity of analysis, the whole byte is often conservatively assumed to be ACE. The ACE byte may be transferred between various levels of the memory hierarchy, but it needs

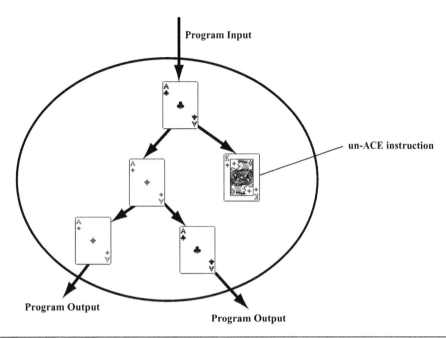

Program Input

un-ACE instruction

Program Output

Program Output

FIGURE 3.3 Architectural ACE versus un-ACE paths in the dynamic dataflow graph of a program. The un-ACE instruction does not affect the final output of a program. The ACE instructions are marked with ACE cards.

to be correct throughout its "journey" through the computer system. Otherwise, by definition, it will result in an SDC event. When a load instruction loads such an ACE byte memory, the load instruction itself becomes ACE. Such notion of ACE-ness can be applied to other objects, such as instruction chunks in the front end of the pipeline or cache blocks moving between processor caches and main memory.

Based on the above definition of ACE and un-ACE objects, it is easy to define whether a bit contains ACE state or not. A bit is in ACE state when ACE values from an object reside in that bit. The fraction of cycles ACE values reside in a bit is its AVF. Hence, a particle strike causing a bit flip during the cycles the bit contains ACE state will cause an SDC event (in the absence of any error detection or correction mechanism). Later in this chapter, three techniques to compute the AVF using such ACE principles are examined.

3.5.2 Point-of-Strike Model versus Propagated Fault Model

The definition of an ACE state has a subtlety with respect to where the particle strike occurred. One must distinguish between whether a storage bit was struck directly by a particle causing a fault (point-of-strike fault) or whether the fault propagated to the bit after it struck a different storage bit (propagated fault). ACE-ness and AVF of a bit are strictly defined for the bit that got struck and hence are based on the point-of-strike model.

It is critical to distinguish between the point-of-strike and propagated fault model when computing the AVF of structures. Otherwise, one may double count the same fault. For example, assume that a system has two structures, A and B, with the same intrinsic FIT/structure of 20 FIT. Assume that 50% of the faults in A propagate to B. That is, one computes A's AVF to be 10%. This does not, however, mean that B's AVF is 50% of 10%, which is 5%. Instead, B's AVF must also be computed using the point-of-strike fault model (let us assume it to be 20%). Then, the total FIT rate of A and B is $(10\% + 20\%) \times 20\,\text{FIT} = 6\,\text{FIT}$.

Later in this chapter, two techniques that use variations of these themes to compute the AVF are described. The first one directly uses a point-of-strike fault model to compute the AVF (see ACE Analysis Using the Point-of-Strike Fault Model, p. 106). The second one uses the propagated fault model to indirectly derive the AVF (see ACE Analysis Using the Propagated Fault Model, p. 114). In the second technique, the definition of the AVF is still based on the point-of-strike model (as it should be), but the calculations are based on a propagated fault model.

■ E X A M P L E

A microprocessor designer is faced with the question about which of the following two structures is more important to protect: a cache controller state table or a processor retire buffer. The designer finds that the AVF and intrinsic FIT for both structures is the same. Nevertheless, the designer argues that all instructions must pass through the instruction retire buffer, whereas only a subset of instructions—specifically load and stores—access the cache controller state table. Hence, the designer concludes that it is much more critical to protect the instruction retire buffer than the cache controller table. Is this observation correct?

SOLUTION The observation is incorrect. This is because the designer is mixing the point-of-strike and propagated fault models. The point-of-strike fault model shows that the FIT rates of both structures is same and hence protecting both structures is equally important. If faults propagate to the retire buffer, then FIT rate contribution of those faults must be ascribed to the structure where the fault originated (i.e., using the point-of-strike model) and not to the retire buffer. Protecting the retire buffer may not help if the fault has already occurred in the instruction entering the retire buffer.

As was discussed earlier in this section, the sources of ACE and un-ACE bits are divided into two general categories: microarchitectural and architectural.

The sources of microarchitectural and architectural un-ACE bits are discussed in the next two sections. How to use such categorization to compute the AVF is discussed later.

3.6 Microarchitectural Un-ACE Bits

Microarchitectural un-ACE bits are those that cannot influence committed architectural state. Microarchitectural un-ACE bits can arise from the following scenarios:

3.6.1 Idle or Invalid State

There are a number of instances in a microarchitecture of a microprocessor or a chipset when a data or a status bit is idle or does not contain any valid information. Such data and status bits are un-ACE bits. Control bits are, however, often conservatively assumed to be ACE bits because a strike on a control bit may cause idle state to be treated as nonidle state.

3.6.2 Misspeculated State

Modern microprocessors often perform speculative operations that may later be found to be incorrect. Examples of such operations are speculative execution following a branch prediction or speculative memory disambiguation. The bits that represent incorrectly speculated operations are un-ACE bits.

3.6.3 Predictor Structures

Modern microprocessors have many predictor structures, such as branch predictors, jump predictors, return stack predictors, and store-load dependence predictors. A fault in such a structure may result in a misprediction and will affect performance but will not affect correct execution. Consequently, all such predictor structures contain only un-ACE bits. An example of a predictor structure in the chipsets is the least recently used (LRU)–state bits that predict the block to be evicted next in a cache. A strike on such an LRU-state bit will cause the processor or chipset to evict a different cache block but may not affect correct execution.

3.6.4 Ex-ACE State

ACE bits become un-ACE bits after their last use. In other words, the bits are dead. This category encompasses both architecturally dead values, such as those in registers, and an architecturally invisible state. For example, after a dynamic instance of an instruction is issued for the last time from an instruction queue, it may still persist in a valid state in the instruction queue, waiting until the processor knows that no further reissue will be needed, but a fault in that instruction will not have any effect on the output of a program.

3.7   Architectural Un-ACE Bits

Architectural un-ACE bits are those that affect correct-path instruction execution and committed architectural state but only in ways that do not change the output of the system. For example, a strike on a storage cell carrying the operand specifier of a NOP instruction will not affect a program's computation. The bits of an instruction that are not necessary for an ACE path are un-ACE instruction bits. Sources of architectural un-ACE bits in processors and chipsets are identified below:

3.7.1 NOP Instructions

Most instruction sets have NOP instructions that do not affect the architectural state of the processor. Fahs et al. [3] found 10% NOPs in the dynamic instruction stream of a SPEC2000 integer benchmark suite compiled with the Alpha instruction set. On the Intel® Itanium® processor, Choi et al. [4] observed 27% retired NOPs in SPEC2000 integer benchmarks. These instructions are introduced for a variety of reasons, such as to align instructions to address boundaries or to fill very long instruction word (VLIW)–style instruction templates. The only ACE bits in a NOP instruction are those that distinguish the instruction from a non-NOP instruction. Depending on the instruction set, this may be the opcode or the destination register specifier. The remaining bits are un-ACE bits.

3.7.2 Performance-Enhancing Operations

Most modern processors and chipsets include performance-enhancing instructions and operations. For example, a nonbinding prefetch operation brings data into the cache to reduce the latency of later loads or stores. A single-bit upset in a nonopcode field of such a prefetch instruction issued by the processor or prefetch operation issued by a chipset cache may not affect the correct execution of a program. A fault may cause the wrong data to get prefetched or may cause the address to become invalid, in which case the prefetch may be ignored, but the program semantics will not change. Thus, the nonopcode bits are un-ACE bits. Fahs et al. [3] reported that 0.3% of the dynamic Alpha processor instructions in SPEC2000 integer suite were prefetch instructions. The Itanium®2 architecture has other performance-enhancing instructions, such as the branch prediction hint instruction.

3.7.3 Predicated False Instructions

Predicated instruction-set architectures, such as IA64, allow instruction execution to be qualified based on a predicate register. If the predicate register is true, the instruction will be committed. If the predicate register is false, the instruction's result will be discarded. All bits except the predicate register specifier bits in a predicated false instruction are un-ACE bits. A corruption of the predicate register specifier bits may erroneously cause the instruction to be predicated true. Hence, those bits are often conservatively referred to as ACE instruction bits. However, if the instruction itself is dynamically dead (see below) and the predicate register is overwritten before any other intervening use, then the predicate register and the corresponding specifier can be considered un-ACE bits. Mukherjee et al. [10] found about 7% of dynamic instructions to be predicated false.

3.7.4 Dynamically Dead Instructions

Dynamically dead instructions are those whose results are not used. Instructions whose results are simply not read by any other instructions are termed first-level dynamically dead (FDD). Transitively dynamically dead (TDD) instructions are those whose results are used only by FDD instructions or other TDD instructions. An instruction with multiple destination registers is dynamically dead only if all its destination registers are unused.

FDD and TDD instructions need to be tracked through both registers and memory. For example, if two instructions A and B successively write the same register R1 without any intervening read of register R1, then A is an FDD instruction tracked via register R1. Similarly, if two store instructions C and D write the same memory address M without any intervening load to M, then C is an FDD instruction tracked via memory address M.

Using the Alpha instruction set running the SPEC2000 integer benchmarks, Butts and Sohi [2] reported that about 9% FDD and 3% TDD instructions tracked only via registers. In contrast, Fahs et al. [3] found that about 14% FDD and TDD instructions tracked via both registers and memory in their evaluation of SPEC2000 integer benchmarks running on an Alpha instruction set architecture (ISA). Evaluation of Mukherjee et al. [10] with IA64 across portions of 18 SPEC2000 benchmarks shows that about 12% FDD and 8% TDD instructions tracked via both registers and memory. Their analysis assumes that memory results produced by FDD and TDD instructions are not used by other I/O devices. Their numbers for dynamically dead instructions are higher than earlier evaluations most likely because of aggressive compiler optimizations, which have been shown to increase the fraction of dead instructions [2].

In general, one needs to count all the opcode and destination register specifier bits of FDD and TDD instructions as ACE bits; all other instruction bits can be un-ACE bits. If the opcode bits get corrupted, then the machine may crash when evaluating those bits. If the destination register specifier bits get corrupted, then an FDD or a TDD instruction may corrupt a nondead architectural register, which could affect the final outcome of a program. This accounting is conservative, as it is likely that some fraction of bit upsets in the opcode or destination register specifier would not lead to incorrect program output.

3.7.5 Logical Masking

There are many bits that belong to operands in a chain of computation whose values still do not influence the computation results. Such bits are said to be logically masked. For example, consider the following code sequence:

1 R2 = R3 OR 0x00FF

2 R4 = R2 OR 0xFF00

3 R3 = 0

4 R2 = 0

5 output R4.

In this case, the lower 16 bits of R4 will be 0xFFFF regardless of the values of R2 and R3. When the value of a bit in an operand does not influence the result of the operation, the phenomenon is called logical masking. In our example, bits 0 to 7 (the low-order bits) of R3 are logically masked in instruction (1), and bits 8 to 15 of R2 are masked in instruction (2). One could identify additional un-ACE bits by considering transitive logical masking, where the effects of logical masking are propagated backward transitively. For a bit in a register to be logically masked (and thus be un-ACE), it must be logically masked for all its uses. In the above code sequence, bits 8 to 15 of R3 contribute only to bits 8 to 15 of R2, since R3 is set to zero and the value of R3 used in instruction (1) is not used anywhere else. Because bits 8 to 15 of R2 are logically masked in instruction (2), via transitive logical masking, bits 8 to 15 of R3 can be considered masked as well.

Logical masking can also arise from compare instructions prior to branches (where the bit may matter only if the value is zero or nonzero but not necessarily every value), bitwise logical operations, and 32-bit operations in a 64-bit architecture (where it is assumed that the upper 32 bits are un-ACE, which may not be true for certain ISAs, such as the Alpha ISA). Typically, all logically masked bits are un-ACE bits and can be factored out of the AVF calculation.

3.8   AVF Equations for a Hardware Structure

Using the classification of ACE and un-ACE bits and assuming the point-of-strike fault model, one can compute the AVF of a hardware structure. As discussed earlier, the AVF of a storage cell is the fraction of time an upset in that cell will cause a visible error in the final output of a program. Thus, the AVF (i.e., SDC AVF) for an unprotected storage cell is the percentage of time the cell contains an ACE bit. For example, if a storage cell contains ACE bits for a billion cycles out of an execution of 10 billion cycles, then the AVF for that cell is 10%.

Although the AVF equations were defined with respect to a storage cell, the AVF is typically computed for an entire hardware structure. The AVF for a hardware structure is simply the average AVF for all the bits in that structure, assuming that all the bits in that structure have the same circuit composition and hence the same raw FIT rate. Then, the AVF of a hardware structure with N bits is equal to

$$AVF_{structure} = \frac{\sum_{i=0}^{N} (\text{bitwise AVF})_i}{N}$$

which can be rewritten as

$$AVF_{structure} = \frac{\sum_{i=0}^{N} \left(\frac{\text{Cycles bit } i \text{ is in ACE state}}{\text{Cycles over which state is observed}} \right)_i}{N} = \frac{\sum_{i=0}^{N} \left(\frac{\text{ACE cycles for bit } i}{\text{Total cycles}} \right)_i}{N}$$

where the cycles bit i in ACE state is denoted as ACE cycles and cycles over which state is observed for all bits as total cycles. Again, rewriting

$$AVF_{structure} = \frac{\sum_{i=0}^{N} \text{ACE cycles}_i}{N \times \text{total cycles}}$$

It will be seen that this equation is very useful to compute the AVF of a structure using a performance simulator. This equation can also be rewritten as

$$AVF_{structure} = \frac{\frac{\sum_{i=0}^{N} \text{ACE cycles}_i}{\text{Total cycles}}}{N} = \frac{\text{Average number of ACE bits in a structure in a cycle}}{\text{Total number of bits in a structure}}$$

The next section shows that this form of the equation can be useful to gain an in-depth understanding of the microarchitectural parameters affecting AVF. Figure 3.4 shows how one can compute the AVF of a structure using this form of the equation and the notion of ACE and un-ACE bits. This figure shows that in a specific cycle, there are two entries with architectural un-ACE instructions, three entries with microarchitectural un-ACE instructions, and one entry with an ACE instruction. The instantaneous AVF of the structure in that cycle is, therefore, equal to $1/6 = 17\%$.

Instruction Queue

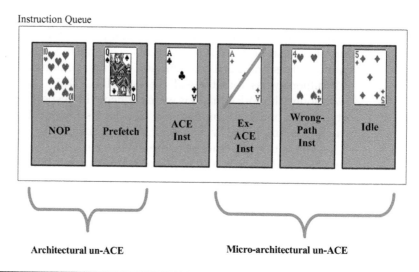

FIGURE 3.4 Identifying ACE and un-ACE bits in an instruction queue in a microprocessor in a particular cycle.

■ E X A M P L E

Compute the AVF of a 4-bit structure over three cycles of execution. In the first cycle, bits 0 and 1 are ACE; in cycle 2, bit 3 is ACE; and in cycle 3, all four bits are ACE.

S O L U T I O N $AVF_{cycle\,1} = 2/4 = 50\%$, $AVF_{cycle\,2} = 1/4 = 25\%$, $AVF_{cycle\,3} = 4/4 = 100\%$. The average AVF over three cycles are then $(50\% + 25\% + 100\%)/3 = 58\%$.

The same ACE and un-ACE analysis can be used to compute the DUE AVFs for bits or structures with error detection. However, this requires specific knowledge of the error detection scheme for specific bits and structures. As discussed earlier, adding error detection to a structure converts its SDC AVF to true DUE AVF but introduces an additional false DUE AVF component. The un-ACE analysis helps compute this false DUE AVF. By identifying the fraction of cycles, un-ACE bits trigger error detection mechanism. This fraction is the false DUE AVF for the specific structure.

3.9 Computing AVF with Little's Law

As was seen in the previous section, a structure's AVF can be expressed as the ratio of the average number of ACE bits in a cycle resident in the structure and the total number of bits in that structure. Little's law [7] is a basic queuing theory equation that enables one to compute the average number of ACE bits resident in a structure. Little's law can be translated into the equation $N = B \times L$, where N is the average number of bits in a box, B is the average bandwidth per cycle into the box, and L is the average latency of an individual object through the box (Figure 3.5a), where none of the objects flowing into the box is lost or removed. Little's law can also be applied to a subset of the bits. Hence, by applying this to ACE bits (Figure 3.5b), one gets the average number of ACE bits in a box as the product of the average bandwidth of ACE bits into the box (B_{ACE}) times the average residence cycles of an ACE bit in the box (L_{ACE}). Thus, one can express the AVF of a structure as

$$AVF_{structure} = \frac{\text{Average number of ACE bits in a structure in a cycle}}{\text{Total number of bits in a structure}}$$

$$= \frac{B_{ACE} \times L_{ACE}}{\text{Total number of bits in a structure}}$$

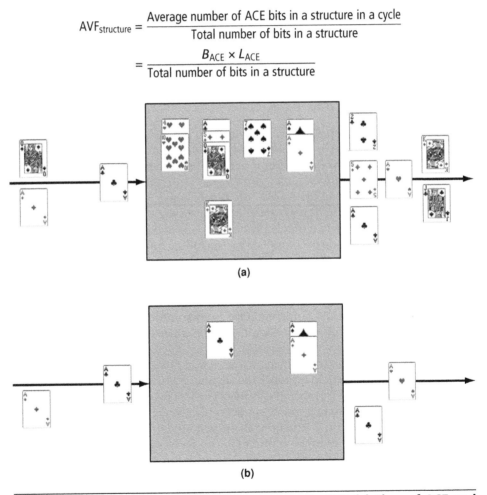

(a)

(b)

FIGURE 3.5 Illustration of Little's law to compute AVF. (a) Flow of ACE and un-ACE instructions through a hardware structure, such as an instruction queue. (b) Flow of only ACE instructions through the structure.

This is a powerful equation that not only allows one to quickly do back-of-the-envelope calculations of AVF but also provides insight into the parameters AVF depends on.

■ EXAMPLE

To quickly compute the approximate AVF of a 32-entry instruction queue, let us categorize instructions into ACE and un-ACE and ignore the ACE bits in un-ACE instructions. Assume that the instruction per cycle (IPC) of ACE instructions is two and average delay of an instruction in the instruction queue is five cycles. What is the approximate AVF of the instruction queue?

SOLUTION $B_{ACE} = 2$ IPCs, $L_{ACE} = 5$ cycles. Then, AVF $= 2 \times 5/32 = 10/32 = 31\%$.

■ EXAMPLE

Compute the AVF of a branch commit table in a microprocessor. At the decode stage, every decoded branch and its associated information are entered into the branch commit table. When the branch commits and is deemed to have been mispredicted, then the information in the commit table is accessed to recover the state of the pipeline and restart the pipeline from the correct-path instruction after the branch. Assume an entire entry in the branch commit table is either ACE or un-ACE, the average IPC of the machine is two, the decode to commit delay (including queueing delay) is 30 cycles, one out of five instructions are branches, the branch misprediction rate is 3%, and the branch commit table has 64 entries.

SOLUTION At any instant, there are four types of entries in the branch commit table: ACE mispredicted branch entries that will be used for recovery, un-ACE branch instructions that are predicted correctly, wrong-path un-ACE branch entries, and idle un-ACE entries. There is one branch per five committed instructions. The branch misprediction rate is 3% so 3 out of 100 branches are mispredicted. In other words, 3 out of 500 instructions are mispredicted. The mispredicted branch IPC is then $2 \times (3/500) = 0.012$. Since the decode to commit delay (including queueing delay) is 30 cycles, the average number of mispredicted branch instructions in the commit table is $0.12 \times 30 = 0.36$. The total number of entries in the commit table is 64. Hence, the AVF $= 0.36/64 = 0.56\%$.

Although Little's law is useful to compute the AVF of hardware structures, it must be applied carefully. Little's law cannot be applied if the ACE objects flowing through a structure change. For example, Little's law cannot be directly applied to an adder, which takes two operands as inputs and produces one output. In this case, Little's law can be applied separately to the input and output datapath latches.

3.9.1 Implications of Little's Law for AVF Computation

Using Little's law to compute the AVF gives one four important insights into the computation of AVF and the factors AVF depends on. First, AVF is a function of the architecturally sensitive area of exposure to radiation. This is expressed through Little's law by multiplying the number of incoming ACE bits into a structure with the delay experienced in the structure. "Sensitive" area in this context refers to the fraction of area that on average is occupied by ACE objects.

Second, IPC alone may not determine the AVF of microprocessor pipeline structures. Let us consider the instruction queue in a processor pipeline. Let us define ACE IPC and ACE latency as the IPC of ACE instructions and latency of ACE instructions through the instruction queue, respectively. The instruction queue usually has the same IPC as the retire unit in a processor pipeline. A benchmark with high IPC can have high ACE IPC but low ACE latency because instructions may be flowing rapidly through the pipeline. Similarly, a benchmark with low IPC can have low ACE IPC, but instructions may be stalled behind cache misses, making ACE latency high. Consequently, both these benchmarks can have very similar AVF for the instruction queue in the pipeline.

Third, it is often not unusual to assume that a structure's AVF decreases if objects move faster through the structure, thereby reducing the exposure time to radiation. However, the AVF may not actually decrease if there is a corresponding increase in the bandwidth of ACE objects flowing into the structure.

Fourth, one can relate AVF of different structures using Little's law. If objects flow from a structure A to a structure B, then in the steady state, the average bandwidth of ACE objects through both A and B will usually be the same. To compute the AVF, however, one needs the average delay through objects A and B and the size of each structure, which may differ. However, in the degenerate case where a sequence of single-bit storage cells with unit delay is connected (e.g., sequence of flow-through latches in a datapath), the AVF of each of these storage cells is the same.

3.10 Computing AVF with a Performance Model

To study the trade-off between performance, power, and soft errors, and to design the appropriate machine, architects need an early AVF estimate for the chip they are designing. Figure 3.6 shows the different steps in the design of a high-performance microprocessor. Typically, this process starts with a performance model—typically

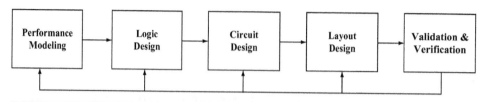

FIGURE 3.6 Typical flow in the design of a microprocessor.

written in C or C++. A performance model is an abstract representation of the machine's timing behavior, which allows architects to predict the performance of the machine under design. This is followed by the actual architectural definition and logic design of the processor. This is often done in a Register Transfer Language (RTL), such as Verilog. Once the RTL is ready, circuit designers convert the logic blocks into circuit blocks using a variety of tools. This is followed by layout design, validation, and verification. The outputs from the validation and verification steps are fed back to the different stages of the design to fix bugs. High-performance chipsets usually follow the same flow. Nevertheless, some application-specific integrated circuits may skip some of the intermediate steps (e.g., they may need to model the performance of the chip).

The two obvious places to model the AVF are in the performance model and in the RTL model. The advantage of the RTL model is that it has the detailed state of microprocessor or chipset structures. Hence, one could use statistical fault injection into different RTL states (e.g., latches) and see if the fault shows up as a user-visible error and thereby compute the AVF. Chapter 4 discusses in detail the advantages and drawbacks of statistical fault injection. AVF evaluation with an RTL model, however, poses two problems for an architect. First, RTL may not be available long after the architecture specification has been defined. By the time RTL is created and the architects find out what the AVFs are, it may be too late to do any major changes to the design because of schedule pressures. It may, however, be possible to use an earlier version of the RTL (from a previous generation chip), but that may still be error prone.

Second, RTL simulations are orders of magnitude slower than performance models. RTL simulations can often only be realistically run for tens of thousands of instructions, but the effect of a fault in a microarchitectural structure may not show up till after tens of millions of instructions. Further, since soft errors are an average and statistical quantity, it is critical to do such simulations over a number of benchmarks, potentially spanning millions to billions of instructions. Further, statistical fault injection into a chip with billions of transistors necessitates an explosive number of simulations to reach a statistical significance.

Nevertheless, for many structures, such as flow-through pipeline latches, the fault-to-error latency is fairly small (e.g., less than 1000 cycles often). Furthermore, some of these structures may not be available in the performance model. For such structures, statistical fault injection into the RTL is a desirable way to measure the

AVF. The section Computing AVFs Using Statistical Fault Injection into RTL, p. 146, Chapter 4, discusses these issues in greater detail.

3.10.1 Limitations of AVF Analysis with Performance Models

The AVF analysis technique using a performance model can provide early estimates of per-structure AVFs, but it also has four limitations that readers should be aware of.

What Is the Scope?

As discussed in Chapter 1 (see Faults, p. 6), the concept of MTTF is fundamentally tied to a "scope." A scope is a domain with which an MTTF value is associated. Because MTTF is inversely proportional to the FIT rate and hence to the AVF, the concept of AVF is also related to the scope for which the AVF is defined. For example, if the scope is defined to be the full system, including I/O devices, then an object is ACE only if its effect shows up at the I/O interface. In contrast, if the scope is defined to be at the cache-to-main memory interface, then one can declare an object arriving at the cache-to-memory interface to be ACE if it shows up at this interface. But what is an ACE bit at the cache-to-memory interface may be un-ACE when the scope is expanded to the full system (e.g., a memory value with an error may never be written back to disk).

Further, the scope at which an error shows up depends on the user's interaction with a program. Normally, a program's outputs are just the values sent by the program via I/O operations. However, if a program is run under a debugger, then the program variables examined via the debugger become outputs and influence the determination of the ACE bits.

For AVF analysis in a performance model, one usually requires a precise definition of what constitutes the scope. It is assumed that the scope extends to an I/O device. In practice, it is hard to track values this far. However, performance models can track values well beyond the point that they are committed to architectural registers or stored to memory to determine whether they could potentially influence the output. Given such a definition of a scope, one can determine in a performance model the ACE and un-ACE bits.

What Is a Correct Output?

Besides the scope, one also needs a precise definition of what constitutes the correct output. This does not necessarily correspond to meeting the precise semantics of the architecture, which can allow multiple correct outputs for the same program. Similarly, in a multiprocessor system, multiple executions of the same parallel program may yield different outcomes due to race conditions. Whether a bit is ACE or un-ACE may depend on the outcome of a race.

To make the methodology precise, ACE analysis computes the AVF for a specific dynamic instance or execution of a program. Given a specific execution of a program, the ACE analysis assumes that the final system output is the correct one. Any bit flip that would have caused the program to generate an output different from the expected system output for that instance constitutes an error. In a multiprocessor system, ACE analysis would use the outcome of the race in the particular execution under study to determine the ACE and un-ACE.

Thus, this style of AVF estimation is a postanalysis method. That is, one runs the programs, collects statistics, and analyzes what the per-structure AVFs would have been. This is, however, not different from how performance simulators estimate performance and power.

ACE Analysis Provides an Upper Bound

As may be obvious by now, proving that a bit is un-ACE is easier than proving that the bit is ACE. For example, if one has two consecutive stores to the same register A without any intervening read, then register A's bits are dynamically dead and hence un-ACE in the interval from the first store to the second store. However, this second store may not occur within the window of simulation or the first store may have been evicted from the AVF analysis window, in which case one cannot determine if register A's bits were ACE or un-ACE, unless the program can be run to completion.

Nevertheless, since a conservative (upper bound) SER and hence AVF estimate are desired, the analysis first assumes that all bits are ACE bits unless it can be shown otherwise. Then it identifies as many sources of un-ACE bits as it can. The analysis does not need (nor claims) to have a complete categorization of un-ACE bits; however, the more comprehensive the analysis is, the tighter the bound will be.

Recently, a couple of studies have examined the preciseness with which AVFs can be computed. Wang et al. [14] computed AVFs using ACE analysis on a relatively less detailed performance model and found that the resulting AVF is two to three times higher than what SFI would predict from an RTL model. The authors argued that this is because it is difficult to add the necessary detail to a performance model to compute the appropriate AVFs. Biswas et al. [1] have shown, however, that such details are not hard to add to a performance simulation model to appropriately compute precise AVFs.

Similarly, Li et al. [8] argue that AVF analysis reaches its limit when one considers tens of thousands of computers or a very high intrinsic FIT rate (significantly higher than what a radiation-induced transient fault would cause). It is unclear what underlying phenomenon is forcing this limit. One possibility is that higher numbers of computers or higher intrinsic FIT rate may be introducing multi-bit faults. In such as case, the basic AVF analysis needs to be extended and the error rate computed differently. For an example of how to compute the SER from double bit errors, the reader is referred to the scrubbing analysis in Chapter 5.

The reader should also note that often a conservative upper bound on AVF is sufficient for what a designer is looking for. If the conservative upper-bound AVF estimates satisfy the soft error budget of a chip, then the designer may not care to further refine them. If not, the designers may continue to refine the AVF analysis for the most vulnerable structures until they are satisfied with the analysis.

Chapter 4 shows how one could extend this analysis to gather best estimate AVF numbers based on the properties of certain structures.

ACE Analysis Approximates Program Behavior in the Presence of Faults

One subtle issue that may not be immediately obvious is that ACE analysis attempts to approximate the behavior of a faulty instance of a program from an analysis of a fault-free execution of the same program. For example, Figure 3.7b shows the dynamic execution of a program in which the program takes the wrong branch direction because of a fault. Currently, ACE analysis assumes that any such control flow inducing instruction is ACE. Nevertheless, it is possible that even if the branch takes the incorrect path, it will eventually produce the correct result, thereby making the result of some branches un-ACE [13]. Such branches are called Y-branches. The impact of this phenomenon is expected to be limited. Nevertheless, this is yet another place where ACE analysis provides an upper-bound estimate. One way to make the analysis more precise for Y-branches is to simulate both paths of a branch during a program's execution and determine if the paths eventually converge with the same architectural state.

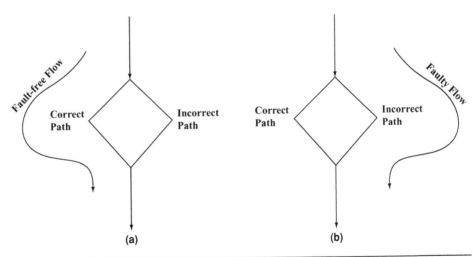

FIGURE 3.7 Examples of fault-free (a) and faulty (b) flows of a program execution. ACE analysis attempts to approximate the behavior of (b) using (a).

CAM structures introduce a similar challenge. Chapter 4 shows how to compute the AVF of such CAM structures.

3.11 ACE Analysis Using the Point-of-Strike Fault Model

As it was discussed earlier, one can use the following equation—based directly on the point-of-strike fault model—to compute the per-structure AVFs:

$$\text{AVF}_{\text{structure}} = \frac{\sum_{i=0}^{N} \text{ACE cycles}_i}{N \times \text{Total cycles}}$$

Thus, the following three terms are needed:

- sum of residence cycles of all ACE bits of objects resident in the structure during program execution

- total execution cycles for which the ACE bits' residence time is observed

- total number of bits in a hardware structure.

Using a performance model, one can compute all the above. This chapter illustrates this technique using objects that carry instruction information along the pipeline (Figure 3.8). Chapter 4 discusses advanced techniques to compute AVF for structures that carry other types of objects, such as cache blocks.

The AVF algorithm can be divided into three parts. As an instruction flows through different structures in the pipeline, the residence time of the instruction in the structure is recorded. Then, before the instruction disappears from the machine, either via a commit or via a squash, the structures through which it flowed are updated with a variety of information, such as the residence cycles, whether the instruction committed, etc. (part 1). Also, if the instruction commits, the instruction

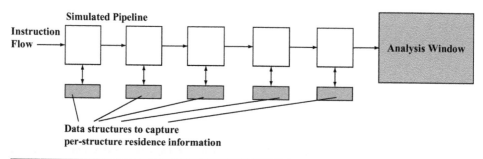

FIGURE 3.8 Pictorial view of data structures for the AVF algorithm used in a performance model.

is put in a postcommit analysis window to determine if the instruction is dynamically dead or if there are any bits that are logically masked (part 2). Finally, at the end of the simulation, using the information captured in parts 1 and 2, one can easily compute the AVF of a structure (part 3).

To compute whether an instruction is an FDD or a TDD instruction and whether any of the result bits has logical masking, one must know about the future use of an instruction's result. The analysis window is used to capture this future use. When an instruction commits, it is entered into the analysis window, linking it with the instructions that produced its operands. At any time, one can analyze the future use of an instruction's results by examining its successors in the analysis window. Of course, because the analysis window must be finite in size, one cannot always determine the future use precisely. An analysis window of a few thousand instructions usually covers most of the needed the future use information, but practitioners have used tens of thousands of instructions for this window.

The analysis window needs three subwindows that compute FDD, TDD, and logical masking information. Each subwindow can be implemented with two primary data structures: a linked list of instructions in commit order and a table indexed via architectural register number or memory address. The linked list maintains the relative age information necessary to compute future use. Each entry in the table maintains the list of producers and consumers for that register or memory location. The FDD, TDD, and logical masking information can all be computed using this list of producers and consumers. Thus, a list with two consecutive producers for a register R and no intervening consumer for the same register R can be used to mark the first producer of R as a dynamically dead instruction.

3.11.1 AVF Results from an Itanium®2 Performance Model

This section presents a case study of AVF evaluations for two structures: an instruction queue and execution units for an Itanium®2 processor pipeline. This case study is based on the evaluations presented by Mukherjee et al. [10]. The evaluation methodology will be discussed first followed by AVF analyses for an instruction queue and latches in input and output datapaths in execution units. This case study demonstrates how ACE analysis can compute both SDC and DUE AVFs.

Evaluation Methodology

The evaluation uses an Itanium®2-like IA64 processor scaled to 2003 technology. Itanium®2 is one of Intel® Corporation's high-end processor architectures. This processor was modeled in detail in the Asim performance model framework [5] developed by a group in the Compaq Computer Corporation and was eventually licensed and developed further by architects in Intel Corporation. In Asim, Red Hat Linux 7.2 was modeled in detail via an OS simulation front end. For wrong paths, the simulator fetched the misspeculated instructions but did not have the correct memory addresses that a load or a store may have accessed. This processor model is augmented with an instrumentation to evaluate per-structure AVFs.

Figure 3.9 lists the skip intervals and input set selected for each of the SPEC 2000 programs used for this analysis. Because it is difficult, if not impossible, to simulate in detail an entire program, simulation models typically simulate only sections of programs or benchmarks to predict performance. The skip intervals shown here were computed using simpoint analysis of Sherwood et al. [12] modified for the IA64 ISA [11]. The numbers presented here are only for the first simpoint. Simpoints provide sections of the program that can best predict the performance of the whole program itself. Although simpoints do not necessarily provide sections that can best predict AVFs, it is probably a fair approximation for the AVFs as well. In this evaluation, each simpoint is simulated for 100 million instructions (including NOPs). The benchmarks were compiled with the Intel® electron compiler (version 7.0) with the highest level of optimization.

Program-Level Decomposition

Figure 3.10 shows a decomposition of the dynamic stream of instructions based on whether the instruction's output affects the final output of the benchmark. An instruction whose result may affect the output is an ACE instruction, while an instruction that definitely does not affect the final output is un-ACE. As the figure shows, on average, about 46% of the instructions are ACE instructions. The rest are

Integer Benchmarks	Instructions Skipped	Floating Point Benchmarks	Instructions Skipped
bzip2-source	48,900 M	ammp	50,900 M
cc-200	16,600 M	applu	500 M
crafty	120,600 M	apsi	100 M
eon-kajiya	73,000 M	art-110	36,400 M
gap	18,800 M	equake	1,500 M
gzip-graphic	2,9000 M	facerec	64,100 M
mcf	26,200 M	fma3d	23,600 M
parser	71,400 M	galgel	5,000 M
perlbmk-makerand	0 M	lucas	123,500 M
twolf	185,400 M	mesa	73,300 M
vortex_lendian3	59,300 M	mgrid	200 M
vpr-route	49,200 M	sixtrack	4,100 M
		swim	78,100 M
		wupwise	23,800 M

FIGURE 3.9 Benchmarks used for AVF studies of the instruction queue and execution unit of the Itanium®2 pipeline under study. M = million. Reprinted with permission from Mukherjee et al. [10]. Copyright ©2003 IEEE.

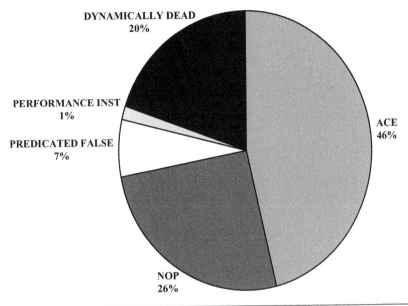

FIGURE 3.10　ACE and un-ACE breakdown of the committed instruction stream in an Itanium®2 pipeline for the benchmarks shown in Figure 3.9.

un-ACE instructions. Some of these un-ACE instructions still contain ACE bits, such as the opcode bits of prefetch instructions. Because this analysis is conservative, there may be other opportunities to move instructions from the ACE to un-ACE category.

NOPs, predicated false instructions, and performance-enhancing prefetch instructions account for 26%, 7%, and 1%, respectively. NOPs are introduced in the IA64 instruction stream to align instructions on three-instruction bundle boundaries. These NOPs are carried through the Itanium®2 pipeline. Finally, dynamically dead instructions account for 20% of the total number of committed instructions.

AVF Analysis for the Itanium®2 Instruction Queue

Figure 3.10 shows the fraction of cycles a storage cell in the instruction queue contains ACE and un-ACE bits. An instruction queue is a structure into which the processor fetches instructions. A separate scheduler usually identifies and dispatches from this queue instructions that are ready to be executed. This calculation assumes each entry of the instruction queue is approximately 100 bits. An IA64 instruction is 41 bits, but the number of bits required in the entry is higher because a large number of bits are required to capture the in-flight state of an instruction in the machine. Out of these 100 bits, five bits are control bits and cannot be derated. Of the remaining 95 bits, seven opcode bits are not derated (i.e., considered

ACE) for any instruction. Additionally, the six predicate specifier bits of falsely predicated instructions or the seven destination register specifier bits for FDD and TDD instructions are not derated.

Figure 3.11a shows that on average, a storage cell in the instruction queue contains an ACE bit about 29% of the time. Thus, the AVF of the instruction queue is 29%. On average, a cell is idle 38% of the cycles and contains a nonidle un-ACE bit

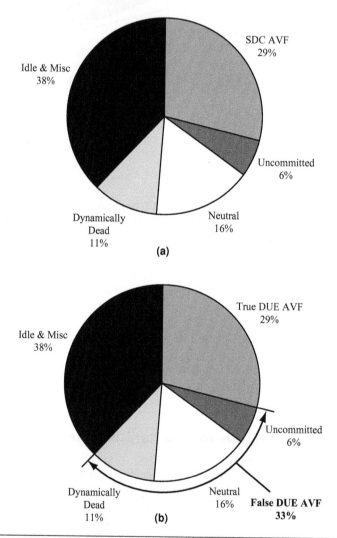

FIGURE 3.11 ACE and un-ACE breakdown of an instruction queue in an Itanium®2 pipeline. (a) The SDC AVF for the unprotected instruction queue is 29%. (b) The DUE AVF of the same instruction queue with parity protection, but no recovery, is 62% (29% true DUE AVF + 33% false DUE AVF).

about 33% of the cycles. This 33% includes 6% uncommitted, such as wrong-path instructions, 16% neutral, such as NOPs, and 11% dynamically dead. Across the simulated portions of the benchmark suite (Figure 3.12), the AVF number ranges between 14% and 47% for the instruction queue. The floating-point programs, in general, have higher AVFs than to integer programs (31% vs. 25%, respectively). This is because floating-point programs usually have many long-latency instructions and few branch mispredictions.

Figure 3.11b shows the false DUE AVF of the instruction queue, assuming that it is protected with parity. The parity bit is checked when an instruction is ready for issue. The entire nonidle un-ACE portion is now ascribed to the false DUE AVF.

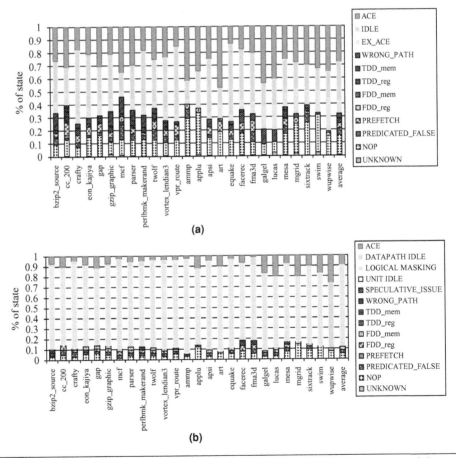

FIGURE 3.12 ACE and un-ACE breakdown for an instruction queue (a) and execution unit (b) for an Itanium®2 pipeline. Reprinted with permission from Mukherjee et al. [10]. Copyright ©2003 IEEE.

Hence, the false DUE AVF of the instruction queue protected with parity is 33%. The total DUE AVF is 29% + 33% = 62%.

Figure 3.13 shows how to approximate AVFs (at the instruction level) for the instruction queue using Little's law. The AVF of the instruction queue can be approximated as the ratio of the average number of ACE instructions in the instruction queue to the total number of instruction entries in the instruction queue, which is 64 in the machine simulated. The number of ACE instructions in the instruction queue in an average cycle, as given by Little's law, is the product of the bandwidth (ACE IPC) and the average number of cycles an instruction in the instruction queue can be considered to be in ACE state (ACE latency). It should be noted that an instruction can persist even after it is issued for the last time. Thus, after an ACE instruction is issued for the last time, the ACE bits holding the ACE instruction become un-ACE. The ACE IPC and ACE latency were obtained from the performance model.

Using the ACE IPC and ACE latency, the AVFs in Figure 3.13 can be computed. This method computes an average AVF of 19%, which is 10% lower than that of actual AVF for the instruction queue reported earlier. This difference can be attributed to the ACE bits of un-ACE instructions, such as prefetch and dynamically dead instructions, whose results do not affect the final output of a program. Figure 3.13 accounts for these as un-ACE bits, but ACE numbers in Figure 3.11 and Figure 3.12 include them. If the Little's law analysis was done at the bit level, instead of instruction level as in Table 2, then the average AVF of 29% in Figure 3.11 would have matched.

Figure 3.13 also explains why lucas has an AVF similar to ammp although lucas has one of the highest ACE IPCs. This is because the AVF depends on both the ACE

Integer Benchmarks	ACE IPC	ACE Latency (cycles)	# ACE Inst	AVF	Floating Point Benchmarks	ACE IPC	ACE Latency (cycles)	#ACE Inst	AVF
bzip2-source	0.55	22	12	19%	ammp	0.23	92	21	33%
cc-200	0.57	18	10	16%	applu	0.82	21	18	27%
crafty	0.37	15	6	9%	apsi	0.31	31	9	15%
eon-kajiya	0.36	20	7	11%	art-110	0.68	37	25	40%
gap	0.78	17	13	21%	equake	0.26	12	3	5%
gzip-graphic	0.60	13	8	12%	facerec	0.41	7	3	5%
mcf	0.25	68	17	26%	fma3d	0.59	11	7	10%
parser	0.49	24	12	19%	galgel	1.10	21	23	35%
perlbmk-makerand	0.38	17	7	10%	lucas	1.23	17	21	33%
twolf	0.30	27	8	13%	mesa	0.47	16	8	12%
vortex_lendian3	0.42	22	9	15%	mgrid	1.28	10	13	21%
vpr-route	0.35	12	4	7%	sixtrack	0.66	20	13	21%
					swim	1.08	16	17	27%
					wupwise	1.60	13	20	31%
average	0.45	23	9	15%	average	0.77	23	14	23%

FIGURE 3.13 AVF breakdown for an instruction queue with Little's law. Number of ACE instructions = ACE IPC × ACE latency. AVF = number of ACE instructions/number of instruction queue entries. Reprinted with permission from Mukherjee et al. [10]. Copyright ©2003 IEEE.

IPC and ACE latency. Although lucas has a high ACE IPC, it has relatively low ACE latency. Consequently, the product of these two terms results in an AVF similar to ammp's, which has a low ACE IPC but a high ACE latency.

AVF for Execution Units

This section describes the AVF numbers of the execution units in the simulated machine model. In this six-issue machine, there are four integer pipes and two floating-point pipes. Integer multiplication is, however, processed in the floating-point pipeline. When integer programs execute the floating-point pipes lie idle. It is also assumed that the execution units overall have about 50% control latches and 50% datapath latches. First, how to derate the entire execution unit is discussed so that the results would apply to both the control and the datapath latches. Then, it will be shown how to further derate the datapath latches.

Figure 3.12b shows that the execution units on average spend 11% of the cycles processing ACE instructions (with a range of 4% to 27%). Thus, the average AVF of a latch in the execution units is 11%. Interestingly, the execution units' AVF is significantly lower than that of the instruction queue. This is due to three effects. First, the instruction queue must hold the state of the instructions until they execute and retire. Thus, ACE instructions persist longer in the instruction queue than the time they take to execute in the execution units.

Second, speculatively issued instructions succeeding cache miss loads must replay through the instruction queue. However, only the last pass through the execution units matters for correct execution. The execution unit state for all prior executions is counted as un-ACE. It should be noted that this is possible in this processor model because a corrupted bit in one of the instructions designated for a replay does not affect the decision to replay. The information necessary to make this decision resides elsewhere in the instruction queue.

Third, the floating-point pipes are mostly idle while executing integer code, greatly reducing their AVFs.

This analysis computes a single AVF for both the control and the datapath latches in the execution units. However, the datapath latches can be further derated based on whether specific datapath bits are logically masked or are simply idle. Here logical masking is applied to data values (and hence datapaths) only; analyzing logical masking for control latches would be a complex task.

Logical masking functions were implemented only for a small but important subset of the roughly 2000 static internal instruction types in the processor model. This subset contains a variety of functions, including logical OR, AND. It was also estimated that another 20% of the instructions (including loads, stores, and branches) will not have any direct logical masking effect. The combination of these two categories covers the vast majority of dynamically executed instructions. Figure 3.12b shows that this logical masking analysis further reduces the AVF by 0.5% (averaged across both control and data latches, although this does not apply to control latches). It is expected that the incremental decrease in AVF due to the remaining unanalyzed instruction types will be small. Transitive logical masking, which would further reduce the AVF number for the datapath latches, was not considered in this analysis.

Datapath latches can be further derated by identifying the fraction of time they remain idle. For example, an IA64 compare instruction produces two predicate values—a predicate value and its complement. However, in the simulated implementation, these two result bits are sent over a 64-bit result bus, leaving 62 of the datapath lanes idle. This effect further reduces the AVF by 1.5%, as shown by DATAPATH_IDLE in Figure 3.12b Depending on the implementation, however, the DATATPATH_IDLE portion can also be viewed as bits that get logically masked at the implementation level. In contrast, UNIT_IDLE in Figure 3.12b refers to the whole execution unit being idle because of the lack of any instruction issued to that unit.

Overall, the average AVF for the execution units is reduced to 9% when logical masking and idle latches are accounted for. Across the simulated portions of our benchmark suite, the AVF for the execution units ranges from 4% to 27%.

It should be noted that Little's law cannot be applied to the entire execution units because the objects flowing through the execution units change. Nevertheless, Little's law can be applied to the input and output datapath latches.

3.12   ACE Analysis Using the Propagated Fault Model

Li et al. [9] developed a tool called *SoftArch*, which makes use of the propagated fault model in a performance simulator to compute the SER. Instead of computing the AVFs directly using the equations shown earlier (see AVF Equations for a Hardware Structure, p. 96), SoftArch evaluates the AVF indirectly by evaluating the derated and nonderated MTTF for a particular structure. The AVF can be obtained by dividing the nonderated MTTF by the derated MTTF. The reader should note that computing AVFs using either the point-of-strike model directly or the propagated fault model indirectly (like SoftArch does) suffers from the same limitations outlined in Limitations of AVF Analysis with Performance Models (p. 103).

Figure 3.14 shows a hypothetical example of how to compute the MTTF using the propagated fault model. Each of the four faults shown in the figure has a certain probability of occurrence and a corresponding nonderated error rate. The second fault gets masked, but the TTF for the first, third, and fourth errors are 1000, 10000, and 20000, respectively. If one assumes that these three errors are completely independent of each other and occur with equal probability (=1/3), then the MTTF would be $[1000 \times (1/3) + 10000 \times (1/3) + 20000 \times (1/3)] = 10230$ cycles. In reality, however, a program may run for longer than 20000 cycles. One cannot assume that the distribution of errors is uniform, although the underlying faults from alpha particle or neutron strikes may occur uniformly.

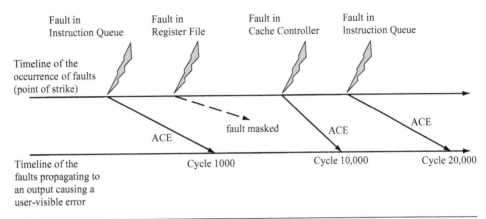

FIGURE 3.14 Hypothetical example showing different TTFs. The first timeline shows the occurrence of four potential faults. Each has a certain probability of occurrence. Some of these will get masked. The second timeline shows when the faults show up as a user-visible error at some output.

To illustrate how SoftArch computes the MTTF, let us consider the following example. Let us assume that

- a chip has only one unit in which the probability of an intrinsic fault (p) is 0.1 in each cycle

- the workload is an infinite loop

- each iteration in the loop has four cycles

- the unit always has ACE values in the first two cycles of the loop and un-ACE values in the last two cycles of the same loop

- the program is fail-stop with respect to this unit, that is, as soon as an error is detected, the program stops

- T is a random variable that designates the TTF, T can assume the values 1, 2, 5, 6, 9, 10, ..., T cannot be 3, 4, 7, or 8 because the unit has un-ACE values in the last two cycles of the loop and the MTTF is the expected value of T or E(T).

Then, $E(T) = 1 \times p + 2 \times (1 - p) \times p + 3 \times 0 + 4 \times 0 + 5 \times (1 - p)^2 \times p + ...$ This is an infinite series. For this particular example, it sums to 18.53 cycles. For such a loop, Li et al. have created a closed form of the equation as

$$MTTF = \frac{LT + 1 \times p + 2 \times p \times (1 - p)}{p \times (2 - p)} - LT$$

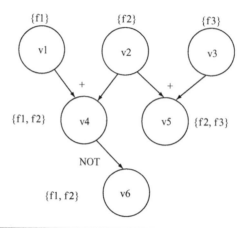

FIGURE 3.15 Example of fault propagation.

where LT is number of cycles in the loop. Using LT = 4, p = 0.1, we get MTTF=18.53 cycles as well.

The above example shows how to compute the MTTF when there is only one unit in a chip. To compute the MTTF in the presence of multiple units across a chip, SoftArch must keep track of how faults propagate through the units during a program's execution. Figure 3.15 shows such an example of fault propagation. Let us assume v6 = NOT (v4), v4 = v1 + v2, v5 = v2 + v3, and f1, f2, and f3 are the faults generated in v1, v2, and v3, respectively, and v6 is an output of the computation. It should be noted that v5 is dynamically dead. In this case, only faults f1 and f2 affect whether output v6 has an error. Hence, v1, v2, v4, and v6 constitute the ACE path. SoftArch tracks propagation of such ACE values through the dataflow graph of the program to compute the probability of an error in an outcome.

Then, the MTTF of an indefinitely executing program can be expressed as

$$\mathrm{MTTF} = \sum_{i=1}^{N} t_i \times \mathrm{Prob}(v_i)$$

where t_i is the time when fault i manifests itself as an error in the output and $\mathrm{Prob}(v_i)$ is the probability that the value v_i corresponding to the fault i has an error, but no prior value $v_1, v_2, \ldots, v_{i-1}$, has an error. As may be obvious, $v_1, v_2, \ldots, v_N$ constitute an ordered set of values, errors in which show up at $t_1, t_2, \ldots, t_N$ times, where $t_1 < t_2 \ldots < t_N$. SoftArch computes the MTTF by keeping track of the v_is, f_is, and t_is. For finite programs, SoftArch proposes to run the same program repeatedly (and therefore indefinitely) and, thereby, compute the MTTF.

Figure 3.16 shows SoftArch's evaluation of AVFs for several processor structures. As described earlier, the AVF for a processor structure can be evaluated as the

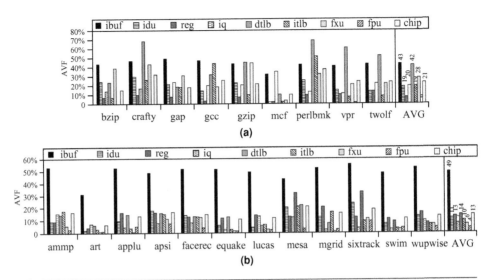

FIGURE 3.16   AVFs from SoftArch's evaluation of a processor roughly resembling IBM's Power4 architecture. (a) AVF for SPECint 2000. (b) AVF for SPECfp 2000. Reprinted with permission from Li et al. [9]. ibuf = instruction buffer, idu = instruction decode unit, reg = architectural register file, iq = instruction queue, dtlb = data TLB, itlb = instruction TLB, fxu = fixed point (integer) execution unit, fpu = floating point unit, and chip = entire chip. Copyright ©2005 IEEE.

ratio of the nonderated and derated MTTF for that particular structure. For this evaluation, SoftArch makes the simplifying assumption that any value stored to memory is an output. Nevertheless, this methodology still provides an upperbound estimate of what the AVF is. Figure 3.16 shows that the AVFs vary from 0% to 49% for different structures and benchmarks.

Unlike ACE analysis that directly computes the AVF, SoftArch first computes the MTTF and then derives the AVFs from them. Theoretically, both are valid methods to compute AVFs. Both suffer from the same disadvantages arising out of using a performance model to compute the AVFs. Both may require significant amounts of memory in the software used to compute the AVFs. ACE analysis may require a significant amount of memory to keep track of dynamically dead instructions. This method attempts to limit the amount of memory needed by using a smaller window of analysis to track dynamically dead instructions, thereby incurring some inaccuracy. SoftArch may require a significant amount of memory to track propagation of ACE values through the dataflow graph. SoftArch reduces this memory requirement by tracking values only through registers but not through memory of the simulated benchmarks, which would also incur some inaccuracy. A systematic evaluation of the merits and demerits of each method, however, is a topic of future work.

3.13 Summary

AVF identifies the fraction of faults that result in a user-visible error for a transistor, bit, or structure. The circuit-level error rate needs to be derated (or multiplied) by the AVF to obtain the overall SER for the component in question. AVFs can vary widely—from 0% to 100%. To reduce the overall SER of the chip under design, designers can choose structures with relatively high bit count and high AVF as candidates for protection.

The SDC FIT of a bit can be expressed as the product of its SDC AVF and circuit-level SER. Similarly, the DUE FIT of a bit can be expressed as the product of its DUE AVF and circuit-level SER. The DUE AVF of a bit is the sum of the true and false DUE AVFs. The true DUE AVF of a bit is its SDC AVF without any error detection. The SDC FIT of a chip can be computed by summing the SDC FIT of all of its constituent transistors, bits, or structures. The same can be done for the DUE FIT.

This formulation of FIT rates and AVFs applies to a single-bit fault model, which is the dominant component of radiation-induced soft errors. Multibit faults can also be analyzed and their formulation will be discussed later in this book. It is also important to note that the AVF is assigned to the bit that was struck by an alpha particle or a neutron and not to a bit where the fault may have propagated.

AVFs can be computed based on ACE principles. A bit is ACE if it needs to be correct for the correct execution of a program. Otherwise, it is un-ACE. The fraction of cycles for which a bit is ACE is its AVF. A structure's AVF is the average AVF of all its constituent bits. Alternatively, a structures's AVF can be expressed as the ratio of the average number of ACE bits in a cycle and the total number of bits in the structure.

Often it is easy to identify the source of un-ACEness and hence AVF can be conservatively computed by analyzing the un-ACE bits during a program's execution. Un-ACE bits can arise from both microarchitectural and architectural components. Examples of microarchitectural un-ACE bits are bits that are idle, bits containing misspeculated states. Examples of architectural un-ACE bits are nonopcode bits of a NOP instruction, nonopcode bits of prefetch and branch hint instructions, and nonopcode and nonregister specifier bits of dynamically dead instructions. Using such an analysis, the AVFs of different structures can be computed using a performance simulation model.

3.14 Historical Anecdote

Prior to the introduction of the term AVF, Intel had widely used the term "logic derating factor" to indicate the same. In 2001, we, a group of four engineers, tried to develop the AVF methodology and frequently talked about cross-purposes because of the term "derating factor." The reason is because the two terms—"derating" and "derating factor"—go in different directions. When we reduce the derating, we *increase* the vulnerability of a bit to soft errors. However, when we reduce the derating factor, we *reduce* the vulnerability of the bit to soft errors. Quite often we would say one thing (e.g., this technique will reduce the derating) when we meant the other (e.g., this technique will reduce the derating factor). To avoid this confusion, we coined the term "vulnerability factor" to replace "derating factor." If you derate more, we increase both the vulnerability and vulnerability factor. Since then, discussions on AVF were more meaningful than in the early days of development of the AVF ideas. Currently, the term AVF is widely used across Intel's soft error groups and management circles.

References

[1] A. Biswas, P. Racunas, J. Emer, and S. S. Mukherjee, "Computing Accurate AVFs using ACE Analysis on Performance Models: A Rebuttal," *Computer Architecture Letters (CAL)*, December 2007.

[2] J. A. Butts and G. Sohi, "Dynamic Dead-Instruction Detection and Elimination," in *10th International Conference on Architectural Support for Programming Languages and Operating Systems (ASPLOS)*, pp. 199–210, October 2002.

[3] B. Fahs, S. Bose, M. Crum, B. Slechta, F. Spadini, T. Tung, S. J. Patel, and S. S. Lumetta, "Performance Characterization of a Hardware Mechanism for Dynamic Optimization," in *34th Annual International Symposium on Microarchitecture (MICRO)*, pp. 16–27, December 2001.

[4] Y. Choi, A. Knies, L. Gerke, and T.-F. Ngai, "The Impact of If-Conversion and Branch Prediction on Program Execution on the Intel Itanium Processor," in *34th Annual International Symposium on Microarchitecture (MICRO)*, pp. 182–191, December 2001.

[5] J. Emer, P. Ahuja, N. Binkert, E. Borch, R. Espasa, T. Juan, A. Klauser, C. K. Luk, S. Manne, S. S. Mukherjee, H. Patil, and S. Wallace, "Asim: A Performance Model Framework," *IEEE Computer*, Vol. 35, No. 2, pp. 68–76, February 2002.

[6] J. L. Hennessy and D. A. Patterson, *Computer Architecture: A Quantitative Approach*, Elsevier Science, 2003.

[7] E. D. Lazowska, J. Zahorjan, G. S. Graham, and K. C. Sevcik, *Quantitative System Performance*, Prentice-Hall, Englewood Cliffs, New Jersey, 1984.

[8] X. Li, S. V. Adve, P. Bose, and J. A. Rivers , "Architecture-Level Soft Error Analysis: Examining the Limits of Common Assumptions," in *International Conference on Dependable Systems and Networks (DSN)*, pp. 266–275, 2007.

[9] X. Li, S. V. Adve, P. Bose, and J. A. Rivers, "SoftArch: An Architecture-Level Tool for Modeling and Analyzing Soft Errors," in *International Conference on Dependable Systems and Networks (DSN)*, pp. 496–505, 2005.

[10] S. S. Mukherjee, C. T. Weaver, J. Emer, S. K. Reinhardt, and T. Austin, "A Systematic Methodology to Compute the Architectural Vulnerability Factors for a High-Performance Microprocessor," in *36th Annual International Symposium on Microarchitecture (MICRO)*, pp. 29–40, December 2003.

[11] H. Patil, R. Cohn, M. Charney, R. Kapoor, A. Sun, and A. Karnunanidhi, "Pinpointing Representative Portions of Large Intel Itanium Programs with Dynamic Instrumentation," in *37th Annual International Symposium on Microarchitecture (MICRO)*, pp. 81–92, 2004.

[12] T. Sherwood, E. Perelman, G. Hamerly, and B. Calder, "Automatically Characterizing Large Scale Program Behavior," in *10th International Conference on Architectural Support for Programming Languages and Operating Systems (ASPLOS)*, pp. 45–57, October 2002.

[13] N. Wang, M. Fertig, and S. Patel, "Y-Branches: When You Come to a Fork in the Road, Take It," in *12th International Conference on Parallel Architectures and Compilation Techniques (PACT)*, pp. 56–67, 2003.

[14] N. Wang, A. Mahesri, and S. J. Patel, "Examining ACE Analysis Reliability Estimates Using Fault-Injection," in *34th International Symposium on Computer Architecture (ISCA)*, pp. 460–469, 2007.

[15] C. Weaver, J. Emer, S. S. Mukherjee, and S. K. Reinhardt, "Techniques to Reduce the Soft Error Rate of a High-Performance Microprocessor," in *31st Annual International Symposium on Computer Architecture*, pp. 264–275, June 2004.

Advanced Architectural Vulnerability Analysis

4.1 Overview

Architectural vulnerability analysis is one of the key techniques to identify candidate hardware structures that need protection from soft errors. The higher is a bit's Architectural Vulnerability Factor (AVF), the greater is its vulnerability to soft errors and hence the need to protect the bit. Besides identifying the most vulnerable structures, the AVF of every hardware structure on a chip is also necessary to compute the full-chip SDC and DUE FIT rates. Hence, a complete evaluation of AVFs of all hardware structures in a chip is critical.

Chapter 3 examined the basics of computing AVFs using ACE analysis. ACE analysis identifies the fraction of time a bit in a structure needs to be correct—that is, ACE—for the program to produce the correct output. This fraction of time is the AVF of the bit. The rest of the time—the time for which the bit does not need to be correct—is called un-ACE.

This chapter extends AVF analysis described in Chapter 3 in three ways. First, this chapter examines how to extend the ACE analysis to address-based structures, such as random access memories (RAMs) and content-addressable memories (CAMs). Chapter 3 focused primarily on ACE analysis of instruction-based structures. ACE analysis of structures carrying instructions is simpler than that for address-based structures. This is because whether a bit is un-ACE or not in a particular cycle depends on whether the corresponding constituent bit of the instruction is un-ACE. For example, for a wrong-path instruction, it can be conservatively assumed that all bits, other than the opcode bits representing the instruction, are un-ACE. Consequently, the bits carrying these un-ACE instruction bits are also un-ACE for the duration they carry the instruction information. In contrast, whether data bits in a cache are un-ACE or not requires a more involved analysis. For example, if a wrong-path load instruction accesses read-only data in a write-through cache, then the corresponding data bits are un-ACE if and only if there is no previous or subsequent access to the same data words by another ACE instruction before the cache block is evicted from the cache. To track whether data bits in a cache, or more generally in RAM and CAM arrays, are ACE or un-ACE, this chapter introduces the concept of *lifetime analysis*.

Second, to track whether bits in a CAM array are ACE or un-ACE, this chapter explains how to augment the lifetime analysis of CAM arrays with a technique called *hamming-distance-one* analysis. The power of ACE analysis to compute AVF arises from its ability to compute the vulnerability of a program from a fault-free execution of the program itself. CAMs, however, make such ACE analysis difficult. A CAM array typically consists of a set of set-associative entries (Figure 4.1). On a CAM match with a set of incoming bits, the corresponding RAM entry is read. ACE analysis with CAMs is difficult because a bit flip in the CAM array can cause an incorrect match against the incoming bits (causing a false-positive match) or a no match when it should have matched (causing a false-negative match). Superficially, it may appear that it will almost be impossible to characterize whether such a bit flip in a CAM array would be ACE or un-ACE without actually flipping a bit in the CAM array and following the subsequent execution of the program. However, the hamming-distance-one analysis can identify the CAM bit or a set of CAM bits that

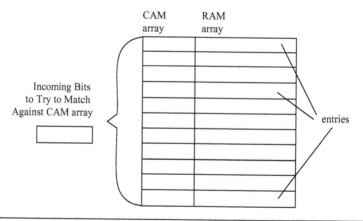

FIGURE 4.1   Mechanics of a CAM array.

need to be ACE to produce the correct output from a program. Once these bits are identified, one can perform the same lifetime analysis, as is done for RAM array bits, to compute the AVF of the CAM array.

Third, this chapter discusses how to compute AVFs using SFI. Unlike ACE analysis that computes AVFs using a fault-free execution of a program, SFI introduces a sample of faults (bit flips) into a hardware model—typically a gate-level representation called an RTL model during a program's execution and observes whether that fault caused a user-visible error. If these bit flips eventually cause a program to produce an incorrect output, then the corresponding state elements are ACE. The difficulty with SFI is that to obtain representative results, one needs to inject faults into a detailed gate-level model, which is significantly slower than a performance model, and carry out a large set of experiments to create a proper statistical representation of the AVF. This allows RTL with SFI simulations to only be run for 1000–10 000 cycles per benchmark, which is often inadequate to decide if these elements are ACE or un-ACE because many microarchitectural and architectural states can live significantly longer than 10 000 cycles. However, SFI into an RTL model can be adequate for latches and flip-flops, whose ACE-ness can often be determined within this window of simulation because the lifetime of data held in these state elements is short (only tens or hundreds of cycles). This chapter discusses the basic principles of SFI and how it can be used to compute AVFs using an RTL model.

4.2 Lifetime Analysis of RAM Arrays

This section explains how to extend the AVF analysis to address-based RAM arrays using *lifetime analysis* and how the differences in properties of different structures and the granularity of the analysis can affect the AVF. Finally, this section illustrates how to compute DUE AVF for RAM arrays that are protected with an error detection mechanism, such as parity.

4.2.1 Basic Idea of Lifetime Analysis

Computing a bit's AVF involves identifying the fraction of time it is ACE. As in Chapter 3, one can focus on identifying un-ACE components in a bit's lifetime since it is typically easier to determine if a bit is un-ACE (as opposed to ACE) in a particular cycle. Subtracting the un-ACE time from total time provides an upper bound on the ACE lifetime of the bit. Lifetime analysis of ACE or un-ACE determination is illustrated using the example in Figure 4.2.

Figure 4.2 shows example activities occurring during the lifetime of a bit in an RAM array, such as in a cache. The bit begins in "idle" state but is eventually filled with appropriate values that could be either ACE or un-ACE. The bit is read and

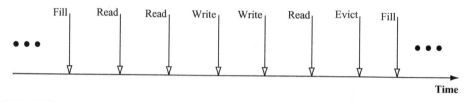

FIGURE 4.2   Lifetime analysis of a bit.

written. Eventually, the state contained in the bit is evicted and refilled. The lifetime of this bit can be divided up into several nonoverlapping components: idle, fill-to-read, read-to-read, read-to-write, write-to-write, write-to-read, read-to-evict, and evict-to-fill. By definition, idle and evict-to-fill are un-ACE since there is no valid state in the bit during those intervals. Read-to-write and write-to-write lifetimes are also un-ACE because a strike on the bit after the read (for read-to-write) or first write (for write-to-write) will not result in an error.

Whether the four other lifetime components—fill-to-read, read-to-read, write-to-read, and read-to-evict—are un-ACE depends on the ACE-ness of the reads and the nature of the architectural structure the bit is a part of. If the read itself in fill-to-read is ACE, then the fill-to-read lifetime is ACE. This can be deduced transitively from the ACE-ness of an instruction. For example, if an instruction reading the register file is ACE, then the read itself is ACE, causing the fill-to-read time to become ACE. However, if the read in fill-to-read is un-ACE (due to an un-ACE read), then one cannot conclude that the fill-to-read time is un-ACE itself until the ACE-ness of the subsequent read is determined. If the first read is un-ACE and the second read is ACE (in the read-to-read lifetime), then both fill-to-read and read-to-read are ACE. However, if both the reads are un-ACE, then one cannot conclude if the fill-to-read and the read-to-read lifetimes are ACE or un-ACE before observing the subsequent write. Once the subsequent write (in read-to-write) is observed, then one can know for sure that the read-to-write lifetime is un-ACE. Then both fill-to-read and read-to-read can be marked as un-ACE.

Finally, whether the read-to-evict is ACE or un-ACE depends on the property of the structure the bit resides in. For example, if the structure is a write-through cache, then the evict operation simply discards the value in the bit. This makes the read-to-evict time un-ACE, independent of the ACE-ness of the read itself. However, in a write-back cache, where the value of a modified bit may be written back to a lower-level cache, the analysis is more complex. To determine whether the read-to-evict is ACE or un-ACE, one has to track the ACE-ness of value through its journey through the computer system until it is overwritten. Often, this interstructure analysis is complicated. Hence, one could conservatively assume that if value in the bit is modified and written back on an evict, then read-to-evict time is ACE. The next subsection further examines how properties of a structure can affect the ACE-ness or un-ACE-ness of a lifetime.

■ E X A M P L E

In Figure 4.2, compute the AVF for the given lifetime from the first fill to the next fill. Assume the following lifetimes are ACE: fill-to-read and write-to-read. Rest is un-ACE. Fill-to-read time is 10 cycles. Write-to-read time is 20 cycles. Fill-to-fill time is 200 cycles.

SOLUTION Total ACE time = 10 + 20 = 30 cycles. Total time = 200 cycles. AVF = 30/200 = 15%.

■ E X A M P L E

A designer is faced with the proposition of evaluating the AVF of a cache and its output latch. Every time the cache is read, the data from the cache are first written to the output latch from where it is read in the next cycle. The designer argues that since every read—hence every ACE read—from the cache is staged through the output latch, the AVF of the output latch should be an upper bound for the AVF of the cache. Is this correct?

SOLUTION This is incorrect. Figure 4.3 shows the lifetime analysis of a counterexample. Consider a one-bit cache with a corresponding output latch. Consider a lifetime sequence of a Fill, 3 ACE Reads, and an Evict. For every ACE Read in the cache, there is a corresponding Write into the output latch followed by an ACE Read in the next cycle after the Write. The AVF of the one-bit cache for this sequence is 12/14 = 86% since Fill to third ACE Read is 12 cycles and total number of cycles is 14. In contrast, the AVF of the output latch is 3/14 = 21%. The AVF of the output latch in this case is significantly smaller than the AVF of the cache. The reason why the AVF of the output latch cannot approximate the AVF of the cache is that the ACE residency time of bits in the cache is significantly longer than that in the output latch. One can easily construct another example to show that the inverse conclusion—whether the AVF of the output latch is a lower bound of the AVF of the cache—is incorrect as well.

4.2.2 Accounting for Structural Differences in Lifetime Analysis

As explained in the previous subsection, the ACE-ness of a bit in a structure depends both on the ACE-ness of the operation on the structure, such as a read, and on certain properties of the structure itself. This section analyzes how ACE-ness can differ

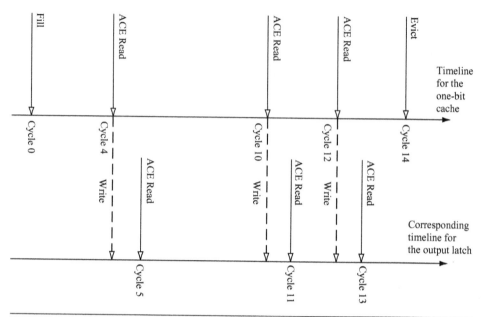

FIGURE 4.3   Lifetime analysis of a one-bit cache and its corresponding output latch.

for four microprocessor structures: a write-through data cache, a write-back data cache, a data address translation buffer (commonly known as the TLB for translation lookaside buffer), and a store buffer. These structures are described briefly below. Readers are referred to Hennessy and Patterson [4] for detailed architectural descriptions of these structures.

Although a data cache is typically protected in a processor, it is a useful structure to explain how lifetime analysis works. Besides, as will be seen later (in Computing the DUE AVF, p. 131), designers often try to optimize the implementation of ECC to reduce the performance degradation by trading off error rate against performance. The AVF analysis of caches is useful for such trade-off analysis.

Data Caches

A processor's closest data cache is a structure that usually sits close to the execution units and holds the processor's most recently and frequently used data. This allows the execution units to compute faster by obtaining data from faster and smaller data caches than from bigger caches or main memory that can be farther away. This section considers a data cache that a processor accesses on every load to read data and on every store to write data. The stores could, however, be routed to the cache via the store buffer discussed later in this section.

Such a data cache can be either a write-through or a write-back. As the name suggests, a write to a write-through data cache writes the data not only to the data cache in question but also to a lower level of the memory hierarchy, which could be a bigger cache or a main memory. Consequently, a write-through data cache never has modified data. In contrast, a write-back data cache keeps modified data and does not propagate the newly written data to a lower level of the memory hierarchy. This is a performance optimization that works well when enough write bandwidth is not available. Eventually, when a cache block needs to be replaced, the modified data are written back and hence the name write-back cache.

As shown in Figure 4.4, the ACE and un-ACE classifications differ for a write-through and a write-back cache (as well as for a data translation buffer and a store buffer). The column "un-ACE" shows the lifetime components that can be definitely identified as un-ACE. The column "Potentially ACE" shows lifetime components, such as fill-to-read, that are possibly ACE, unless one can show through further analysis that these components are un-ACE. For example, if the read and all subsequent reads before eviction can be shown to be un-ACE, then the fill-to-read component becomes un-ACE. Hence, this column is marked potentially ACE. Finally, the column "Unknown" shows lifetime components that remain unresolved because the simulation ends before the resolution could be achieved. The section Effect of Cooldown in Lifetime Analysis, p. 138, examines how a technique called cooldown can reduce this unknown component.

Processor Structure	Lifetime Classification		
	un-ACE	ACE	
		Potentially ACE	Unknown
Write-through data cache	idle, fill-to-write, fill-to-evict, read-to-write, read-to-evict, write-to-write, write-to-evict, evict-to-fill	fill-to-read, read-to-read, write-to-read	fill-to-end, read-to-end, write-to-end
Write-back data cache	idle, fill-to-write, fill-to-evict, read-to-write, read-to-evict, write-to-write, evict-to-fill	fill-to-read, read-to-read, write-to-read, write-to-evict, write-to-end	fill-to-end, read-to-end
Data Translation Buffer	idle, read-to-evict, evict-to-fill	fill-to-read, read-to-read	fill-to-end, read-to-end
Store Buffer	idle, fill-to-write, fill-to-evict, read-to-write, read-to-evict, write-to-write, write-to-evict, evict-to-fill	fill-to-read, fill-to-evict, fill-to-end, read-to-read, read-to-evict, read-to-end	none

FIGURE 4.4 Lifetime classification of RAM arrays in four processor structures. The sections Accounting for Structural Differences in Lifetime Analysis, p. 125, and Effect of Cooldown in Lifetime Analysis, p. 138, explain components of the table. End here denotes the end of simulation.

In Figure 4.4, the most striking difference in the lifetimes for a write-through and a write-back cache is the write-to-evict component, which is un-ACE for a write-through cache but could be ACE for a write-back cache. The write-to-evict is un-ACE in a write-through cache because on eviction, the data are thrown away. So an upset on any of those bits would not matter. In contrast, in a write-back cache, modified data generated by a write will be written back to the lower levels of the memory hierarchy on eviction. Consequently, those data will be used and could potentially be ACE.

To track whether write-to-evict and other lifetime components are ACE or not, one needs to track the values through the memory hierarchy and hence perform interstructure ACE analysis. For practical purposes, industrial design teams could assume conservatively that any write-to-evict in a write-back cache is ACE, unless the AVF is too high and calls for a more precise analysis. Figure 4.4 bins the lifetime components into ACE or un-ACE based on this conservative assumption. Some of the ACE components could, however, be categorized as un-ACE with further information from interstructure analysis.

Data Translation Buffer

The data translation buffer is a processor structure that caches virtual-to-physical address translations and associated page protection information from the page table. A page table is a common OS structure maintained in software. It allows a virtual user process to map its address space to the physical address space of the machine, thereby supporting virtual memory. Every load and store instruction performs a CAM operation (Figure 4.1) on the data translation buffer with its virtual address. On a CAM hit, a load or a store obtains the corresponding physical address and the associated protection information.

ACE analysis of the data translation buffer RAM is relatively straightforward, particularly because there are no "writes" to a data translation buffer besides the "fill" that writes an entry when it is brought into the buffer. Fill-to-read and read-to-read are again ACE, by default, unless the corresponding load or store that initiated the read is un-ACE and there is no subsequent ACE read on this entry.

Store Buffer

The store buffer is a staging buffer for stores before the store data are written into a coalescing merge buffer. From the merge buffer, the data are written to the cache. For dynamically scheduled processors, a store buffer is particularly important to ensure memory ordering within a program. Like the data cache, the store buffer is written into by store instructions but read by loads. Unlike the data cache, however, each store creates a new entry in the store buffer. Thus, the store buffer can concurrently hold multiple stores to the same address. Also, the store buffer typically has per-byte mask bits to identify the bytes that have been modified. As soon as a store instruction retires, it becomes a candidate for eviction. When a store is evicted, the pipeline moves it to a coalescing merge buffer from where the data are eventually written into the cache hierarchy. The residency times of entries in the store buffer are much shorter than corresponding ones in the data cache or data translation buffer.

Let us now examine which components are ACE and un-ACE. The evict-to-fill time in a store buffer is un-ACE because there is no valid entry in a store buffer once the entry is evicted. Read-to-fill time may be ACE, unlike in a write-through cache. This is because even if the read in read-to-evict is initiated by an un-ACE load, once the entry is evicted, the corresponding data are written back to the cache and could be potentially ACE.

Further, the store buffer is somewhat unique in that a write to the data bit of one entry can change the ACE status of a data bit in a completely different entry. Consider two stores that write to the same byte of a data address. These stores will occupy different entries in the store buffer. In a single-processor system, the bits representing this byte in the store buffer entry associated with the older store become un-ACE as soon as the younger store in the store buffer retires. This is because any subsequent loads to this address will receive their value from the younger store buffer entry.

4.2.3 Impact of Working Set Size for Lifetime Analysis

Besides the nature of an address-based structure, the working set size of a program resident in such structures can have a significant impact on a structure's AVF. For example, if in a 128-entry data translation buffer, only one entry is ever used, then the AVF will never exceed 1/128. Similarly, if a structure's miss rate is very high, then the AVF is likely to be low because part of the ACE lifetime gets converted to un-ACE time. For example, an intervening eviction between a write and a read—arising possibly from reduction in a structure's size or forced eviction—can convert ACE time to un-ACE, thereby reducing that entry's contribution to overall AVF (Figure 4.5).

The AVF is harder to predict, however, if the working set size of the structure is just around the size of the structure itself. If the working set size fits just in the structure, then the AVF could be potentially high. But if the structure's size is reduced slightly, the AVF could go down significantly because of evictions and turnover experienced by the structure.

◼ E X A M P L E

Consider two scenarios for a 64-entry RAM structure. In both cases, there is a miss followed by one or more ACE reads, followed by a miss again. Then, the pattern repeats. However, in the first one, the following sequence plays out: a

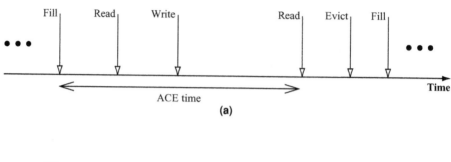

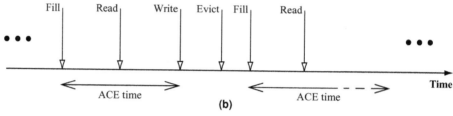

FIGURE 4.5   Evictions can reduce the ACE time of a bit or a structure. (a) The original flow. (b) The reduction in ACE time due to an early eviction.

miss-to-last read time is 99 cycles followed by a read-to-miss time of one cycle. In the second one, the following sequence occurs: a miss-to-read is one cycle and read-to-miss is one cycle. Consider the ACE reads. Compute the AVF in the two different scenarios.

SOLUTION In the first case, the AVF = $99/(99+1) = 99\%$. In the second case, the AVF is $1/(1+1) = 50\%$. The first case experiences a low miss rate, whereas the second case has a fairly high miss rate.

4.2.4 Granularity of Lifetime Analysis

The granularity at which lifetime information is maintained can have a significant impact on the lifetime analysis of certain structures, such as a cache. This relates to accurate accounting in ACE analysis, unlike the previous two issues that relate to the property of a structure and the working set size of a program. Let us consider a cache to understand this point. A cache data (or RAM) array is divided into cache blocks, whose typical size ranges between 32 and 128 bytes. When a byte in a cache block A is accessed, the entire cache block A is fetched into the cache. However, not all the remaining bytes in the block A will be read or written by the processor. When a new cache block B replaces the cache block A, the remaining bytes in block A become un-ACE. This is reflected as fill-to-evict time (see Figure 4.4), which represents the bytes of the data never used before the line is evicted or used only by the initial access. For a write-through cache, fill-to-evict time could be as high as 45% of its total un-ACE time [3].

The write-back cache also needs special consideration. Consider the following scenario in the data array. Two consecutive bytes A and B in the same cache block are fetched into the data array. If A is only read and B is never read prior to eviction, then the fill-to-evict time for B is un-ACE. In contrast, if A is written into, then the fill-to-evict time for B becomes potentially ACE because the entire block (including B) could now be written back into the next level of the cache hierarchy. Thus, a fault in B could propagate to other levels of the memory hierarchy. To handle a write-back cache, the lifetime breakdown must be modified. Bytes of a modified line that have not themselves been modified are ACE from fill-to-evict, regardless of what may happen to them in the interim. The bytes that have been modified are ACE from last write-to-evict and any from earlier write-to-read time. The bytes of an unmodified line work identically to those of a write-through cache. Hence, two of the un-ACE components of a write-back cache (as shown in Figure 4.4), fill-to-evict and read-to-evict, can be conditionally ACE at certain times. This extra ACE component could potentially be reduced by adding multiple modified bits, each representing a portion of the cache line.

For the data translation buffer, it is sufficient to maintain the ACE and un-ACE components on a per-entry basis. For the store buffer data array, however, it may be necessary to maintain the information on a per-byte basis because of the per-byte masks.

EXAMPLE

Compute the AVF of a structure in which the average residence time is 100 cycles and 40% of the bytes in a block are never touched once they are brought in. There are on average two ACE reads to the other 60% of the bytes in the block. Average fill-to-read time is 40 cycles and read-to-read time is 30 cycles. Read-to-evict time is un-ACE.

SOLUTION The AVF would be $40\% \times 0 + 60\% \times (40 + 30)/100 = 42\%$.

4.2.5 Computing the DUE AVF

All prior discussions in this section focused on determining the SDC AVF, which assumes no protection for a specific structure. Instead, if these structures had fault detection (e.g., via parity protection) and no recovery mechanism, then the corresponding AVF is called DUE AVF. As described in False DUE AVF, p. 86, Chapter 3, one can derive the DUE AVF by summing the original SDC AVF and the resulting false DUE AVF.

In the structures referred to in this section, false DUE AVF from parity protection arises only for a write-back cache and the store buffer. On detecting a parity error, the write-through cache and the data translation buffer can refetch the corresponding entry from the higher-level cache and page table, respectively. That is, with parity and an appropriate recovery mechanism, the DUE AVF of both a write-through data cache and a data translation can be reduced to zero.

In both the write-back cache and the store buffer RAM false DUE arises from dynamically dead loads. When a dynamically dead load reads an entry in the cache or the store buffer RAM array, it can check for errors by recomputing the parity bit. If there is a mismatch between the existing and the computed parity bit, then the cache or the store buffer will signal an error, resulting in a false DUE event.

■ E X A M P L E

Assume a processor has a few architecturally visible scratch registers, which are used only in a special mode. Nevertheless, the processor initializes them to zero every time it boots and the OS saves and restores them on every context switch. What is the SDC AVF of these scratch registers? What would be the DUE AVF if the registers were protected with parity and checked for errors every time they are read?

S O L U T I O N   Since the registers are rarely used, the SDC AVF is probably close to zero. If the registers are protected with parity, then whenever the OS saves them, it will be forced to read the registers and declare an error on a parity check violation. Hence, the DUE AVF for these registers are probably close to 100% (Figure 4.6a). Note, however, if the register is actively read and written, the false DUE AVF goes down, even if the OS saves and restores the registers (Figure 4.6b).

■ E X A M P L E

To protect against transient faults, processors often have their caches protected with SECDED ECC, where SECDED is for single-error correction and double-error detection (see Chapter 5 for details on how SECDED ECC works). Although SECDED ECC can correct single-bit errors, it can incur an extra cycle of penalty in a high-frequency processor pipeline. To avoid this extra cycle of penalty, a processor designer decided to use the ECC for in-line fault detection, which will not incur this penalty, but out-of-band error recovery to ensure that errors in the cache are eventually corrected. In-line fault detection ensures that when a load accesses the cache line with a fault, it will always detect the error but will not be able to correct it. A background scrubber wakes up periodically, scans one cache line at a time, and corrects any resident error in the cache using

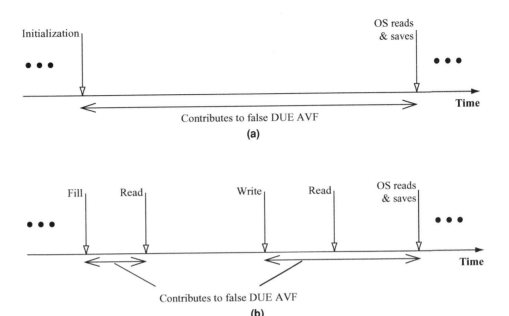

FIGURE 4.6   False DUE exposure in two different cases. (a) For a bit that is hardly written or read but saved by the OS. (b) For a bit that is actively read and written as well as saved by the OS.

the ECC. This is called out-of-band correction since the scrubber is not in the critical path of a load access. AMD's Opteron^{TM} processor uses a somewhat similar scheme [2].

Let us consider a 16-kilobyte write-through cache, with each ECC covering 8 bytes of data. Also, assume that load accesses to the cache are uniformly distributed, and each ECC-protected word undergoes the following lifetime sequence: fill, ACE read, ACE read, evict. Assume both fill-to-ACE read and ACE read-to-evict time are negligible. ACE read-to-ACE read time per word on average is 1000 cycles. Assume that the scrubber wakes up every 20 cycles (ignore the absence of free read port in the cache), finds the next cache block (size = 64 bytes), and corrects any existing error in all the eight words in the block. For simplicity, assume that the scrubber typically accesses a word halfway between the two ACE reads. How much will the DUE AVF of the cache reduce from this scrubbing scheme?

S O L U T I O N   In the absence of any scrubbing, a load accessing a faulty word will incur a DUE event. Whenever a scrubbing event occurs between two ACE reads, the ACE time from ACE read-to-scrub is converted to un-ACE, which causes the reduction in the DUE AVF (Figure 4.7). However, the cache itself has 256 cache blocks, so the DUE AVF reduction would be roughly 500 cycles

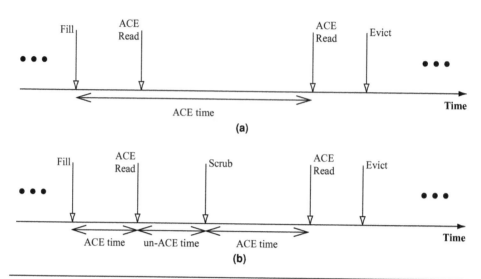

FIGURE 4.7 **Effects of scrubbing. (a) Fill-to-read and read-to-read are both ACE. (b) Fill-to-read is ACE, but read-to-scrub is un-ACE due to the intervening scrub. It should be noted that the scrub can happen anywhere between the two ACE Reads, and it will still convert the read-to-scrub time to un-ACE.**

(half of the ACE read-to-ACE read time, as per the example) for every 256 × 20 = 5120 cycles. The total time to sequentially scrub the cache is 5120 cycles. Consequently, the DUE AVF reduction in this specific scenario would be ~10% (500/5120).

Using simulation, Biswas et al. [3] showed that the DUE AVF could be reduced by 42% by scrubbing with a 16-kilobyte cache, a 2-GHz processor, and a scrubbing interval of 40 ns (which was the best scrubbing interval for the Opteron™ processor). This analysis scrubs only on idle cache cycles to minimize any disruption in the processor's performance. The reduction in DUE AVF from scrubbing, however, is highly dependent on the interaccess time of a word, the size of the cache, and the frequency of scrubbing. Consequently, the advantage of scrubbing must be carefully computed based on these design parameters.

4.3 Lifetime Analysis of CAM Arrays

The lifetime analysis of CAMs has both similarities to and differences from that of RAM arrays. Like RAM arrays, CAMs are common hardware structures used in a silicon chip design. For example, tags in data caches are designed with CAMs. As in Figure 4.2, the lifetime analysis of CAMs also involves monitoring activities on the CAM array and identifying the un-ACE portion of a CAM bit's lifetime. If

Should Have	Actual Outcome	Potential Error?			Scenario
		Write-through Cache	Data Translation Buffer	Store Buffer	
Mismatched	Matched	Yes	Yes	Yes	False Positive
Matched	Mis-matched	No	No	Yes	False Negative

FIGURE 4.8 Un-ACE CAM lookup scenarios with single-bit faults.

the soft error contribution of a CAM array is deemed high from its bit count or its AVF computed via the lifetime analysis, then a designer can choose to add to it a protection mechanism, such as parity, ECC bits (see Chapter 5), or radiation-hardened circuits (see Chapter 2).

Unlike RAM arrays, however, one needs to handle false-positive and false-negative matches in CAM arrays. As discussed before, a CAM, such as the tag store, operates by simultaneously comparing the incoming match bits (e.g., an address) against the contents of each of several memory entries (Figure 4.1). Such a CAM can give rise to two types of mismatches, as shown in Figure 4.8. First, incoming bits can match against a CAM entry in the presence of a fault when it should really have mismatched (the *false-positive* case). If the RAM entry corresponding to the CAM entry is read, then this will cause the RAM array to deliver incorrect data, potentially causing incorrect execution. Similarly, there is a potential for incorrect execution if the RAM entry is written into.

Alternatively, incoming bits may not match any CAM entry, although they should have really matched (the *false-negative* case). For a write-through cache or a data translation buffer, this would result in a miss, causing the entry to be refetched without causing incorrect execution. However, for a store buffer that holds modified data, this may cause an error because the incoming load would miss in the store buffer and obtain possibly stale data from the cache. An incoming write to a write-back cache will have a similar problem. A false-negative match would refetch an incorrect block to which the write will deposit its data. Methods to handle these scenarios are discussed in more detail below.

4.3.1 Handling False-Positive Matches in a CAM Array

Figure 4.9 shows an example of the false-positive match. The incoming match bits, 1001, would not have matched against the existing CAM entry, 1000, unless the fourth bit is flipped to 1. One can use a technique called *hamming-distance-one* analysis to compute the AVF of CAM entries. Two sets of bits are said to be apart by hamming distance of one, if they differ in only one bit position. Thus, 1001 and 1000 are apart by a hamming distance of one (in the third bit position, assuming

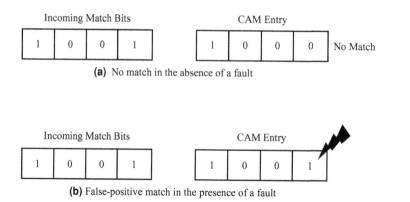

FIGURE 4.9   False-positive match for a CAM entry.

bit positions are marked from zero to three). Similarly, 0001 and 1000 are apart by a hamming distance of two because they differ in the zero-th and third bit positions.

Assuming a single-bit fault model, an incoming set of bits can cause a false-positive match in the CAM array if and only if there exists an entry in the CAM array that differs from the incoming set of bits in one bit position. In other words, false positives are introduced in the CAM entries that are at a hamming distance one from the incoming set of bits. The example in Figure 4.10 shows that only two of the CAM entries are at a hamming distance of one from the incoming bits.

Once the bits that may cause a mismatch have been identified, one can perform the lifetime analysis on these bits. However, because the false-positive case is caused by one particular bit in a tag entry, the ACE analysis of the tag array must be done on a per-bit basis, rather than on a per-entry or a per-byte basis as in the RAM arrays. That is, when a match is found, the bit is marked as potentially ACE. All other bits in the same entry remain un-ACE.

■ E X A M P L E

Compute the AVF of a write-through CAM array with 10 4-bit-wide entries over 10 cycles. Assume that only false-positive matches (not false-negative ones) contribute to the AVF. In each cycle, there is an incoming address CAM-ing against the CAM array. During these 10 cycles (marked 0 through 9), there is only one hamming-distance-one match in bit position 1 of entry 1 in cycle 7. Entry 1 gets evicted in cycle 9.

S O L U T I O N   Because the CAM array is write-through, the structure does not hold modified data in the corresponding RAM array. Bit 1 of entry 1 is potentially ACE for eight cycles (cycles 0 through 7) because any fault in bit position

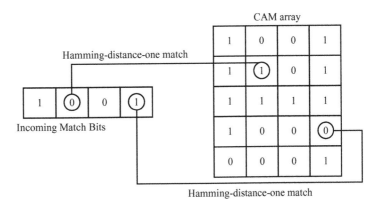

FIGURE 4.10   Hamming-distance-one match in a CAM array.

1 between cycles 0 and 7 would deliver a false-positive match in cycle 7. The rest of the bits are un-ACE throughout. The AVF of bit 1 of entry 1 is $8/10 = 80\%$. The AVF of the rest of the bits throughout these 10 cycles is zero. So the average AVF of the CAM entry is $(80\% \times 1 + 0 \times 39)/40 = 2\%$.

4.3.2 Handling False-Negative Matches in a CAM Array

The false-negative case is easier to track than the false-positive case. The false-negative case occurs when the incoming match bits do not find a match when they should have really matched the bits in a CAM entry. A single-bit fault in any bit of the CAM entry would force a mismatch. On a false-negative match, therefore, all bits in the CAM entry are marked either ACE or un-ACE depending on whether a false miss in the structure would cause incorrect execution. In a data translation buffer or a write-through cache, such a false miss will not cause an incorrect execution, but this would cause an incorrect execution in a write-back cache or a store buffer. In the write-back cache, a false miss could fetch stale data, causing incorrect execution. In the store buffer, an incoming load address CAM-ing against the store buffer will miss and obtain stale data from the data cache, potentially causing incorrect execution.

Interestingly, there is a subtle difference between the RAM and CAM analyses. On a single-bit fault in the RAM array, the actual execution does not necessarily change because the effect of the single-bit fault is localized. In contrast, on a false-negative match in the CAM array, one may not get an actual fault (e.g., as in the write-through cache or the data translation buffer). But the fault can alter the flow of execution because the hardware would potentially bring in a new entry in the

CAM array. Biswas et al. [3], however, verified that this effect did not alter the AVF of a microprocessor data translation buffer in any significant way.

4.4   Effect of Cooldown in Lifetime Analysis

As should be apparent by now, properties of structures (e.g., write-back vs. write-through) and time of occurrence of events, such as write, read, or evictions, may have a significant impact on the AVF of a structure. In a similar way, *when* the AVF simulation ends can also have a significant impact on the AVF of structures. This is termed as an "edge effect."

The edge effects arise as an artifact of not running a benchmark to completion in a performance model. For example, in Figure 4.11a, if the simulation ended after the fill, then one would not know if the fill-to-end time is ACE or un-ACE and therefore must be marked as unknown. If there were an ACE read after the simulation ended, the fill-to-end time should really have been ACE. Conversely, if there were an eviction after the simulation ended (Figure 4.11b), then the ACE read-to-end time should have been un-ACE instead of unknown. These can have significant impacts on the AVF numbers because designers may have to conservatively assume that AVF = (ACE time + unknown time)/(total time).

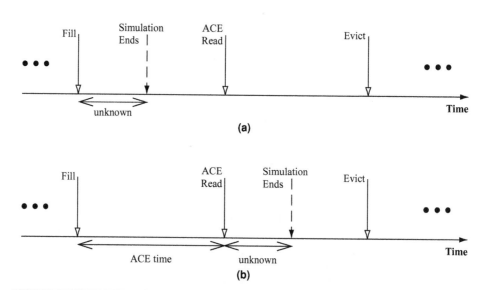

FIGURE 4.11   Edge effect in determining ACE and un-ACE time in a write-through structure. In (a), fill-to-end time is unknown when the simulation ends but should really be ACE. In (b), ACE read-to-end time is unknown when the simulation ends but should really be un-ACE.

To tackle these edge effects, Biswas et al. [3] introduced the concept of *cooldown*, which is complementary to the concept of warm-up in a performance model. A processor model faces a problem at start-up in that initially all processor states, such as the content of the cache, are uninitialized. If simulation begins immediately, the simulator will show an artificially high number of cache misses. This problem can be solved by warming up the caches before activating full simulation. In the warm-up period, no statistics are gathered, but the caches and other structures are warmed up to reflect the steady-state behavior of a processor.

Cooldown is the dual of warm-up and follows the actual statistics-gathering phase in a simulation. During the cooldown interval, one only needs to track events that determine if specific lifetime components, such as fill-to-end or read-to-end, should be ACE or un-ACE. If after the end of the cooldown interval, one cannot precisely determine if the specific lifetime components are ACE or un-ACE, they can be marked as unknown (Figure 4.4).

Figure 4.12 shows the effect cooldown has in reducing the unknown component. The y-axis of the graph shows the average AVF for each structure over all benchmarks. There are two bars associated with each structure. The first bar represents the structure's AVF without cooldown, and the second represents the AVF with cooldown. The gray section of each bar represents the fraction of AVF that is unknown at simulation end. For every structure other than the tags (CAM) of the data translation buffer, the cooldown period reduces the unknown component by over 50%. Less effect is seen in the data translation buffer because an unknown

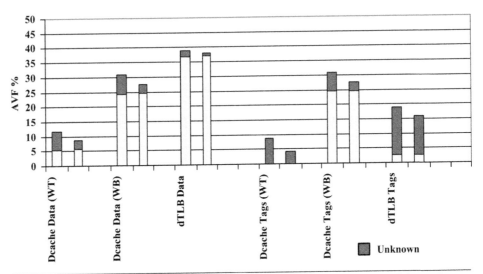

FIGURE 4.12 Effect of cooldown with a 10-million instruction cooldown interval. For each structure, there are two bars. The first bar shows the AVF without cooldown. Reprinted with permission from Biswas et al. [3]. Copyright © 2005 IEEE.

entry can only be classified after it is evicted, and the data translation buffer has a much lower turnover rate than the other structures. Increasing the cooldown period, however, progressively reduces the unknown component without raising the ACE component, suggesting that asymptotically the unknown component may become negligible.

4.5 AVF Results for Cache, Data Translation Buffer, and Store Buffer

This section discusses AVF results for a write-through and a write-back cache, a data translation buffer, and a store buffer. These structures and their differences have been described earlier in this chapter (see Accounting for Structural Differences in Lifetime Analysis, p. 125). The evaluation methodology, performance simulator, and benchmark slices are same as described in Evaluation Methodology, p. 107, Chapter 3. Both the data caches studied are 16-kilobyte four-way set associative. The data translation buffer is fully set associative with 128 entries. The store buffer has 32 entries. All simulations were run for 10 million instructions followed by a cooldown of 10 million instructions. Further details of these experiments are described in Biswas et al. [3].

As per the ACE methodology, the AVF can be divided into two components: *potentially ACE* and *unknown* components (Figure 4.4). Potentially ACE components are those that are possibly ACE unless later analysis proves they are un-ACE. Unknown lifetime components are those that are unknown because the simulation ended before resolving whether the components are unknown. As it was seen in Effect of Cooldown in Lifetime Analysis, p. 138, typically these unknown components can be reduced significantly using cooldown techniques. Based on this observation, one can use two AVF terms: upper-bound AVF and best-estimate AVF. Upper-bound AVF includes both potentially ACE and unknown components. In contrast, best-estimate AVF includes only potentially ACE components under the assumption that if one ran the programs to completion, the unknowns would resolve and become mostly un-ACE.

4.5.1 Unknown Components

From the graphs in Figures 4.13–4.16, it is seen that the RAM arrays have an average unknown component of 3% and data cache and store buffer CAM arrays have an average of 4%. The data translation buffer CAM array has a significantly higher unknown component of 13%. This is because the data translation buffer has a significantly lower turnover rate than the data cache. That is, entries tend to stick around in the data translation buffer for long durations—even beyond the simulation and cooldown periods. Thus, all the CAM bits in an entry that do not hamming-distance-one match with a memory operation remain in the unknown state until that entry is evicted from the translation buffer.

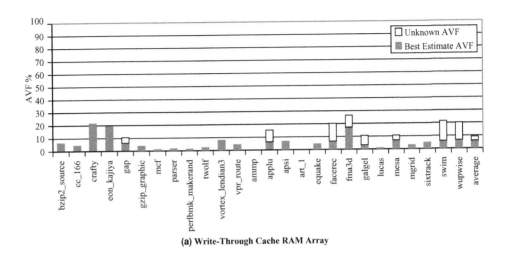

(a) Write-Through Cache RAM Array

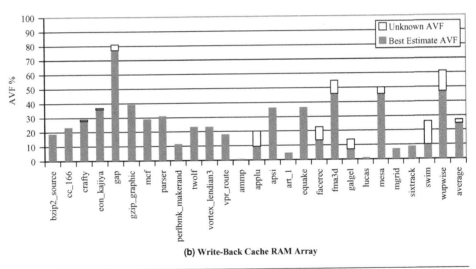

(b) Write-Back Cache RAM Array

FIGURE 4.13 AVFs of the RAM arrays of a write-through and a write-back cache.

Fortunately, hamming-distance-one matches are rare, and each match only adds ACE time to a single bit of the matched tag in the CAM array. Further, separate experiments done by Biswas et al. (not shown here) show that nearly all these bits will eventually resolve to the un-ACE state. Similarly, the unknown lifetime components for the data arrays also resolve mostly to un-ACE if the cooldown period is extended further. Hence, the rest of this section primarily discusses the best-estimate AVF numbers.

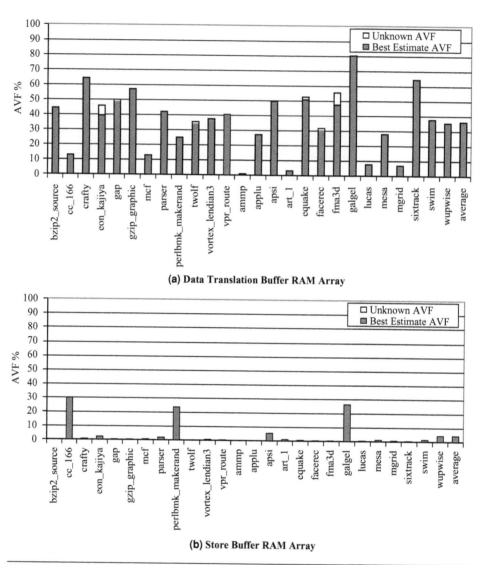

FIGURE 4.14   **AVFs of the RAM arrays of a data translation buffer and a store buffer.**

4.5.2 RAM Arrays

The best estimate of SDC AVFs varies widely across the RAM arrays: from 4% for the store buffer (Figure 4.14) to 6% for the write-through data cache (Figure 4.13) to 25% for the write-back cache (Figure 4.13) to 36% for the data translation buffer (Figure 4.14). If unknown time is included, these rise to 4%, 9%, 28%, and 38%, respectively.

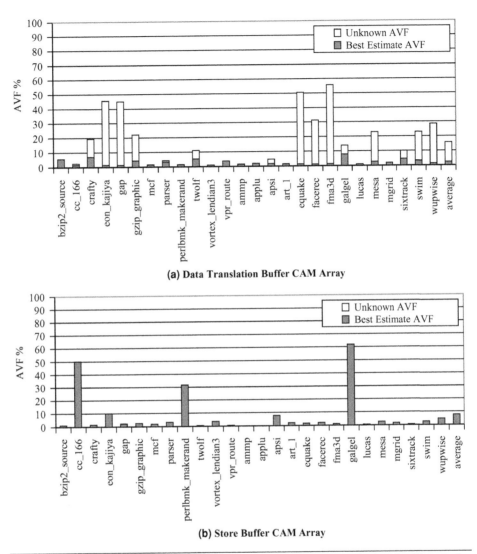

(a) Data Translation Buffer CAM Array

(b) Store Buffer CAM Array

FIGURE 4.15   AVFs of the tag (CAM) array of a data translation buffer and a store buffer.

The store buffer's low SDC AVF arises from its bursty behavior and lower average utilization in most benchmarks. Additionally, the store buffer has per-byte mask bits that identify which of the 16 bytes of an entry is written. Entries that are not written remain un-ACE and do not contribute to the AVF. In the average in-use store buffer entry, only 6 out of the 16 bytes were written.

The data translation buffer's RAM array has an SDC AVF of 36%, the highest among the RAM arrays discussed in this chapter. This is due to its read-only status and relatively low turnover rate.

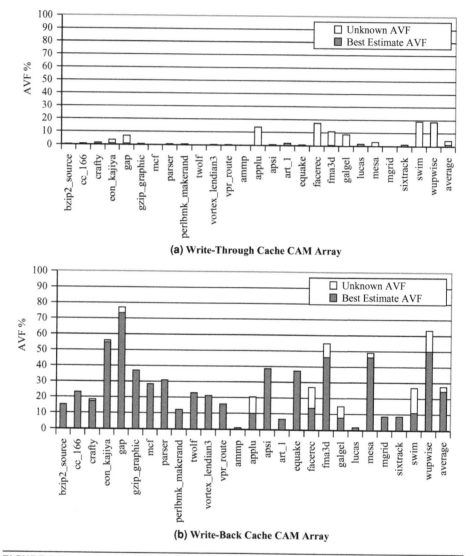

FIGURE 4.16 AVFs of the tag (CAM) arrays of a write-through and a write-back cache.

The write-through cache's SDC AVF is relatively low for three reasons. First, on average, over 45% of the bytes read into the cache are accessed only on that initial access or not at all. A read or write to a specific byte or word brings in a whole cache block full of bytes, but many of these other bytes are never accessed. Second, read-to-evict constitutes a significant fraction of the average overall lifetime (over 20%). These results agree with observations made by others (e.g., Lai et al. [6],

Wood et al. [10]). Read-to-evict is un-ACE for a write-through cache. It is also un-ACE for a write-back cache, assuming that no write preceded the read. Last, unlike the data translation buffer, a data cache line is modified by stores, indicating that any bytes overwritten by a store are un-ACE in the time period from the last useful read until the write. On the graph, this component of un-ACE time accounts for a little less than 5% of the overall lifetime.

The write-back cache's SDC AVF is 25% compared to 6% of the write-through cache. This is because a write to a byte of a cache line makes all unmodified bytes in the cache line ACE from the time of the initial fill until the eviction of the line. This is true regardless of the intervening access pattern to that line. Per-byte mask bits (as in the store buffer) would help avoid the write-backs of bytes never modified, thereby reducing the write-back cache's AVF.

It should be noted that there is a wide variation among the AVFs across benchmarks. Further, AVFs of the different structures for the same benchmark are not necessarily correlated, that is, a high AVF on one structure may not necessarily imply a high AVF on a different structure for the same benchmark. This is true for both the RAM and CAM arrays.

4.5.3 CAM Arrays

The best-estimate SDC AVFs of the CAM arrays of the write-through cache and the data translation buffer, which do not include unknown time, are quite low: 0.41% and 3%, respectively. These values are considerably lower than the same values for the corresponding data arrays: 6% and 36%, respectively. The low AVF in these CAM arrays arises because the false-negative case—mismatch when there should have been a match—forces a miss and refetch in these structures but does not cause an error. In contrast, the false-positive case—match when there should have been a mismatch—causes an error, but it only affects the ACE state of the single bit that causes the difference. Consequently, there are significantly fewer ACE bits on average in the tag arrays of these structures compared to the data arrays. The write-through tag AVF is particularly low because a hamming-distance-one match would have to occur between the four members of a set to contribute to ACE time.

When including unknown time, SDC AVFs of the CAM arrays of the write-through cache and data translation buffer increase to 4.3% and 16%, respectively. The lower numbers are more representative of the actual AVFs of these two structures because hamming-distance-one matches are so rare, and each hamming-distance-one match makes only 1-bit ACE. If the higher numbers were used, one would be classifying all bits of the tag as ACE at the end point of the simulation. Also, it is likely that these structures would be effectively flushed by a context switch before the end of the cooldown phase (unless the structures have per-process identifiers).

In contrast, the SDC AVF of the store buffer is 7.7%, which is, as per our expectation, higher than that of the corresponding RAM arrays. This is because the store buffer tags are always ACE from fill to evict. The only contributor to un-ACE time for the store buffer tags is the idle lifetime component. The low AVF implies that the store buffer utilization is on average quite low. The CAM AVF is higher than the RAM AVF for the store buffer because all the bytes in the store buffer entry are not written by each store. On average, an in-use store buffer entry contains only 6 valid bytes out of 16 total bytes. Only the valid bytes contribute ACE time.

The SDC AVF of the CAM array of the write-back cache is 25% but is not directly correlated with that of its RAM array. This is because two of the un-ACE components for a write-through cache—fill-to-evict and read-to-evict—may become potentially ACE in a write-back cache. A write-back cache's CAM entry is always ACE from the time of first modification of its corresponding RAM entry until that entry is evicted.

4.5.4 DUE AVF

A DUE occurs when a fault in a structure can be detected but cannot be recovered from it. Putting parity on the RAM arrays of a write-back cache or a store buffer allows one to detect a fault in a dirty block but not recover from it. To compute the DUE AVF, one can sum the original SDC AVF (same as the true DUE AVF with parity) and false DUE AVF arising from dynamically dead loads. Analysis shows that the false DUE AVF is, on average, an additional 0.2% and 0.5% arising out of dynamically dead loads and stores. Hence, the total DUE AVF for the RAM arrays of a store buffer and write-back cache is 4.2% and 25.5%, respectively. It should be noted that the false DUE AVF component for these structures is significantly less than what was found for an instruction queue (Figure 3.11).

The DUE AVF of the CAM arrays of both a write-back cache and a store buffer is the same as their corresponding SDC AVF (in the absence of parity) since the CAM bits are required to be (conservatively) correct even if the store is dynamically dead. An incorrect CAM entry could result in data being written to a random memory location when the entry is evicted.

Putting parity on a RAM or CAM array of a write-through cache or a data translation buffer allows one to recover from a parity error by refetching the corresponding entry from the higher-level cache or page table, respectively. Consequently, DUE AVF of these arrays can be reduced to zero in the presence of parity.

4.6 Computing AVFs Using SFI into an RTL Model

This chapter so far has discussed how to compute the AVF of RAM and CAM arrays using ACE analysis of fault-free execution. This section discusses how to compute the AVF and assess the relative vulnerabilities of different structures using SFI. First, the equivalence of fault injection and ACE analyses, as well as their advantages and disadvantages, is discussed. Then, two key aspects of the SFI methodology are described. Finally, a case study on SFI done at the University of Illinois at Urbana–Champaign is discussed.

4.6.1 Comparison of Fault Injection and ACE Analyses

Figure 4.17 shows how ACE analysis and SFI can compute the same AVF. Figure 4.17a shows a sequence of operations: a Fill, an ACE Read, an ACE Read, an un-ACE Read, and an Evict. Let us assume that the ACE time is from the Fill to the second ACE Read and the un-ACE time is from the second ACE Read till Evict. Hence, the AVF = ACE Time/(Fill to Evict Time) = 6/15 = 40%.

Figure 4.17b shows how the same can be achieved via fault injection. In this experiment, the same program is executed 15 times (instead of once as done in ACE analysis). In each of the executions, a fault is injected into a different cycle. For example, during the first execution, the fault is injected into cycle 0; during the second execution, the fault is injected into cycle 1 etc. When a fault is injected, the bit's value is flipped. That is, it changes to one if it were zero and zero if it were one. The bit's AVF in this case is defined as the number of errors divided by the number of faults injected. A fault results in an error only when the bit is ACE. Hence, the AVF = 6/15 = 40%. In other words, both fault injection and ACE analyses will ideally yield the same AVF for a bit.

In reality, however, both ACE analysis and fault injection suffer from a number of shortcomings, which limit the scope of their use. ACE analysis relies on precise identification of ACE and un-ACE components of a bit's lifetime and hence requires executing the program through tens of millions of instructions as well as by using cooldown. Typically, in a microprocessor, a performance model is able to run such a huge number of instructions; hence, ACE analysis is reasonable to apply in a

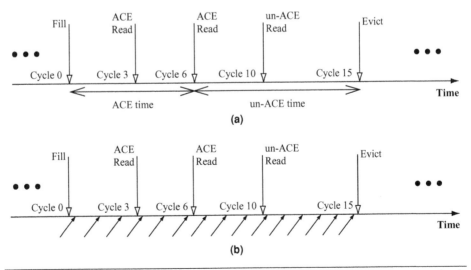

FIGURE 4.17 (a) ACE analysis of an example lifetime. (b) Analysis using fault injection in the same lifetime period. The arrows with the solid head show the injection of a fault into a bit.

performance simulator. However, a performance simulator does not capture all the detailed microarchitectural state of a machine. For example, typically latches are not present in a performance simulator. Hence, it is difficult to compute the AVF of latches in a performance simulator.

Also, ACE analysis may not be able to track bit flips that cause a control-flow change but do not change the final output of the program (see ACE Analysis Approximates Program Behavior in the Presence of Faults, p. 105). Fault injection can, however, compute the AVF of latches and track control-flow changes and consequent masking because it simulates the propagation of the injected fault through the program execution. Nonetheless, this may not have a large impact on the AVF numbers. For example, using microbenchmarks, Biswas et al. [3] found that both SFI into an RTL model and ACE analysis in a performance model of a commercial-grade microprocessor (built by Intel Corporation) yielded a very similar AVF of the Data Translation Buffer (DTB) CAM and RAM.

Fault injection requires executing a program repeatedly for each injected fault to see if the injected fault results in a user-visible error for the specific instance of the fault. An exhaustive search of this space is almost impossible because it requires an explosive number of experiments spanning the total state of a silicon chip (e.g., upward of 200 million bits in a current microprocessor), potential space of benchmarks running on the chip, and the cycles in which faults can be injected. Instead, practitioners use statistical sampling, which is covered in the next subsection. Because of the statistical nature of this methodology, it is called statistical fault injection.

SFI into a performance simulator may be less meaningful because a performance simulator does not precisely capture the logic state of a silicon chip. Instead, SFI is typically done into a chip's RTL model consisting primarily of logic gates. Although an RTL model exposes the detailed operation of a processor, it also makes the SFI simulations orders of magnitude slower than those of a performance simulator. This is because large number of simulations must be run to evaluate the AVF of each structure and because the RTL model is itself slow. Consequently, such simulations are often run for 1000–10 000 simulated processor cycles, which is often insufficient to determine if a bit flip is ACE or un-ACE. Hence, SFI may result in large unknowns and hence a conservative estimate of AVF because of lack of knowledge about the ACE-ness of a latch or a bit.

Further, because an RTL model cannot run a program to completion, SFI typically runs two copies of the same program on two RTL models: one faulty and one clean (with no fault). For each fault injection experiment, a fault is injected into one of the RTL models denoted as the faulty model. The clean copy runs the original program without any injected fault. After several cycles—typically, 1000–10 000 cycles—the architectural states of the two models are compared. If the architectural states do not match, then it is often assumed that there is an error and the state in to which the fault was injected is assumed to be ACE. This is, however, not strictly correct since the mismatch could have resulted from the fault injected into an un-ACE state. For example, if the fault was injected into a dynamically dead state, which is un-ACE, the architectural states of the two copies can mismatch, but this may produce no error in the final outcome of the program. This issue is discussed in greater detail later in this chapter.

Finally, an RTL model is often not available during the early design exploration of a processor or a silicon chip, which makes it hard to compute AVFs using SFI. However, performance models are typically created early in the design cycle of a high-performance microprocessor (but not necessarily for chipsets). This leaves the designer with two options to compute AVFs: ACE analysis in a performance model or SFI into an RTL model for an earlier generation of the processor or the silicon chip, if available.

4.6.2 Random Sampling in SFI

To determine a structure's AVF using fault injection, one can use the following algorithm: pick a bit in the structure, pick a benchmark, and pick a cycle among the total number of cycles the benchmark will run. Start the benchmark execution, and at the predetermined cycle, flip the state of the bit. Then continue running the program until one can determine if the bit flip results in a user-visible error (i.e., bit was ACE at the point of fault injection) or not. Repeat this procedure for the selected list of benchmarks, for every cycle the benchmark executes, and for every bit or state element in the silicon chip. As the reader can easily guess, this results in an explosive number of experiments. For example, with 30 benchmarks, each running a billion cycles, and 200 million state elements, one ends up with 6×10^{18} experiments. If each experiment took 10 hours to run and one had a thousand computers at one's disposal to do the experiments, this would still take about 7×10^{12} years to complete all the experiments. Clearly, this is infeasible.

To reduce this space of experiments, one can randomly select a set of benchmarks, a set of cycles to inject faults into, and a set of bits in each structure. If the random samples are selected appropriately, the computed AVF should asymptotically approach that of the full fault injection or ACE analysis. It should be noted that each fault injection represents a Bernoulli trial with the bit's AVF as the probability that the specific experiment will cause an error. Then, the minimum number of experiments necessary or n can be computed as

$$n = \frac{4z_{\alpha/2}{}^2 \times \text{AVF}\,(1 - \text{AVF})}{w^2}$$

where $z_{\alpha/2}$ denotes the value of the standard normal variable for the confidence level $100 \times (1 - \alpha)\%$ and w is the width of the confidence interval at the particular confidence level [1]. In a layman's terms, an experimental result that yields an X% confidence interval at a Y% confidence level indicates that if the experiment was repeated 100 times, then on average, Y of the experiments would return a result within X% of the true value of the random variable. The values of z can be obtained from statistical tables (e.g., table 5 in Appendix A in Allen [1]). The AVF is the sample AVF computed as the ratio of user-visible errors and the number of fault injection experiments.

■ EXAMPLE

Compute the minimum number of experiments necessary for an AVF of 30% for a 95% confidence level with a 10% error in the AVF in either direction. Repeat the calculation for a 99% confidence level.

SOLUTION For the 95% confidence level, $z_{\alpha/2} = 1.96$. For 10% error in the AVF in either direction, $w = 2 \times 0.1 \times 0.3 = 0.06$. Hence, $n = 4 \times 1.96^2 \times 0.3(1 - 0.3)/0.06^2 = 896.37$. So the minimum of fault injection experiments necessary is 897. For the 99% confidence level, $z_{\alpha/2} = 2.576$, so $n = 4 \times 2.576^2 \times 0.3(1 - 0.3)/0.06^2 = 1548.35$. Hence, the minimum number of fault injections necessary is 1549.

Figure 4.18 shows that the number of experiments required increases with the increase in confidence level and decrease in the error bound. For example, with a 10% error margin in one direction, the number of experiments required increases 4-fold when the confidence level is raised from 90% to 99.9%. Similarly, for a confidence level of 90%, the number of experiments increases 25-fold when the error bound in one direction is tightened from 50% to 10%.

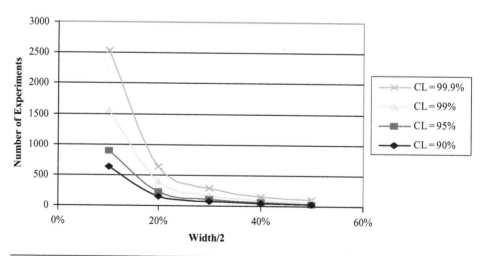

FIGURE 4.18 Number of fault injection experiments necessary to achieve the appropriate confidence level (CL) and error margin (or width) for a sample AVF of 30%. The $z_{\alpha/2}$ values for 99.9%, 99%, 95%, and 90% CL are 3.291, 2.576, 1.96, and 1.645, respectively. Using these values, the same graph can be reconstructed for different values of AVF.

4.6.3 Determining if an Injected Fault Will Result in an Error

One of the key questions in SFI into an RTL model is to determine if an injected fault actually will result in an error. In ACE terminology, one needs to determine if the bit or the state element is ACE when the fault is injected. Because of the thousands of simulations that must be run for each AVF evaluation for each structure and because the RTL model is usually orders of magnitude slower than a performance simulator, each SFI experiment is typically run for a small number of cycles—typically between 1000 and 10 000 cycles. Determining ACE-ness in this short interval is often a challenge.

SFI will typically run two copies of the RTL simulation: one with a fault and one without a fault (called the clean copy). At the end of the simulation, the faulty model is compared with the clean copy to determine if the injected fault resulted in an error. There are two choices for state to compare: architectural state and microarchitectural state. Architectural state includes register files and memory, and microarchitectural state represents internal state that has not yet been exposed outside the processor (e.g., instruction queue state).

If there is a mismatch in an architectural state between the faulty and the clean copies, then the fault could be potentially ACE. However, just because architectural state is incorrect, it does not necessarily mean that the final outcome of the program will be altered. For example, a faulty value could propagate to an architectural register but later could be overwritten without any intervening reads. That is, the faulty value is dynamically dead. In such an instance, the fault is actually un-ACE and gets masked. Hence, the mismatch in the architectural state only provides a conservative estimate about whether the injected fault is ACE or not.

In contrast, a mismatch in the microarchitectural state, but no mismatch in architectural state, indicates that the fault is latent and may result in an error later in the program after the simulation ends. This is often labeled as unknown since the ACE-ness cannot be determined when the simulation ends. However, if there is no mismatch in either architectural or microarchitectural state, then the fault is masked and therefore is un-ACE.

Many microarchitectural structures, such as a data translation buffer or a cache, can have latent faults long after the simulation ends. This makes it difficult to compute the AVF of such structures using SFI due to the short simulation interval. However, faults in flow-through latches that only contain transient data in the pipeline and affect architectural state can fairly quickly show up as errors in the architectural state if the injected fault is ACE. Hence, computing the AVF of latches using SFI into an RTL model is well suited.

To reduce the simulation time for each fault, the state comparison between the faulty and clean copies is often done periodically, instead of at the end. This is because if the fault is masked completely from both the architectural and the microarchitectural states, then the simulation can end earlier than the predesignated number of cycles the simulation was supposed to run for. Since significant fraction of faults is masked, this could significantly improve simulation time.

4.7   Case Study of SFI

This section will describe the Illinois SFI study conducted by Wang et al. [9]. The Illinois SFI study not only investigates the absolute vulnerability of various processor structures but also delves into the relative vulnerabilities of different structures and latches used in their processor core. For further studies of SFI into an RTL model, readers are referred to Kim and Somani's work on SFI into a picoJava™ core [5] and study of SFI into the RTL model of an Itanium® processor by Nguyen, et al. [7].

4.7.1 The Illinois SFI Study

The Illinois SFI study shows the masking effects of injected faults in a dynamically scheduled superscalar processor using a subset of the Alpha ISA. For this study, the authors wrote the RTL model in Verilog from scratch. The authors' goal was to have a latch-accurate model to study the masking effects from injected faults both in latches as well as in microarchitectural structures.

Processor Model

The processor model used in the Illinois SFI study is a dynamically scheduled superscalar pipeline. Figure 4.19a shows a diagram of this pipeline. Figure 4.19b shows the key processor parameters. To understand the details of this pipeline, the readers are referred to Hennessy and Patterson's book on computer architecture design [4]. This processor model uses the Alpha ISA but does not execute floating-point instructions, synchronizing memory operations, and a few miscellaneous instructions. The processor resembles a modern, dynamically scheduled superscalar processor.

The processor has a 12-stage pipeline with up to 132 instructions in flight. The 32-entry dynamic scheduler can issue up to six IPCs. To support dynamic scheduling, the pipeline supports a load queue (LDQ), store (store queue), speculative RAT (register allocation table), memory dependence predictor (mem dep pred 0 and mem dep pred 1), and a 64-entry ROB (ReOrder Buffer).

4.7.2 SFI Methodology

The Illinois SFI study divides the fault injection experiments into two varieties: those targeting both RAM arrays and latches and those targeting only latches. Each experiment requires repeatedly injecting faults and determining if the fault would result in a user-visible error. Each such fault injection is called a trial of

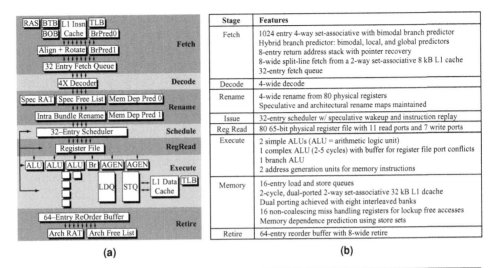

Stage	Features
Fetch	1024 entry 4-way set-associative with bimodal branch predictor Hybrid branch predictor: bimodal, local, and global predictors 8-entry return address stack with pointer recovery 8-wide split-line fetch from a 2-way set-associative 8 kB L1 cache 32-entry fetch queue
Decode	4-wide decode
Rename	4-wide rename from 80 physical registers Speculative and architectural rename maps maintained
Issue	32-entry scheduler w/ speculative wakeup and instruction replay
Reg Read	80 65-bit physical register file with 11 read ports and 7 write ports
Execute	2 simple ALUs (ALU = arithmetic logic unit) 1 complex ALU (2-5 cycles) with buffer for register file port conflicts 1 branch ALU 2 address generation units for memory instructions
Memory	16-entry load and store queues 2-cycle, dual-ported 2-way set-associative 32 kB L1 dcache Dual porting achieved with eight interleaved banks 16 non-coalescing miss handling registers for lockup free accesses Memory dependence prediction using store sets
Retire	64-entry reorder buffer with 8-wide retire

(a) (b)

FIGURE 4.19 The processor model used in the Illinois SFI study. (a) The pipeline. (b) The key processor parameters. Reprinted with permission from Wang et al. [9]. Copyright © 2004 IEEE.

the experiment. In each trial, the authors injected faults into randomly chosen bits from the 31 000 RAM bits and 14 000 latch bits. To achieve statistical significance, fault injection trials were repeated 25 000–30 000 times for each experiment. Each fault injection trial consisted of running a clean copy and a faulty copy for up to 10 000 cycles. However, for simplicity, instead of injecting faults at any randomly chosen cycle, the authors injected the faults on a set of 250–300 start points in the designated faulty copy.

Each trial was continuously monitored for one of the four outcomes: microarchitectural state match (*μArch Match*), incorrect program output (*SDC-Output*), premature termination of the workload (*SDC-Termination*), and none of the above or unknown (*Gray Area*). *μArch Match* occurs when the entire microarchitectural state of the processor model (i.e., every bit of state in the machine) is equivalent to that of the clean copy. If a trial results in a microarchitectural state match with no previous architectural state inconsistencies, then it is safe to declare that the injected fault's effects have been masked by the microarchitectural layer. These trials are placed in the *μArch Match* category. Since checking for microarchitectural state matches is relatively expensive requiring a full microarchitectural state comparison between the faulty and clean copies, this study only performed the check periodically, taking advantage of the fact that once a microarchitectural state match occurs, the remainder of the simulation is guaranteed to have a consistent microarchitectural state. The check was performed on an exponential time scale: at 1, 10, 100, 1000, and 10 000 cycles after the injection.

The faulty copy's architectural state (i.e., program-visible state such as memory, registers, and program counter) is compared with that of the clean copy in every cycle. If the architectural state comparison fails, then the transient fault has corrupted architectural state, and the trial is put in the *SDC-Output* or *SDC-Termination* bin. Trials that result in register and memory corruptions are conservatively placed into the *SDC-Output* category, along with those that result in TLB misses. Trials in the *SDC-Termination* category are those trials that resulted in pipeline deadlock or resulted in an instruction generating an exception, such as memory alignment errors and arithmetic overflow. Strictly speaking, all errors in the *SDC-Termination* category are SDCs, unless one can definitely prove that the program terminated before corrupting the data that are visible to the user. For example, if corrupted data are written to a database on disk before the program terminates, then the SDC-Terminated error definitely falls in the SDC bin.

If a trial does not result in a *μArch Match*, *SDC-Output*, or *SDC-Termination*, within the 10 000-cycle simulation limit, the trial is placed into the *Gray Area* category. Either the fault is latent within the pipeline or it is successfully masked, but the timing of the simulation is thrown off such that a complete microarchitectural state match is never detected. Of those that are latent, some will eventually affect architectural state, while others propagate to portions of the processor where they will never affect correct execution.

4.7.3 Transient Faults in Pipeline State

Figure 4.20 shows the results of fault injection experiments into the pipeline state. Each bar in the graph represents a different benchmark application from the

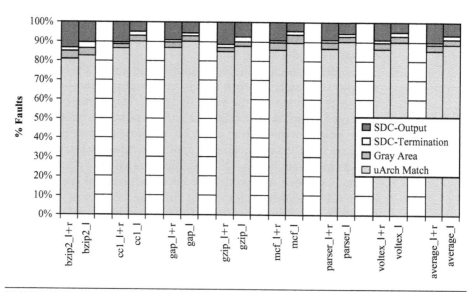

FIGURE 4.20 Results from fault injection into pipeline state. *l* + *r* denotes fault injection into both latch and RAM. l denotes fault injection only into latches. Reprinted with permission from Wang et al. [9]. Copyright © 2004 IEEE.

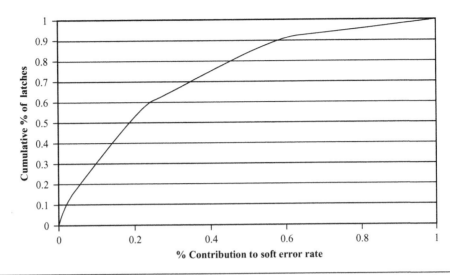

FIGURE 4.21 Variation in latch vulnerability. Reprinted with permission from Wang and Patel [8]. Copyright © 2006 IEEE.

SPEC2000 integer benchmark suite. The data from fault injection into latches and RAMs are labeled with an $l + r$ suffix, while data from injection into only latches are labeled with an l suffix. The different benchmarks represent different workloads on the processor, which affect the masking rate of the microarchitecture. The average results are presented in the rightmost bar in the graph I.

Examining the aggregate bars of both graphs, one can observe that approximately 85% of latch+RAM faults and about 88% of latch-based faults are successfully masked. The fraction of trials in the *Gray Area* accounts for another 3% for both experiments; the study was not able to determine conclusively whether these faults are masked or not. The remaining 12% of latch+RAM trials and 9% of latch trials are known errors that are either SDC-Output or SDC-Termination.

Figure 4.20 shows the average vulnerability of a latch or a processor state element for different benchmarks. However, previously, it was seen that the AVF of processor structures varies from structure to structure within a benchmark. Figure 4.21 shows a similar variation for latches. This study was done by some of the authors of the Illinois SFI study using the same experimental framework [8].[1] In particular, Figure 4.21 shows that 30% of the latches account for 70% of the total soft error contribution from latches. Consequently, these 30% of the latches should be the first target for soft error protection mechanisms.

---

[1]Although the original paper [8] reports the variation for all processor states, Figure 4.21 shows the variation only for the latches in the design. The authors graciously extracted the latch vulnerability data for this book.

4.7.4 Transient Faults in Logic Blocks

To understand the vulnerability of logic blocks, each latch or RAM cell in the processor was categorized based on the general function provided by that bit of state. For example, latches and RAM cells that hold instruction input and output operands were placed into a data category. Figure 4.22 lists the various categories of logic blocks used in the Illinois SFI study and provides a brief description for each, as well as the number of bits of latches and RAM cells within that category.

The results of the fault injection experiments (for latches and latches+RAMs) were then categorized by the logic block of the bit of state that the fault was injected into and the resulting outcome of the trial. Figure 4.23 shows the results of these experiments categorized by the functional block.

Figure 4.23a shows that the architectural register alias table (*archrat*) and the physical register file (*regfile*) are especially vulnerable to soft errors. This is not surprising since these structures contain the software-visible program state. The speculative register alias table (*specrat*) and the speculative free list (*specfreelist*) also appear to be particularly vulnerable. In order to bolster the overall reliability of our microarchitecture, it would be sensible to protect these structures.

Category	Description	Bits of Latches	Bits of RAMs
addr	64-bit address field for memory operations.	384	3584
archfreelist	Architectural register free list.	0	336
archrat	Architectural register alias table.	0	224
ctrl	Miscellaneous control state such as decoded instruction bundle control words and state machines.	2502	1916
data	Instruction input and output operands.	5899	2820
insn	Parts of the instruction word passed along with each instruction.	1525	2016
pc	62-bit program counter fields.	1984	12480
qctrl	Control state associated with queues.	176	0
regfile	65-bit register file entries and scoreboard bits.	80	5200
regptr	7-bit physical register file pointers.	978	1852
robptr	6-bit ROB tags.	352	444
specfreelist	Speculative register free list.	0	336
specrat	Speculative register alias table.	0	224
valid	Valid bits throughout the pipeline.	263	124

FIGURE 4.22 Description of different categories of state. Reprinted with permission from Wang and Patel [8]. Copyright IEEE © 2006.

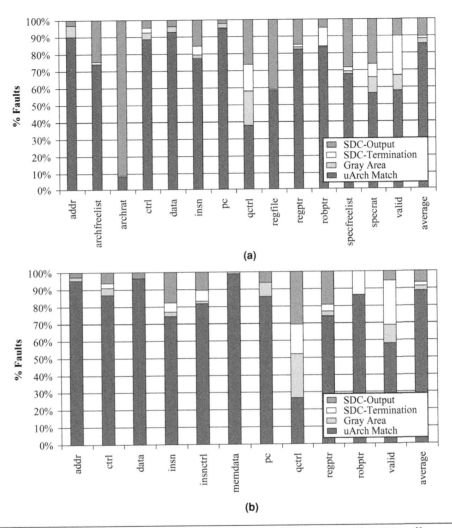

FIGURE 4.23 Results of fault injection into (a) latches and RAM cells and (b) latches alone. Reprinted with permission from Wang and Patel [8]. Copyright © 2006 IEEE.

Both the latch+RAM injections and the latch-only injections show high vulnerability for the bits categorized as *qctrl* and *valid*. Their impact on the overall fail rate is small, however, since they constitute only a small fraction of the total state of the machine. Also, it is interesting to note that the fail rate of the *data* category is the lowest due to a combination of low utilization rate, speculation, and logical masking.

4.8   **Summary**

AVF of a hardware structure expresses the fraction of faults that show up as user-visible errors. The intrinsic SER of a structure arising from circuit and device properties must be multiplied by the AVF to obtain the SER of the structure. Both the absolute and relative AVFs are useful. The absolute value of AVF is used to compute the overall chip-level SER. The relative values of AVFs are used to identify structures in need of protection. This chapter extends the AVF analysis described in Chapter 3 to three types of structures: address-based RAM arrays, CAM arrays, and latches.

Computing the AVF of address-based RAM arrays requires an extension called lifetime analysis. Lifetime analysis of a bit or a structure divides up the lifetime of a bit or a structure as a program executes into multiple nonoverlapping segments. For example, for a byte in the cache, the lifetime can be divided into fill-to-read, read-to-read, read-to-write, write-to-read, read-to-evict, etc. Each of these individual lifetime components can be categorized into potentially ACE, un-ACE, or unknown. For example, if the read in fill-to-read is an ACE read (implying the read value affects the final outcome of the executing program), then the fill-to-read component is ACE as well. Similarly, the read-to-evict component could be un-ACE in a write-through cache. Unknown components arise when the analysis cannot determine how the lifetime of a component should be binned (perhaps because the simulation ends prior to the determination of ACE-ness or un-ACE-ness of a component). By summing up all the potentially ACE and unknown lifetime components of a structure and dividing by the total simulation time, one can obtain an upper bound on the structure's AVF.

The AVF analysis of CAM arrays is slightly more involved. A CAM array, such as a tag store, operates by simultaneously comparing a set of incoming match bits (e.g., an address) against the contents of each of several memory entries. A single-bit fault in the CAM array results in two scenarios that could cause incorrect execution: a false-positive match or a false-negative match. A false-positive match occurs when a bit flip in the CAM array causes a match when there should not have been a match. A false-negative match occurs when a bit flip in the CAM array causes a mismatch when there should have been a match. False-positive matches can be tracked using hamming-distance-one analysis. The incoming match bits and bits of an entry in the CAM array are said to be Hamming distance one apart if they differ in one bit. When a fault occurs, only this bit can give rise to a false-positive match. On encountering a false-positive match and by tracking the lifetime of the bit causing the match, one can compute the AVF arising from such a match. The false-negative match is easier to track. By tracking the lifetime of the CAM entry associated with a false-negative match (i.e., the entry that should have matched), one can compute the corresponding AVF component.

Computing latch AVFs using ACE analysis in a performance model is difficult because a performance model does not typically model the behavior of latches.

Instead, it is easier to do in an RTL model that captures the actual behavior of the hardware being built. However, RTL models are usually 100 to 1000 times slower than performance models, making it difficult to run RTL models for more than tens of thousands of cycles to determine AVFs. This is too short for ACE analysis to determine ACE-ness of many structures, and it may be somewhat cumbersome to implement the ACE analysis of a low-level gate-centric RTL model. Instead, it is easier to use SFI to compute the AVF of latches. In SFI, random bit flips simulating faults are introduced in an RTL model to create a faulty copy. Simultaneously, a clean copy is run separately. If after some number of the cycles, the architectural states of the faulty and clean copies differ, then one can assume that the fault is likely to show up as an error. This is not strictly correct but used in practice due to limitations imposed by the RTL model. Such experiments are repeated thousands of times. Each time a different bit may be flipped at a different point in the simulation. By repeating these fault injection experiments, one can obtain a desired level of statistical significance in the AVF numbers.

4.9   Historical Anecdote

I was the lead soft error architect of an Itanium® processor design that was eventually canceled. Itanium® processors are, in general, aggressive about soft error protection. During the design process, we assumed, based on established design guidelines (and not a proper AVF analysis), that the data translation buffer CAM was highly vulnerable to soft errors. Protecting this CAM with parity was difficult because of Itanium architecture's support of multiple page sizes.

A simple scheme such as adding a parity bit to the CAM and then augmenting the incoming address with the appropriate parity bit to enable the CAM operation did not work. This is because the number of bits of the incoming address over which we needed to compute the parity was determined by the page size, whose information was embedded in the RAM array of the data translation buffer. Thus, we could not compute the parity of the incoming address without looking up the data translation buffer RAM array first.

To tackle this problem, we came up with a novel scheme that simultaneously computed the parity for multiple page sizes and then CAM-ed the data translation buffer appropriately. Details of this scheme are described in the US Patent Application 20060150048 filed to the US patent office on December 30, 2004. This was even coded into the RTL model of the Itanium processor we were building.

Later when we formulated the AVF analysis of address-based structures and computed the AVF of the data translation buffer CAM, we found it to be very small, around 2–3%. This immediately made the soft error contribution of the data translation buffer CAM noncritical to the processor we were building. We would not have protected this CAM, if we had these data prior to making the decision about protecting the CAM array.

References

[1] A. O. Allen, *Probability, Statistics, and Queue Theory with Computer Science Applications*, Academic Press, 1990.

[2] AMD, "BIOS and Kernel Developer's Guide for AMD Athlon™64 and AMD OpteronTM Processors." Publication #26094, Revision 3.14, April 2004. Available at: http://www.amd.com/us-en/assets/content_type/white_papers_and_tech_ docs/26094.PDF.

[3] A. Biswas, P. Racunas, R. Cheveresan, J. Emer, S. S. Mukherjee, and R. Rangan, "Computing Architectural Vulnerability Factors for Address-Based Structures," in *32nd Annual International Symposium on Computer Architecture (ISCA)*, pp. 532–543, June 2005.

[4] J. L. Hennessy and D. L. Patterson, *Computer Architecture: A Quantitative Approach*, Morgan Kaufmann Publishers, 2003.

[5] S. Kim and A. K. Somani, "Soft Error Sensitivity Characterization for Microprocessor Dependability Enhancement Strategy," in *International Conference on Dependable Systems and Networks (DSN)*, pp. 416–425, June 2002.

[6] A. Lai, C. Fide, and B. Falsafi. "Dead-Block Prediction and Dead-Block Correlating Prefetchers," in *28th International Symposium on Computer Architecture*, pp. 144–154, June 2001.

[7] H. T. Nguyen, Y. Yagil, N. Seifert, and M. Reitsma, "Chip-Level Soft Error Estimation Method," *IEEE Transactions on Device and Materials Reliability*, Vol. 5, No. 3, pp. 365–381, September 2005.

[8] N. Wang and S. J. Patel, "ReStore: Symptom-Based Soft Error Detection in Microprocessors," *IEEE Transactions on Dependable and Secure Computing*, Vol. 3, No. 3, pp. 188–201, July–September 2006.

[9] N. Wang, J. Quek, T. M. Rafacz, and S. J. Patel, "Characterizing the Effects of Transient Faults on a High-Performance Processor Pipeline," in *International Conference on Dependable Systems and Networks (DSN)*, pp. 61–70, June 2004.

[10] D. Wood, M. Hill, and R. Kessler. "A Model for Estimating Trace-Sample Miss Ratios," in *1991 SIGMETRICS Conference on Measurement and Modeling of Computer Systems*, pp. 79–89, May 1991.

5

Error Coding Techniques

5.1   Overview

Architectural vulnerability analysis, which was described in Chapters 3 and 4, identifies the most vulnerable hardware structures that may need protection. This chapter discusses how to protect these vulnerable structures with error coding techniques. The theory of error coding is a rich area of mathematics. However, instead of delving deep into the theory of error coding techniques, this chapter describes coding techniques from a practitioner's perspective with simple examples to illustrate the basic concepts. Examples of common error codes used in computer systems and covered in this chapter include parity codes, single-error correct double-error detect (SECDED) codes, double-error correct triple-error detect (DECTED) codes, and cyclic redundancy check (CRC) codes. This chapter also discusses advanced error codes, such as AN codes, residue codes, and parity prediction circuits, which protect execution units.

For practitioners, implementation overhead of such codes is an important metric, which is also discussed in this chapter. Protecting structures from multibit faults becomes important as the structure grows in size. This chapter describes a technique called scrubbing, which can reduce the SER without incurring the overhead of the larger multibit fault detectors, and discusses how architecture-specific knowledge can reduce the overhead of error detection, help identify false errors, and help create hardware assertions to detect faults in a processor or a chipset.

Finally, this chapter discusses the role of a machine check architecture (MCA), which is invoked when the hardware detects a fault or corrects an error.

5.2   Fault Detection and ECC for State Bits

This section describes some simple, yet powerful, fault detection and error correction schemes. First, basics of error coding are described with simple examples. Then a number of coding schemes, parity codes, single error correction (SEC) codes, SECDED codes, and CRC codes, are discussed. There are other and more complex codes available, but the reader is referred to Peterson and Weldon [15] for further reading on error coding theory.

5.2.1   Basics of Error Coding

Coding schemes are one of the most powerful and popular architectural error protection mechanisms used in computing systems today. Coding schemes can be used to detect or correct single-bit or multibit error. If a fault in a bit is always detected by a code, then the bit's SDC AVF is zero, but the DUE AVF can still be nonzero. In contrast, if a fault in a bit is not only detected but also the corresponding error is corrected by a code, then both its SDC and DUE AVFs can be made zero. First, this section illustrates the basics of single-bit fault detection and error correction using simple examples. Second, it elaborates on how the concept of Hamming distance relates to the number of errors that can be detected or corrected. Finally, it discusses the computation of the minimum number of code bits needed to correct a given number of errors in a set of bits.

Simple Examples to Illustrate Single-Bit Fault Detection and Error Correction

The basic idea of error codes can be explained with a simple example. Assume that one *data bit* needs to be protected against single-bit errors during its residence in a buffer (Figure 5.1). The value of this data bit can be either zero or one. Let us add

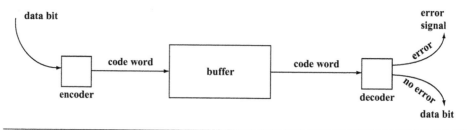

FIGURE 5.1   Example of encoding and decoding processes.

one *code bit* to the data bit to form a *code word*, such that the code word is a tuple: <data bit, code bit>. Each entry of the buffer will hold such a code word. Let us define an *encoding scheme* that sets the code bit as follows:

$$\text{code bit} = \text{data bit}$$

Before writing the data bit into the buffer, the encoder creates the code bit, appends it to the data bit, and writes the code word into the buffer. When the data bit needs to be read out of the buffer, the whole code word is read out and decoded. The valid code words must be either 00 or 11 since the code bit must equal the data bit. In such a case, there is no error, and the data bit is returned. However, if the code word read out is either 01 or 10, then the value of the code word changes while it was resident in the buffer. One of the causes of this bit flip could be due to an alpha particle or a neutron strike either on the data bit or on the code bit.

The decoding process is easier for code words that are *separable*. A code word is separable if the data bits are distinct from the code bits within a code word. For example, in the above example, the code word = <data bit, code bit>. If the code word is valid, the decoder simply extracts the data bit from the code word and returns it. Examples of nonseparable codes are covered later in this chapter.

The code word representation of 00 and 11 can only detect a single-bit fault but cannot correct it. For example, the data bit could be struck, so that the code word changes from a valid 00 word to invalid 10 code word. However, looking at 10 without any prior knowledge of what the data bit was, the decoder cannot determine if code word changed from 00 to 10 (data bit flipped) or 11 to 10 (code bit flipped).

Expanding the set of code bits can help identify the bit in error. For example, let us define a new code word as a three-tuple: <data bit, code bit 1, code bit 2>. Then, let us define the code bits as:

$$\text{code bit} = \text{data bit}$$
$$\text{code bit 2} = \text{NOT (data bit)}$$

The valid code words are 001 and 110. Based on this code, one can easily identify the bit position that got struck. For example, if the code word is 101, then one can say that the data bit was struck. This is because the code bit 1 = NOT (code bit 2), which is the correct relationship. But the relationship between data bit and code bit 1 and that between data bit and code bit 2 are inconsistent. This inconsistency can arise only if the data bit experienced a fault. It should be noted that code bit 2 could be set to data bit itself (without the inversion).

▮ E X A M P L E

For the code word defined above, determine which bit is in error when the code word is 111.

S O L U T I O N   Code bit 2 is in error. This is because the code bit 1 = data bit = 1. But code bit 2 is not the inverse of either data bit or code bit 1.

The examples in this section illustrate the detection and correction of single-bit errors. The same basic concepts can be expanded to cover the detection and/or correction of multibit errors. Such codes are covered later in this section.

Determination of Number of Bit Errors That Can Be Detected and Corrected

As must be evident by now, code words are often divided into two distinct spaces: words that are fault-free and words that have faults in them. In our example for fault detection earlier in this section, the fault-free code words were 00 and 11 and erroneous code words were 10 and 01. Similarly, in our error correction example, the error-free code words were 001 and 110 and any other bit combination is erroneous. When one or more errors occur, the code word moves from an error-free to an erroneous space, which makes it possible to detect or correct the error.

There is, however, a limit to the number of errors a code word can detect or correct. For example, the example code word for error detection can only detect one error. However, if two bits are in error, then this code word may not be able to detect the double bit error. For example, if the first bit of the code 00 gets struck, then it can change to 10 (erroneous word). But if the second bit is struck now, it can change to 11, making it a valid code word. Consequently, this code cannot detect double-bit errors. The number of bits a coding scheme can detect or correct is determined by its minimum Hamming distance.

The Hamming distance between two words or bit vectors is the number of bit positions they differ in. For example, consider the following two words: $A = 00001111$ and $B = 00001110$. They only differ in the last bit position, so the Hamming distance between A and B is 1. If B were 11110011, then A and B differ the in the first six bit positions, hence the Hamming distance between A and B would be 6. If A and B were the nodes of a binary n-cube, then the Hamming distance would be the minimum number of links to be traversed to get from A to B.

Given a code word space, the minimum Hamming distance of the code word is defined as the minimum Hamming distance between any two valid (fault-free) code words in the space. For example, for our example code word that only detects faults, the valid code words are 00 and 11. The minimum Hamming distance for this code space is 2. Similarly, the valid code words for our error correction example are 001 and 110. The minimum Hamming distance for this code word space is 3.

■ E X A M P L E

Consider the following set of code words: 000, 011, 101, 110. What is the minimum Hamming distance for this set of code words?

S O L U T I O N Assume HD(x, y) = Hamming distance between x and y. Then, HD(000, 011) = 2, HD(000, 101) = 2, HD(000, 110) = 2, HD(011, 101) = 2, HD(011, 110) = 2, HD(101, 110) = 2. Hence, the minimum Hamming distance = 2.

The minimum Hamming distance of a code word determines the number of bit errors that can be detected and/or corrected by the code word. There are three key results:

- *The minimum Hamming distance of a code word must be $\alpha + 1$ for it to detect all faults in α or fewer bits in the code word.* In our error correction example, the minimum Hamming distance of the code word was 3, with the valid code words being 001 and 110 ($\alpha = 2$). Consequently, a single error or a double-bit error will convert a valid error-free code word into a code word with error. A third error, however, can potentially bring the code word back into the error-free space, thereby avoiding detection. Hence, this code word can only detect either single-bit or double-bit errors but not triple-bit errors.

- *The minimum Hamming distance of a code word must be $2\beta + 1$ for it to correct all errors in β or fewer bits in the code word.* Again, in our error correction example, the minimum Hamming distance of the code word was 3, so $\beta = 1$. Any fault in β or fewer bits will still be at least $\beta + 1$ Hamming distance from the nearest valid code word. Thus, given a valid code word 001 (in our error correction example), a bit flip in the first bit would convert it to 101, which is still at a Hamming distance of 2 away the other valid code word of 110. Thus, no other single-bit error in any bit other than the first one can reach this erroneous code word 101. Hence, the bit position in error can be precisely identified and hence corrected.

- *The minimum Hamming distance of a code word must be $\alpha + \beta + 1$, where $\alpha \geq \beta$, for it to detect all errors in α or fewer bits and correct all errors in β or fewer errors.* This result follows from the prior two results about error detection and correction. Thus, if the minimum Hamming distance of a code word is 4, then it can correct single-bit errors and detect double-bit errors, if $\alpha = 2$ and $\beta = 1$. Such a coding scheme is referred to as SECDED codes, which are covered later in this section. Figure 5.2 shows the number of bit errors that can be detected or corrected, given a minimum Hamming distance for a code word. It should be noted that different numbers of bit errors can be detected and corrected depending on the values of α and β.

Minimum Hamming Distance	1	2	3	3	4	4	5	5	5
Number of bits in which error is detected (α)	0	1	1	2	2	3	2	3	4
Number of bits in which error is corrected (β)	0	0	1	0	1	0	2	1	0

FIGURE 5.2 Relationship between minimum Hamming distance and number of bit errors that can be detected and corrected.

◼ E X A M P L E

What is the minimum Hamming distance of a code that can detect two faults and correct one?

S O L U T I O N From Figure 5.2, such a code must have a minimum Hamming distance of 4.

Determination of the Minimum Number of Code Bits Needed for Error Correction

There is an alternate formulation that computes the minimum number of code bits needed, given the number of bit errors that need to be corrected. In contrast, the formulation above shows the number of bits that can be corrected, given the minimum Hamming distance of a code word. Given k data bits and r code bits (where $n = k + r$), the r code bits must be able to precisely determine the bit position or positions in error. To correct a single-bit error, the 2^r combinations arising from the r code bits must be able to determine where the error occurred in the n bits. This combination must also be able to represent that no bit position is in error. This results in the equation

$$2^r \geq k + r + 1$$

If $k = 1$, then r must be at least 2 to satisfy the above inequality. Thus, in our error correction example above, the number of code bits chosen (two) was optimal.

The number of error correction bits used determines the code's storage overhead. Figure 5.3 shows how this overhead grows with the number of data bits for single-bit error correction. The overhead of single-bit error correction decreases with the increase in the number of data bits. For example, for a single data bit, the overhead of error correction is 200%. In contrast, for 64 data bits, only 7 code bits are necessary for single-bit correction, which results in an overhead of only 11%. However, the number of data bits that can be covered in a single code word depends on a variety of implementation issues, such as the number of available data bits, timing. Implementation issues for ECC are discussed later in this section.

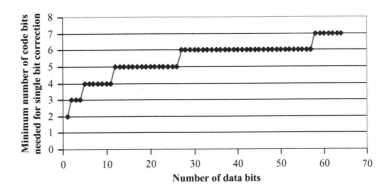

FIGURE 5.3 The minimum number of code bits needed to correct a single-bit error for a given number of data bits.

The minimum number of code bits required to correct multiple bit errors can be computed in a similar fashion. To correct a double-bit error, for example, the number of code words must cover the following three cases: no error (1), single-bit errors $(k + r)$, and double-bit errors $^{(k+r)}C_2$.[1] Hence, one has the inequality:

$$2^r \geq 1 + k + r + ^{k+r}C_2$$

Thus, to correct double-bit errors in a 64-bit data word (i.e., $k = 64$), r or the number of code bits must be at least 12. In general, the minimum number of code bits (r) needed to correct m bit errors in a data word with k bits is given by

$$2^r \geq \sum_{i=0}^{m} {}^{k+r}C_i$$

■ EXAMPLE

Using the above equation, compare the increase in storage overhead to correct a single-bit error, a double-bit error, and a triple-bit error in a 64-bit data word.

SOLUTION For all three correction schemes, $k = 64$. For a single-bit correction $(m = 1)$, the minimum $r = 7$. For a double-bit correction $(m = 2)$, the minimum $r = 13$. For a triple-bit correction $(m = 3)$, the minimum $r = 20$. Consequently, the overheads for single-bit, double-bit, and triple-bit corrections are 11%, 22%, and 32%, respectively.

---

[1] $^aC_b = (a!)/((a - b)!b!)$, where $a! = a^*(a - 1)^*(a - 2)^*...^*2^*1$, *represents multiplication, and $a > b$.

A designer usually carefully weighs these overheads in error correction against the performance degradation the chip may experience. For example, a processor cache often occupies a significant fraction of the chip. For a single-error correction (SEC), the overhead is about 11%, whereas for a triple-error correction, the overhead is as high as 32%. An overhead of 11% indicates that about 10% (= 11/111) of the bits available for a cache are used for error correction. Similarly, a 32% overhead indicates that about 24% (= 32/132) of the bits available for a cache are used for error correction. Thus, going from a single-bit correction to a triple-bit correction, about 14% more bits are needed. Instead of using these bits for error correction, these bits could be used to increase the performance of the chip itself by increasing the size of the cache. The designer must carefully weigh the advantage of a triple-error correction against increasing the performance of the chip itself.

The next few sections describe different fault detection and error correction strategies and the overheads associated with them.

5.2.2 Error Detection Using Parity Codes

Parity codes are perhaps the simplest form of error detection. In its basic form, a parity code is simply a single code bit attached to a set of k data bits. Even parity codes set this bit to 1 if there is an odd number of 1s in the k-bit data word (so that the resulting code word has an even number of 1s). Alternatively, odd parity codes set this bit to 1 if there is an even number of 1s in the k-bit data word (so that the resulting code has an odd number of 1s). Given a set of k bits (denoted as $a_0 a_1 \ldots a_{k-1}$), one can compute the even parity code corresponding to these k bits using the following equation:

$$\text{Even parity code} = a_0 \oplus a_1 \oplus K \oplus a_{k-1}$$

where $\oplus$ denotes the bit-wise XOR operation.[2] For example, given a set of four bits 0011, the corresponding even parity code is 0. Parity codes are separable since the parity bit is distinct and separate from the data bits.

The minimum Hamming distance of any parity code is two, so a parity code can always detect single-bit faults. Such a code can also detect all odd numbers of faults since every odd number of faults will put the code word back into the erroneous space. Parity codes cannot, however, detect even numbers of faults since two faults will put the word back into fault-free space.

Parity codes can be made to detect spatially contiguous multibit faults using a technique called *interleaving*. Figure 5.4 shows an example of two interleaved code

---

[2]Bitwise XOR operation: 0 XOR 0 = 0, 0 XOR 1 = 1, 1 XOR 0 = 1, 1 XOR 1 = 0.

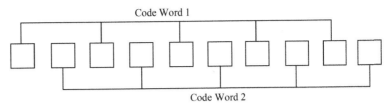

FIGURE 5.4 **Example of interleaving. Bits (represented by squares) of code words 1 and 2 are interleaved. The interleaving distance is 2.**

words. If two contiguous bits are upset by a single alpha particle or a neutron strike, then the particle strike will corrupt both code words 1 and 2. The parity codes for the individual code words will detect the error. Interleaving distance is defined as the number of contiguous bit errors the interleaving scheme can catch. The interleaving distance in Figure 5.4 is 2. Alternatively, for example, if three code words are interleaved, such that any three contiguous bits are covered by three different parity codes, then the interleaving distance is 3. The greater is the interleaving distance for a given number of code word bits, the greater is the distance the XOR tree for the parity computation has to be typically spread out. Hence, a greater interleaving distance will typically require a longer time to compute the parity tree. Whether this affects the timing of the chip under design depends on the specific implementation.

The number of bits a parity code can cover may depend on either the implementation or the architecture. To meet the timing constraints of a pipeline, the parity code may have to be restricted to cover a certain number of bits. Then, the set of bits that need protection may have to be broken up into smaller sized chunks—with each chunk covered by a single parity bit.

In some cases, the architecture may dictate the number of bits that need be covered with parity. Many instruction sets allow reads and writes to both byte-sized data and word-sized data. A word in this case can be multiple bytes. Per-byte parity, instead of per-word parity, allows the architecture to avoid reading the entire word to compute the appropriate parity for a read to a single byte. For example, a word could be composed of the following two bytes: 00000001 and 00000001. If parity is computed over the whole word, then the parity bit is 0 (assuming even parity). But if the first byte and the word-wide parity code are read, one may erroneously conclude that there has been a parity error, when there was not a parity error to begin with. Hence, to determine if the first byte had an error, the entire word along with its parity bit needs to be read out. Instead, having per-byte parity allows the architecture to only read the first byte and its corresponding parity bit to check for the error in the first byte.

The next section describes how to extend the concept of parity bits to correct bit errors.

5.2.3 Single-Error Correction Codes

Given a set of bits, a conceptually simple way to detect and correct a single-bit error is to organize the data bits in a two-dimensional array and compute the parity for each row and column. This is referred to as a product code. In Figure 5.5, 12 data bits are arranged in a 4 × 3 two-dimensional array. Then one can compute the parity for each row to generate the horizontal parity bits and for each column to generate the vertical parity bits. In such an arrangement, if an error occurs in one of the data bits, then the error can be precisely isolated to a specific row and column. The combination of the row and column indices will point to the exact bit location where the error occurred. By flipping the identified bit, one can correct the error. These codes are called product codes.

■ E X A M P L E

In the example in Figure 5.5, the horizontal and vertical parity bits are read out as 110 and 0100, respectively. Was there a fault? If so, which bit encountered the fault?

S O L U T I O N The vertical parity bits are all correct, but the second horizontal parity bit is incorrect. This implies that there was no fault in the data bits. Hence, one can conclude that the second horizontal parity bit encountered a fault. By flipping the bit to zero, we can correct it.

Product codes using horizontal and vertical parity bits are, however, not optimal in the number of code bits used. For example, to detect a single-bit error in 12 data bits, the product code in Figure 5.5 uses seven code bits. However, as Figure 5.3 shows, the optimal number of code bits necessary to correct a single-bit error in 12 bits is five. More sophisticated ECC can reduce the number of code bits necessary for SEC. To describe how ECC works, the concept of a parity check matrix is now introduced.

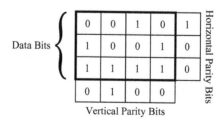

FIGURE 5.5 Single-error correction with a product code using horizontal and vertical parity bits.

A parity check matrix consists of r rows and n columns ($n = k + r$), where k is the number of data bits and r is the number of check bits. Each column in the parity check matrix corresponds to either a data bit or a code bit. The positions of the 1s in a row in the parity check matrix indicate the bit positions that are involved in the parity check equation for that row. For example, the parity check matrix in Figure 5.6a defines the following parity check equations:

$$C_3 = D_2 \oplus D_3 \oplus D_4$$
$$C_2 = D_1 \oplus D_3 \oplus D_4$$
$$C_1 = D_1 \oplus D_2 \oplus D_4$$

where $\oplus$ denotes the bit-wise XOR operation. If $D_1 D_2 D_3 D_4 = 1010$, then $C_3 = 1$, $C_2 = 0$, $C_1 = 1$. The corresponding code word $C_1 C_2 C_3 D_1 D_2 D_3 D_4$ is 1011010. If E represents the code word vector, then in matrix notation, $E = [1\,0\,1\,1\,0\,1\,0]$.

The parity check matrix is created carefully to allow the generation of the syndrome, which identifies the bit position in error in a given code word. If P is the parity check matrix and in E is the code word vector, then the syndrome is expressed as

$$S = P \cdot E^{\mathsf{T}}$$

where E^{T} is the transpose of E. For example, if P is the parity check matrix in Figure 5.6a and $E = [1\,0\,1\,1\,0\,1\,0]$, then one can derive S as

$$S = \begin{bmatrix} 0 & 0 & 1 & 0 & 1 & 1 & 1 \\ 0 & 1 & 0 & 1 & 0 & 1 & 1 \\ 1 & 0 & 0 & 1 & 1 & 0 & 1 \end{bmatrix} \cdot \begin{bmatrix} 1 \\ 0 \\ 1 \\ 1 \\ 0 \\ 1 \\ 0 \end{bmatrix} = \begin{bmatrix} 0 \\ 0 \\ 0 \end{bmatrix}$$

C_1	C_2	C_3	D_1	D_2	D_3	D_4
0	0	1	0	1	1	1
0	1	0	1	0	1	1
1	0	0	1	1	0	1

(a)

C_1	C_2	D_1	C_3	D_2	D_3	D_4
0	0	0	1	1	1	1
0	1	1	0	0	1	1
1	0	1	0	1	0	1

(b)

FIGURE 5.6 Parity check matrices for the code word: $D_1 D_2 D_3 D_4$ with corresponding code bits $C_1 C_2 C_3$. (a) Parity check matrix where code bit columns are contiguous. (b) Parity check matrix where the code bit columns are in power of two positions. The syndrome (discussed in the text) for (b) points to the bit position in error, if the count of the column is started from 1. The syndrome zero would mean no error.

If S is expressed as $[S_3 \; S_2 \; S_1]$, then $S_3 = (0 \text{ AND } 1) \oplus (0 \text{ AND } 0) \oplus (1 \text{ AND } 1) \oplus$ $(0 \text{ AND } 1) \oplus (1 \text{ AND } 0) \oplus (1 \text{ AND } 1) \oplus (1 \text{ AND } 0) = 0$, where the dot represents the bit-wise AND operation and $\oplus$ represents the bit-wise XOR operation. S_2 and S_1 can be computed in a similar fashion. The syndrome in this example is zero, indicating that there is no single-bit error in the code word E.

Any nonzero syndrome would indicate an error in E. By construction, $S_1 = C_1 \oplus C_1'$, $S_2 = C_2 \oplus C_2'$, and $S_3 = C_3 \oplus C_3'$, where C_i is the parity code computed before the code word was written into the buffer and C_i' is the parity code computed after the code word is read out of the buffer. If any of C_i (resident parity) and C_i' (recomputed parity) do not match, then the code word had a bit flip while it was resident in the buffer. Hence, S will be nonzero. For example, if D_1 flips due to an alpha particle or a neutron strike (denoted in bold in the vector E^T shown below, then one can obtain S as

$$S = \begin{bmatrix} 0 & 0 & 1 & 0 & 1 & 1 & 1 \\ 0 & 1 & 0 & 1 & 0 & 1 & 1 \\ 1 & 0 & 0 & 1 & 1 & 0 & 1 \end{bmatrix} \cdot \begin{bmatrix} 1 \\ 0 \\ 1 \\ 0 \\ 0 \\ 1 \\ 0 \end{bmatrix} = \begin{bmatrix} 0 \\ 1 \\ 1 \end{bmatrix}$$

The reader can easily verify that the bit error in each code word position creates a different syndrome value that can help identify and correct the bit in error. More interestingly, if the check bits are placed in power-of-two positions (1, 2, 4, ...) and the column count starts with 1 (zero indicates no error), then the syndrome identifies exactly which bit position is in error. Such a coding scheme is referred to as a Hamming code. The parity check matrix in the prior example is not organized as such. Thus, the syndrome is 3 although the error is in bit position 4. In contrast, the parity check bits are placed in power-of-two positions in Figure 5.6b. Using this parity check matrix and code word matrix E rewritten as $[C_1 C_2 D_1 C_3 D_2 D_3 D_4]$, one can get the syndrome as

$$S = \begin{bmatrix} 0 & 0 & 0 & 1 & 1 & 1 & 1 \\ 0 & 1 & 1 & 0 & 0 & 1 & 1 \\ 1 & 0 & 1 & 0 & 1 & 0 & 1 \end{bmatrix} \cdot \begin{bmatrix} 1 \\ 0 \\ 0 \\ 1 \\ 0 \\ 1 \\ 0 \end{bmatrix} = \begin{bmatrix} 0 \\ 1 \\ 1 \end{bmatrix}$$

Thus, the syndrome is three, which is also the bit position in error. This is also referred to as a (7,4) Hamming SEC code. In general, a Hamming code is referred to as (n, k) Hamming code.

EXAMPLE

Using the above parity check matrix, find the bit position in error in the code word 1010010.

SOLUTION $S_3 = 1, S_2 = 0, S_1 = 0$, indicating the bit position in error is 100 or in the fourth position.

From Figure 5.2, it follows that the minimum Hamming distance of SEC codes is 3. This also implies that no two columns in the parity check matrix of a SEC code can be identical (that is, each column is unique). This is because if two columns were identical, then one can create two valid code words that only differ in E^T's bit positions that correspond to the identical columns. For example, if the last two columns are identical in the parity check matrix shown above, then one could create two valid code words, both of which have identical bits except for the last two bits. The two code words could have 00 and 11 as the last two bits. This would imply that the code word's minimum Hamming distance is 2 and, hence, it will not be able to correct any single-bit error.

EXAMPLE

In the parity check matrix shown below, the last two columns are identical. Create two code words that when multiplied with the parity check matrix yield zero syndrome but are only apart by a Hamming distance of 2.

$$H = \begin{bmatrix} 0 & 0 & 0 & 1 & 1 & 1 & 1 \\ 0 & 1 & 1 & 0 & 0 & 1 & 1 \\ 1 & 0 & 1 & 0 & 1 & 1 & 1 \end{bmatrix}$$

SOLUTION The following two code words—1001100 and 1001111—yield a syndrome of zero with the above parity check matrix but are only apart by a Hamming distance of 2. Hence, this parity check matrix will not be able to correct single-bit errors.

As may be clear by now, a SEC ECC is simply a combination of multiple parity trees (e.g., for C_1, C_2, and C_3). By creating and combining the parity check equations carefully, one is able to detect and correct a single-bit error. Further, the code words protected with ECC can be interleaved to correct multibit errors in the same way as the code words protected with parity bits can be interleaved to detect multibit errors (as discussed in the previous section, Error Detection Using Parity Codes). For further reading on SEC and the theory of ECC spanning the use of Galois fields and generator matrices, readers are referred to Peterson and Weldon [15].

5.2.4 Single-Error Correct Double-Error Detect Code

A SEC code can detect and correct single-bit errors but cannot detect double-bit errors. As suggested by Figure 5.2, a SECDED code's minimum Hamming distance must be 4, whereas this SEC code's minimum Hamming distance is 3 and hence it cannot simultaneously correct single-bit errors and detect double-bit errors. For the parity check matrix for the Hamming SEC code in the previous subsection, consider the following code word with errors in bit positions 3 and 8: [1 0 **0** 1 0 1 **1**]. The Hamming SEC code syndrome is 100, which suggests a single-bit error in bit position 4, which is incorrect. The problem here is that a nonzero syndrome cannot distinguish between a single- and a double-bit error.

A SEC code's parity check matrix can be extended easily to create a SECDED code that can both correct single-bit errors and detect double-bit errors. Today, such SECDED codes are widely used in computing systems to protect memory cells, such as microprocessor caches, and even to register files in certain architectures. To create a SECDED code from a SEC code, one can add an extra bit that represents the parity over the seven bits of the SEC code word. The syndrome becomes a 4-bit entity, instead of a 3-bit entity, for SEC codes. Then, one can distinguish between the following cases:

- If the syndrome is zero, then there is no error.
- If the syndrome is nonzero and the extra parity bit is incorrect, then there is a single-bit correctable error.
- If the syndrome is nonzero, but the extra parity bit is correct, then there is an uncorrectable double-bit error.

The parity check matrix of this SECDED code can be represented as

$$
\begin{bmatrix}
1 & 1 & 1 & 1 & 1 & 1 & 1 & 1 \\
0 & 0 & 0 & 1 & 1 & 1 & 1 & 0 \\
0 & 1 & 1 & 0 & 0 & 1 & 1 & 0 \\
1 & 0 & 1 & 0 & 1 & 0 & 1 & 0
\end{bmatrix}
$$

The additional first row now represents the extra parity bit. The additional last column places the extra parity bit in the power-of-two position to ensure it is a Hamming code. Hence, the code word with the double-bit error would look like [1 0 0 1 0 1 1 0], where the last 0 represents the extra parity check bit. Then, $H \times E^{\mathrm{T}}$ gives us:

$$
S = \begin{bmatrix} 1 & 1 & 1 & 1 & 1 & 1 & 1 & 1 \\ 0 & 0 & 0 & 1 & 1 & 1 & 1 & 0 \\ 0 & 1 & 1 & 0 & 0 & 1 & 1 & 0 \\ 1 & 0 & 1 & 0 & 1 & 0 & 1 & 0 \end{bmatrix} \cdot \begin{bmatrix} 1 \\ 0 \\ 0 \\ 1 \\ 0 \\ 1 \\ 1 \\ 0 \end{bmatrix} = \begin{bmatrix} 0 \\ 1 \\ 0 \\ 0 \end{bmatrix}
$$

Here the extra bit (last bit in code word) is correct, but the syndrome 0100 is nonzero. Therefore, this SECDED code can correctly identify a double-bit error. However, if the code word is [1 0 0 1 0 1 0 0], then the extra bit is zero and incorrect, which suggests a single-bit error. The resulting syndrome is 1011, the lower three bits of which indicate the bit position in error.

The parity check matrix of a SECDED code has important properties that can be used to optimize the design of the code. Like the parity check matrix for a SEC code, all the columns in SECDED code's parity check matrix must be unique for it to correct single-bit errors. Also, since the minimum Hamming distance of a SECDED code is 4, no three or fewer columns of its parity check matrix can sum to zero (under XOR or modulo-2 summation). This also implies that no two columns of a SECDED code's parity check matrix can sum to a third column. These properties can be used to optimize the implementation of a SECDED ECC.

The number of 1s in the parity check matrix has two implications on the implementation of the SECDED code. First, each entry with a 1 in the parity check matrix causes an XOR operation. Hence, typically, greater the number of 1s in the parity check matrix, greater is the number of XOR gates. Second, greater the number of 1s in a single row of the parity check matrix, the greater is the time taken or complexity incurred to compute the SECDED ECC. For example, the first row of the SECDED ECC shown above has all 1s and hence requires either a long sequence of bit-wise XOR operations or a wide XOR gate. Consequently, reducing the number of 1s in each row of the parity check matrix results in a more efficient implementation. For example, the following parity check matrix (with the code word sequence $[C_1 C_2 D_1 C_3 D_2 D_3 D_4 C_4]$)

$$
\begin{bmatrix} 0 & 0 & 1 & 0 & 1 & 1 & 0 & 1 \\ 0 & 0 & 0 & 1 & 1 & 1 & 1 & 0 \\ 0 & 1 & 1 & 0 & 0 & 1 & 1 & 0 \\ 1 & 0 & 1 & 0 & 1 & 0 & 1 & 0 \end{bmatrix}
$$

has only a total of sixteen 1s compared to the earlier one that has twenty 1s. Further, the maximum number of 1s per row is four compared to eight in the earlier one.

This optimized parity check matrix is based on a construction proposed by Hsiao and is referred to as the *Odd-weight column SECDED code* [5]. Hsiao's construction satisfies the properties of the parity check matrix (e.g., no three or fewer columns can sum to zero) yet results in a much more efficient implementation. Hsiao's construction imposes the following restrictions on a parity check matrix's columns:

- no column has all zeros
- every column is distinct
- every column contains an odd number of 1s.

The parity check matrix above satisfies these properties. A zero syndrome indicates no error, but the condition to distinguish between a single- and a double-bit error is different from what was used before. If the number of 1s in the syndrome is even, then it indicates a double-bit error; otherwise it is a single-bit error. Thus, with the code word [1 0 0 1 0 1 0 0], one gets a syndrome of 1011. Since the number of 1s is odd, it indicates a single-bit error. The lower three bits specify the bit position in error. Hsiao [5] describes how to construct Odd-weight column SECDED codes for an arbitrary number of data bits.

■ E X A M P L E

For the parity check matrix that uses Hsiao's formulation, determine whether the code word 10110111 is correct, has a single-bit error, or has a double-bit error.

S O L U T I O N The syndrome is 1111. The number of 1s is even, so this indicates a double-bit error.

5.2.5 Double-Error Correct Triple-Error Detect Code

As seen earlier, SEC or SECDED codes can be extended to correct specific types of double-bit errors. Interleaving independent code words with an interleaving distance of 2 allows one to correct a spatially contiguous double-bit error. A technique called scrubbing, described later in this chapter, can help protect against a temporal double-bit fault, which can arise due to particle strikes in two separate bits protected by the same SEC or SECDED code. The scrubbing mechanism would attempt to read the code word between the occurrences of the two faults, so that the SEC or SECDED code could correct the single-bit fault before the fault is converted into a double-bit fault.

The applicability of interleaving or scrubbing, however, is limited to specific types of double-bit faults and may not always be effective (e.g., if the second error occurs before the scrubber gets a chance to correct the first error). In contrast, a DECTED code can correct any single- or double-bit error in a code word. It can also detect triple-bit faults. To understand the theory of DECTED codes, one needs a background in advanced mathematics. The theory itself is outside the scope of this book. Nevertheless, using an example, this section illustrates how a DECTED code works. The construction of this example DECTED code follows from a class of codes known as BCH codes named after the inventors R. C. Bose, D. K. Ray-Chaudhuri, and A. Hocquenghem.

There are some similarities between how DECTED and SECDED codes work. Like in a SECDED code, a DECTED code can construct a parity check matrix, which when multiplied by the code word gives a syndrome. By examining the syndrome, one can identify if there was no fault, a single-bit fault, a double-bit fault, or a triple-bit fault. The minimum Hamming distance of a DECTED code is 6 (see Determination of Number of Bit Errors That Can Be Detected and Corrected, p. 164). This also implies that any linear combination of five or fewer columns of the parity check matrix of such a code is not zero.

A parity check matrix for a $(N-1, N-2m-2)$ DECTED code is usually expressed in a compact form as

$$
\begin{bmatrix}
1 & 1 & 1 & K & 1 \\
1 & X & X^2 & K & X^{N-2} \\
1 & X^3 & X^6 & K & X^{3(N-2)}
\end{bmatrix}
$$

where X is the root of a primitive binary polynomial $P(x)$ of degree m, $N = 2^m$, the number of data bits is $N - 2m - 2$, and number of check bits is $2m + 1$. The design may have to incur the overhead of extra bits if the number of data bits is fewer than $N - 2m - 2$.

As an example, one can define a (31, 20) DECTED code with $m = 5$ and $P(X) = 1 + X^2 + X^5$. In a binary vector format, X can be expressed as a vector [0 1 0 0 0], where the entries of the vector are the coefficients of the corresponding polynomial. Similarly, X^4 can be expressed as [0 0 0 0 1]. X^5 will be computed as X^5 mod $P(X)$, which equals $1 + X^2$ and is expressed in a binary vector format as [1 0 1 0 0]. Expanding the Xs for the (31, 20) DECTED code, one can obtain the following parity check matrix:

$$
\begin{bmatrix}
1&1&1&1&1&1&1&1&1&1&1&1&1&1&1&1&1&1&1&1&1&1&1&1&1&1&1&1&1&1&1\\
1&0&0&0&0&1&0&0&1&0&1&1&0&0&1&1&1&1&0&0&0&1&1&0&1&1&1&0&1&0&\\
0&1&0&0&0&0&1&0&0&1&0&1&1&0&0&1&1&1&1&0&0&0&1&1&0&1&1&1&0&1&\\
0&0&1&0&0&1&0&1&1&0&0&1&1&1&1&0&0&0&1&1&0&1&1&1&0&1&0&1&0&0&\\
0&0&0&1&0&0&1&0&1&1&0&0&1&1&1&1&0&0&0&1&1&0&1&1&1&0&1&0&1&0&\\
0&0&0&0&1&0&0&1&0&1&1&0&0&1&1&1&1&0&0&0&1&1&0&1&1&1&0&1&0&1&\\
1&0&0&0&0&1&1&0&0&1&0&0&1&1&1&1&0&1&1&1&0&0&0&1&0&1&0&1&1&0&\\
0&0&1&1&1&1&1&0&1&1&1&0&0&0&1&0&1&0&1&1&0&1&0&0&0&0&1&1&0&0&1\\
0&0&0&0&1&1&0&0&1&0&0&1&1&1&1&0&1&1&1&0&0&0&1&0&1&0&1&1&1&0&1\\
0&1&1&1&1&1&0&1&1&1&0&0&0&1&0&1&0&1&1&0&1&0&0&0&0&1&1&0&0&1&0\\
0&0&0&1&0&1&0&1&1&0&1&0&0&0&0&1&1&0&0&1&0&0&1&1&1&1&1&0&1&1&1
\end{bmatrix}
$$

The vector X embedded in this matrix is shaded. Column 2 of this matrix is $[1\ 0\ 1\ 0\ 0\ 0\ 0\ 0\ 0\ 1\ 0]$, which is the same as $[1\ X\ X^3]$.

The syndrome can be obtained by multiplying the parity check matrix with the code word vector. The syndrome is a binary vector with $2m+1$ entries and can be expressed as $[S0\ S1\ S2]$, where $S0$ consists of one bit and $S1$ and $S2$ both consist of m bits each. The syndrome can be decoded as follows:

- If the syndrome is zero, then there is no fault.

- If $S0 = 1$, then there is a single-bit fault and the error position is the root of the linear equation $y + S1 = 0$. The error can be corrected by inverting the bit position.

- If $S0 = 0$, then there is a double-bit error. If the bit positions in error are $E1$ and $E2$ (expressed as numbers less than 2^m), then it turns out that $S1 = E1 + E2$ and $S2 = E1^3 + E2^3$. It can be shown that $E1$ and $E2$ are roots of the quadratic equation $S1\ y^2 + S1^2 y + (S1^3 + S2) = 0$. It should be noted that for a single-bit error (say with $E1 > 0$ and $E2 = 0$), this equation degenerates into the linear equation $y + S1 = 0$. This is because $S1^3 + S2 = E1^2 E2 + E1\ E2^2 = E1\ E2 S1$. If $E2 = 0$, then $S1^3 + S2 = 0$.

- If $S0 = 0$ and the quadratic equation has no solution or if $S0 = 1$ and the quadratic equation does not degenerate into a linear equation, then there is an uncorrectable error.

For example, if the 31-bit code word is $[0\ 0\ 0\ ...\ 0]$, then the syndrome is zero and there is no fault. If the code word is $[1\ 0\ 0\ ...\ 0]$, then the syndrome is $[1\ 1\ 0\ 0\ 0\ 0\ 1\ 0\ 0\ 0\ 0]$. That is, $S0 = 1$, $S1 = [1\ 0\ 0\ 0\ 0]$, and $S2 = [1\ 0\ 0\ 0\ 0]$. Since $S0 = 1$, one would expect a single-bit fault. The solution to the equation $y + S1 = 0$ is $y = 1$, indicating that the bit position 1 is in error. If the 31-bit code word is $[1\ 1\ 0\ ...\ 0]$, then the syndrome is $[0\ 1\ 1\ 0\ 0\ 0\ 1\ 0\ 0\ 1\ 0]$. $S0 = 0$, $S1 = [1\ 1\ 0\ 0\ 0]$, $S2 = [1\ 0\ 0\ 1\ 0]$. $S1$ can be interpreted as the number 3, and $S2$ can be interpreted as the number 9. The reader can verify that the first and second bit positions are in error. Then, $S1 = E1 + E2 = 1 + 2 = 3$. $S2 = 1^3 + 2^3 = 9$. Instead of solving the quadratic equation in hardware to find the bit positions in error, researchers have proposed to solve the quadratic equation for all possible values of the syndrome and store the values in a table for use by the hardware.

5.2.6 Cyclic Redundancy Check

CRC codes are a type of *cyclic code*. An end-around shift of a cyclic code word produces another valid code word. Like the Hamming codes discussed so far in this chapter, CRC codes are also linear block codes. CRC codes are typically used to detect burst errors, errors in a sequence of bits. Typically, such errors arise in transmission lines due to coupling and noise but not due to soft errors.

Soft errors usually do not affect transmission lines because metal lines transferring data can typically recover from an alpha particle or a neutron strike. Nevertheless, CRC codes can provide soft error protection for memory cells as a by-product of protecting transmission lines. For example, if there is a buffer into which the data from a transmission line are dumped before the data are read out and the CRC is decoded, then the buffer is protected against soft errors via the CRC code. CRC codes by themselves only provide error detection, so the CRC code reduces the buffer's SDC AVF to close to zero. However, often when a CRC error is detected by the receiver, it sends a signal back to the sender to resend the data. In such a case, the faulty data in the buffer are recovered, so both the SDC and DUE AVFs of the buffer reduce to almost zero.

The underlying principle of CRC codes is based on a polynomial division. CRC codes treat a code word as a polynomial. For example, the data word 1011010 would be represented as the polynomial $D(x) = x^6 + x^4 + x^3 + x$, where the coefficients of x^i are the data word bits. In general, given a k-bit data word, one can construct a polynomial $D(x)$ of degree $k-1$, where x^{k-1} is the highest order term. Further, the sender and receiver must agree on a generator polynomial $G(x)$. For example, $G(x)$ could be $x^4 + x + 1$. If the degree of $G(x)$ is r, then k must be greater than r.

The CRC encoding process involves dividing $D(x) \cdot x^r$ by $G(x)$ to obtain the quotient $p(x)$ and the remainder $R(x)$, which is of degree $r-1$. This results in the equation $D(x) \cdot x^r = G(x) \cdot p(x) + R(x)$. Then the transmitted polynomial $T(x) = G(x) \cdot p(x) = D(x) \cdot x^r - R(x)$. Because the addition of coefficients is an XOR operation and the product is an AND operation, it turns out that the coefficients of $T(x)$ can simply be expressed as a concatenation of coefficients of $D(x)$ and $R(x)$. Figure 5.7 shows an example of how to generate the coefficients of $T(x)$, given the data bits 1011010 and the corresponding generator polynomial $x^4 + x + 1$. The bits corresponding to the transmitted polynomial $T(x)$ are the ones sent on the transmission lines.

```
                                    p(x) = 1  0  1  0  1  0  1
          G(x) = 1  0  0  1  1 | 1  0  1  1  0  1  0  0  0  0  0
                                1  0  0  1  1
                                ─────────────
                                   1  0  1  1  0
                                   1  0  0  1  1
                                   ────────────
                                      1  0  1  0  0
                                      1  0  0  1  1
                                      ────────────
                                         1  1  1  0  0
                                         1  0  0  1  1
                                         ────────────
                          R(x) =            1  1  1  1
```

FIGURE 5.7 Polynomial division to generate the CRC code. The data bits represented by *D(x)* are 1011010. The bits corresponding to the generator polynomial *G(x) = x^4 + x + 1* are 10011, which is the divisor. The corresponding remainder is 1111. Hence, the transmitted bits = original data bits concatenated with remainder bits = 10110101111.

The CRC decoding process involves simply dividing $T(x)$ by $G(x)$. If the remainder is zero, then $D(x)$'s coefficients have been transmitted without error and can be extracted from $T(x)$. However, if the remainder is nonzero, then $T(x)$ was in error during transmission. For example, it arrives as $T'(x)=T(x)+E(x)$, where $E(x)$ is the error polynomial. Each 1 bit in $E(x)$ corresponds to 1 bit that has been inverted. $T(x)/G(x)$ will always be zero but $E(x)/G(x)$ may not be. Errors that cause $E(x)/G(x)$ to be zero will not be detected by the CRC code.

The challenge of a CRC code is, therefore, to minimize the number of errors that cause $E(x)/G(x)$ to be zero. If a single-bit error occurs, then $G(x)$ will never divide $E(x)$ as long as $G(x)$ contains two or more terms. Similar properties can be derived for double-bit errors. For example, on a double-bit error, one can have $E(x)=x^i+x^j$, where $i>j$. $E(x)=x^j \cdot (x^{i-j}+1)$. So, as long as $E(x)$ is not divisible by x and x^m+1 is not divisible by $G(x)$, where $i-j \leq m$, $G(x)$ can detect double-bit errors m distance apart. However, more importantly, a CRC code with r check bits can detect all burst errors of length $\leq r$, as long as $G(x)$ contains x^0 (i.e., 1) as one of its polynomial terms. $E(x)$ in this case can be represented as $x^i \cdot (x^{l-1}+x^{l-2}+1)$, where i is the bit position where the burst starts and l is the length of the burst error. $G(x)$ will not divide $E(x)$ in this case as long as it contains x^0 as a term and $r \geq l$.

CRC is particularly attractive for error detection because of the simplicity of its encoding and decoding implementations. The CRC encoding scheme can be implemented simply as a chain of XORs and flip-flops. Figure 5.8 shows the CRC encoding circuit corresponding to the generator polynomial x^4+x+1. In every clock iteration, one bit of $D(x) \cdot x^r$ is fed into the circuit. In the example used earlier, the corresponding coefficient would be 10110100000. After $k+r-1$ iterations, the data flip-flops will contain the coefficients of the remainder, which when concatenated with the coefficients of $D(x)$ will provide the word to be transmitted.

The key to the construction of a CRC encoding circuit is the placement of the XOR gates. Whenever the coefficient is 1 (except for the highest order 1), one can insert an XOR gate before the data flip-flop. For example, in Figure 5.8, the two XOR gates to the right correspond to the two rightmost 1s in the generator polynomial

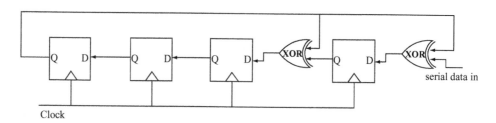

Clock

FIGURE 5.8　CRC encoding circuit corresponding to the generator polynomial x^4+x+1 (with corresponding coefficients 10011). "Serial data in" is the port through which the $D(x)$ polynomial enters its bit stream from the most significant bit. The square boxes are data flip-flops with D as the clocked input and Q as the clocked output.

coefficient 10011. The leftmost one (or one in the most significant bit) does not need an XOR gate by itself. Instead, the output Q of the leftmost flip-flop is fed back into the XOR gates to enable the division operation that the CRC encoding performs. It will be left as an exercise to the reader to perform the clocking steps to be convinced that indeed this circuit will generate the remainder 1111 (Figure 5.7) for the data word 1011010 and generator polynomial coefficients 10011.

Today, CRC is widely used in a variety of interconnection networks and software protocols. Three of the most commonly used generator polynomials are:

- CRC-12: $G(x) = x^{12} + x^{11} + x^3 + x^2 + x^1 + 1$

- CRC-16: $G(x) = x^{16} + x^{15} + x^2 + 1$

- CRC-CCITT[3]: $G(x) = x^{16} + x^{12} + x^5 + 1$.

■ EXAMPLE

Compute the transmitted polynomial for $D(x) = 10011$ when the generator polynomial $G(x) = x + 1$.

SOLUTION Dividing $D(x) \cdot x$ by $G(x)$ gives $p(x) = 1100$ and $R(x) = 1$. So, the transmitted bits are $D(x) \cdot x + R(x) = 100111$. Interestingly, $R(x)$ is also the even parity code for the data bits 1011. It turns out that an even parity code is the same as a 1-bit CRC code.

5.3 Error Detection Codes for Execution Units

In a microprocessor pipeline, execution units, such as adders and multipliers, are in general less vulnerable to soft errors than structures that hold architectural state or are stall points in the pipeline. This is because of two reasons. First, execution units are largely composed of logic circuits, which have high levels of logical, electrical, and latch-window masking (see Masking Effects in Combinatorial Logic Gates, p. 52, Chapter 2). In contrast, structures holding architectural state or stall points are composed of state bits that do not have many of these masking properties.

The second reason is more subtle but can be easily explained using Little's law formulation for AVF (see Computing AVF with Little's Law, p. 98, Chapter 3). Using Little's law, one can show that the AVF of a structure is proportional to $B_{ACE} \times L_{ACE}$, where B_{ACE} is the throughput of ACE instructions into the structure and L_{ACE} is the latency or delay of ACE instructions through the structure. On average, B_{ACE} remains the same through similar structures in the pipeline (e.g., ones that process all ACE instructions).

---

[3]CCITT stands for International Telephone and Telegraph Consultative Committee.

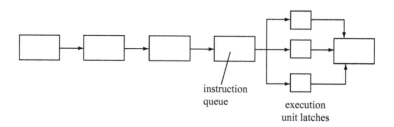

instruction
queue

execution
unit latches

FIGURE 5.9   An example processor pipeline. Instruction queue is a stall point.

However, L_{ACE} varies significantly across the pipeline (Figure 5.9). Stall points in a pipeline, such as an instruction queue, have a significantly higher L_{ACE} than the input or output latches of execution units. This is because the pipeline can back up due to a cache miss or a branch misprediction. In contrast, L_{ACE} can be much lower for execution unit latches because execution units typically do not hold stalled instructions. Hence, AVF of stall points in a pipeline is typically significantly higher than that of execution units. The window of exposure of structures containing architectural state is also usually higher than that of execution unit latches. Hence, both structures with architectural state and stall points in a pipeline are more prone to soft errors than execution units. Consequently, for soft errors, structures with architectural or stalled state are the first candidate for protection. Nevertheless, once these more vulnerable structures are protected, execution units could potentially become targets for protection from soft errors. The techniques described in this chapter can also be used to protect the execution units from other kinds of faults described in Chapter 1.

This section discusses three schemes to detect transient faults in execution unit logic and latches. These are called AN codes, residue codes, and parity prediction circuits. The first two are called arithmetic codes. Also, AN codes are not separable, whereas residue codes and parity prediction circuits are separable codes. These codes are cheaper than full duplication of functional units, which makes them attractive to designers. Some of these codes ensure important properties, such as *fault secureness*. These properties are not discussed in this book because these are not as relevant for soft errors. The readers are referred to Pradhan [16] for a detailed discussion on these properties.

5.3.1 AN Codes

AN codes are the simplest form of arithmetic codes. Arithmetic codes are invariant under a set of arithmetic operations. For example, given an arithmetic operation • and two bit strings a and b, then C is an arithmetic code if $C(a \cdot b) = C(a) \cdot C(b)$. $C(a)$ and $C(b)$ can be computed from the source operands a and b, whereas $C(a \cdot b)$ can be computed from the result. Thus, by comparing $C(a \cdot b)$ obtained from the result and $C(a) \cdot C(b)$ obtained from the source, one can determine if the operation incurred an error.

Specifically, an AN code is formed by multiplying each data word N by a constant A (hence, the name AN code). AN codes can be applied for addition or subtraction operations since $A(N_1 + N_2) = A(N_1) + A(N_2)$ and $A(N_1 - N_2) = A(N_1) - A(N_2)$. The choice of A determines the extra bits that are needed to encode N. A typical value of A is 3 since $3N = 2N + N$, which can be derived by a left shift of N followed by an addition with N itself. However, the same does not hold for multiplication and division operations. The next subsection examines another class of arithmetic codes called residue codes that do not have this limitation.

5.3.2 Residue Codes

Residue codes are another class of arithmetic codes. Unlike AN codes, they are separable codes and applicable to a wide variety of execution units, such as integer addition, subtraction, multiplication, and division, as well as shift operations. This section discusses the basic principles of residue codes for integer operations. Iacobovici has shown that residues can be extended to shift operations [6]. Extending residue codes to floating-point and logical operations is still an active area of research [8]. IBM's recent mainframe microprocessor—codenamed z6—incorporates residue codes in its pipeline [21].

Modulus is the operation under which residue codes are invariant for addition, subtraction, multiplication, and division and used as the underlying principle to generate residue codes. For addition, $(N_1 + N_2) \bmod M = ((N_1 \bmod M) + (N_2 \bmod M)) \bmod M$. For example, $N_1 = 10$, $N_2 = 9$, $M = 3$. Then $19 \bmod 3 = 1$ (left-hand side of equation), and $((10 \bmod 3) + (9 \bmod 3)) \bmod 3 = (1 + 0) \bmod 3 = 1$ (right-hand side of equation). The same is true for subtraction. Figure 5.10 shows the block diagram of residue code logic for an adder.

For multiplication, $(N_1 \times N_2) \bmod M = ((N_1 \bmod M) \times (N_2 \bmod M)) \bmod M$. Using the same values for the addition example, one gets the left-hand side of this equation $= (10 \times 9) \bmod 3 = 0$. Similarly, right-hand side $= ((10 \bmod 3) \times (9 \bmod 3))$

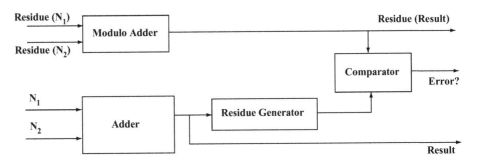

FIGURE 5.10 Block diagram of residue code logic for an adder.

mod $3 = (1 \times 0)$ mod $3 = 0$. For division, the following equation holds: $D - R = Q \times I$, where D = dividend, R = remainder, Q = quotient, and I = divisor. Since subtraction and multiplication are individually invariant under the modulus operation, one can have $((D \bmod M) - (R \bmod M)) \bmod M = ((Q \bmod M) \times (I \bmod M)) \bmod M$. Signed operations in 2's complement arithmetic cause some additional complications but can be handled easily.

Residue codes for addition can be implemented using a logarithmic tree of adders if the numbers are represented in binary arithmetic and the divisor M is of the form $2^k - 1$. The process is known as "casting out Ms," so when $M = 3$, the process is called "casting out 3s." To understand how this works, let us assume that a binary number is represented as a sequence of n concatenated values: $a_{n-1}a_{n-2}$ ... a_1a_0, where each a_i is represented with k bits. For example, if $N = 1111001$, $k = 2$, $M = 3$, $n = 4$, then one can have $a_3 = 1$, $a_2 = 11$, $a_1 = 10$, and $a_0 = 01$. N can be expressed as

$$N = a_{n-1} \cdot 2^{k \times (n-1)} + a_{n-2} \cdot 2^{k \times (n-2)} + L + a_1 \cdot 2^k + a_0$$

Then, $N \bmod M$ becomes

$$N \bmod M = ((a_{n-1} \cdot 2^{k \times (n-1)}) \bmod M + (a_{n-2} \cdot 2^{k \times (n-2)}) \bmod M + L \\ + (a_1 . 2^k) \bmod M + (a_0) \bmod M) \bmod M$$

Since $(A \times B) \bmod M = ((A \bmod M) \times (B \bmod M)) \bmod M$ and $(2^{k \times i} \bmod (2^k - 1)) = 1$, one can have

$$N \bmod M = ((a_{n-1}) \bmod M + (a_{n-2}) \bmod M + L + (a_1) \bmod M + (a_0) \bmod M) \bmod M$$

Or,

$$N \bmod M = (a_{n-1} + a_{n-2} + L + a_1 + a_0) \bmod M$$

In other words, the modulus of N with respect to M is simply the modulo sum of the individual concatenated values. So, for $N = 121 = 1111001$ in binary, one can have $((1) + (11) + (10) + (01)) \bmod 3 = 7 \bmod 3 = 1$.

Low-cost residue codes with low modulus, such as 3, are particularly attractive for soft errors. But higher the modulus, the greater is the power of residue codes to detect single and multi-bit faults. For more on residue code implementation, readers are referred to Noufal and Nicolaidis [11].

■ E X A M P L E

An adder has a residue code generator associated with it. The source operands are 8 and 9. The final result is 17 with a residue of 1 computed from the source operands. Assume the modulus is 3. Was there an error in the addition operation?

SOLUTION 8 in binary representation is 1000 and 9 is 1001. The modulus
$= (10 + 00 + 10 + 01) \bmod 3 = 5 \bmod 3 = 2$. The residue computed from the result
was 1. Hence, there was an error. It should be noted that the hardware cannot
tell if the addition operation or the residue computation was in error in this
case.

The next subsection examines a different style of fault detection that extends the
concept of parity to execution units.

5.3.3 Parity Prediction Circuits

Parity prediction circuits use properties of carry chains in addition and multiplication operations. Like residue codes, parity prediction circuits can be made to
work for subtraction and division. Like AN and residue codes, parity prediction
computes the parity of the result of an operation in two ways: first from the source
operands and second by computing the parity of the result value itself. Although
the term "prediction" suggests that the parity of the result of an operation is "predicted" from the source operands, in reality it is computed accurately. The term
"prediction" here is not used in the sense speculative microprocessors use it today
to refer to, for example, branch prediction logic. Parity prediction circuits are currently used in commercial microprocessors, such as the Fujitsu SPARC64 V microprocessor [2].

To understand how parity prediction works, let us consider the following
addition operation: $S = A + B$. Let A_c, B_c, and S_c be the parity bits for A, B,
and S, respectively. Further, let $A = a_{n-1}a_{n-2} \ldots a_1a_0$, $B = b_{n-1}b_{n-2} \ldots b_1b_0$, and
$S = s_{n-1}s_{n-2} \ldots s_1s_0$, where a_i, b_i, and s_i are individual bits representing A, B, and
S, and n is the total number of bits representing each of these variables. Now, S_c
can be computed in two ways (Figure 5.11). First, by definition, $S_c = s_{n-1} \text{ XOR } s_{n-2}$
XOR ... XOR s_1 XOR s_0, which can be computed from the result S. Second, it turns

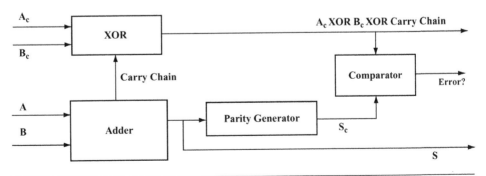

FIGURE 5.11 Parity prediction circuit.

out that S_c can also be computed from A_c, B_c, and the carry chain. Let c_i be the carry out from the addition of a_i and b_i. Then, $S_c = A_c$ XOR B_c XOR $(c_{n-1}$ XOR c_{n-2} XOR ... XOR c_1 XOR $c_0)$. Thus, by computing S_c in two different ways and by comparing them, one can verify whether the addition operation was performed correctly. For example, if in binary representation, if $A = 01010$, $B = 01001$, then $S = A + B = 10011$. $A_c = 0$, $B_c = 0$, and $S_c = 1$. The carry chain is 01000. So, S_c computed as an XOR of A_c, B_c, and the carry chain is 1.

The implementation of a parity prediction circuit must ensure that a strike on the carry chain does not cause erroneous S and S_c values (or its AVF will be higher). This could happen because the carry chain feeds both S (from which one version of S_c is computed) and S_c, which can be computed from A_c, B_c, and the carry chain. If the same error propagates to both versions, then the error may elude detection by this circuit. To avoid this problem, parity prediction circuits typically implement dual-rail redundant carry chains to ensure that a strike on a carry chain does produce erroneous S_c on both paths.

Parity prediction circuits for multipliers can be constructed similarly since a multiplication is a composition of multiple addition operations. Nevertheless, parity prediction circuits may differ depending on whether one can use a Booth multiplier or a Wallace tree. Readers are referred to Nicolaidis and Duarte [13] and Nicolaidis et al. [14] for detailed description of how to construct parity prediction circuits for different multipliers.

Given that both residue codes and parity prediction circuits are good candidates to protect execution units against soft errors, it is natural to ask which one should a designer choose. It is normally accepted that parity prediction circuits are cheaper in area than residue codes for adders and small multipliers. But for large multipliers (e.g., 64 bits wide), it turns out that the residue codes are cheaper. For further discussion on the implementation of residue codes and parity prediction circuits, readers are referred to Noufal and Nicolaidis [11] and Nicolaidis [12].

■ EXAMPLE

Compute the parity for the addition operation of the two source operands 01111 and 00001 directly and through the parity prediction circuit.

SOLUTION The sum is 10000, so the parity of the sum is 1. The parity of the first operand is 0 and the second operand is 1. The carry chain is 01111, whose parity is 0. Consequently, the parity from the parity prediction circuit is 0 XOR 1 XOR 0 = 1.

As should be obvious by now, different protection schemes have different overheads and trade-offs. The next section discusses some of the implementation overheads and issues that arise in the design of error detection and correction codes.

5.4 Implementation Overhead of Error Detection and Correction Codes

Implementors of error detection and coding techniques typically worry about two overheads: number of logic levels in the encoder and decoder and the overhead in area incurred by the extra bits and logic. This section elaborates on each of these.

5.4.1 Number of Logic Levels

Figure 5.12 shows the logic diagram of a SEC encoder and decoder described in Single-Error Correction Codes earlier in this chapter. For example, before writing the data bits to a register file, the check bits must be generated and stored along with the data bits (the encoding step). Similarly, before reading the data bits, the syndrome must be generated and checked for errors (the decoding step).

As Figure 5.12 shows, encoding is simpler and takes fewer logic levels than decoding. Nevertheless, even the simple encoding logic may be hard to fit into the cycle time of a processor. For example, assume that a processor pipeline has 12 stages of logic per pipeline stage and the critical path for a stage is already 12 logic levels. Then adding ECC encoding to the pipeline stage may increase the critical path to 13 stages. This reduces the processor frequency by $(1/12) = 8.3\%$ by stretching the cycle time, which may be unacceptable to the design.

Decoding can pose an even greater performance penalty because it requires several levels of logic to decode the error code associated with the data. Hence, like encoding, it may not fit into a processor's or chipset's cycle time. Such a style of error code decoding is referred to as *in-line* error detection and/or correction, in which the error code is decoded and the error state of the data is identified before the data are allowed to be used by the next pipeline stage. Alternatively, one can allow *out-of-band* decoding in which the data are allowed to be read by intermediate stages of the pipeline but eventually the error is tracked down before it is allowed to propagate beyond a certain boundary. Three ways in which the performance degradation associated with error code decoding can be reduced with both in-line and out-of-band decoding are discussed subsequently.

First, like Hsiao's formulation for odd-weight column SECDED codes (see Single-Error Correct Double-Error Detect Code, p. 174), one can try to reduce the average number of 1s in a row of the parity check matrix. The number of 1s in the parity check matrix determines the binary XOR operations one has to do in the critical path. Hence, reducing the average number of 1s across the rows of the parity check matrix would reduce the height of the logic tree necessary to decode the ECC. This would still be in-line ECC decoding but with less overhead in time.

Second, if part of the error decoding logic fits into the cycle time, but not the full logic, then one may consider alternate schemes. For example, the AMD's Opteron™ processor uses an in-line error detection mechanism but an out-of-band probabilistic ECC correction scheme for its data cache [1]. For every load access to

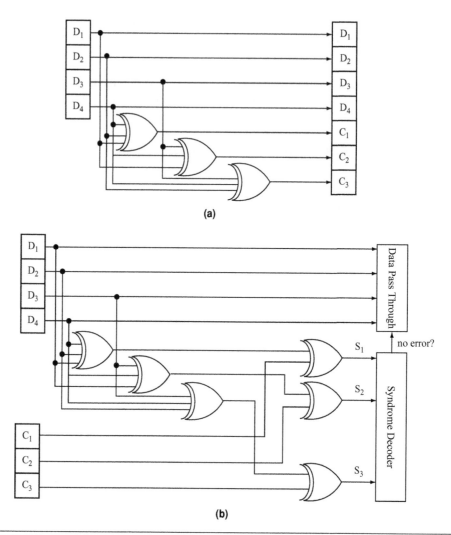

FIGURE 5.12   (a) SEC Encoder. (b) SEC Decoder. (7,4) code. D_1–D_4 are data bits, C_1–C_3 are check bits, and S_1–S_3 are syndrome bits.

the cache, the ECC logic checks whether there is an error in the cache. If there is no error, the load is allowed to proceed. In the background, a hardware scrubber wakes up periodically and examines the data cache blocks for errors. If it finds errors, it corrects them using the ECC correction logic. Thus, if there is a bit flip in the cache and the hardware scrubber corrects it before a load accesses it, then there will not be any error. As discussed in Computing the DUE AVF, p. 131, Chapter 4, the effectiveness of such a scheme is highly dependent on the frequency with which loads access the cache, the size of the cache, and frequency of scrubbing.

Third, if the error decoding logic does not fit in the cycle time, then one can try out-of-band error decoding and correction. For example, the error signal from the parity prediction logic in the Fujitsu SPARC64 V execution units is generated after the result moves to the next pipeline stage [2]. The error signal is propagated to the commit point and associated with the appropriate instruction before the instruction retires. The commit logic checks for errors in each instruction. This allows the pipeline not only to detect the error before the instruction's result is committed but also helps to recover from the error by flushing the pipeline from the instruction that encountered the error. Then, by refetching and executing the offending instruction, the processor recovers from the error.

5.4.2 Overhead in Area

Error coding techniques incur two kinds of area overhead: number of added check bits and logic to encode and decode the check bits. The cost of the error coding and decoding logic is typically amortized over many bits. For example, one could have a single-error encoder and a decoder for a 32-kilobyte cache, which would amortize the cost of the encoding logic and decoding logic.

The greater concern for area is the additional check bits. Figure 5.3 shows that the number of extra bits necessary to correct a single bit of error in a given number of data bits grows slowly with the number of data bits. Consequently, the greater the data bits a set of code bits can cover, the lower is the overhead incurred in an area. However, there is a limit to the number of data bits that can be covered with a given code. For example, a cache block is typically 32 or 64 bytes. But the ECC granularity is usually 64 bits (8 bytes) or fewer. This is because a typical instruction in a microprocessor can only work on a maximum of 64 bits. If the ECC were to cover the entire 32 bytes (four 64-bit words), then the ECC encoding step would be a read-modify-write operation, that is, first read the entire cache block, write the appropriate 64-bit data word in there, recompute ECC for the whole block, and then write the ECC bits into the appropriate block. This read-modify-write can be an expensive operation and can cause performance degradation by stretching the critical path of a pipeline stage. Hence, today's processors typically limit the ECC to 64 bits. For a SECDED code, the overhead is 8 bits, so a typical word consists of 72 bits: 64 bits for data and 8 bits for the code.

It is noteworthy though the same read-modify-write concern may not exist for parity codes. This is because the parity of the whole cache block can be recomputed using only the parity of the word just changed and the parity of the whole block. The parity of the whole cache block still has to be read but not the words untouched by the specific operation.

The overhead of ECC increases if double-bit errors need to be corrected. There are two kinds of double-bit errors: spatial and temporal. Spatial double-bit errors are those experienced by adjacent bits (typically caused by the same particle strike). Temporal double-bit errors are experienced by nonadjacent bits (typically caused by two different particle strikes). One way to protect against both spatial and temporal double-bit errors is to use double-error correction (DEC) codes. DEC codes, however, need significantly more check bits than SEC codes. For example, for 64 data bits, SEC needs only 7 bits but DEC needs 12 bits (see Determination of the Minimum Number of Code Bits Needed for Error Correction, p. 166).

To avoid incurring the extra overhead for double-bit correction, computer systems often use two schemes: interleaving to tackle spatial double-bit errors and scrubbing to tackle temporal double-bit errors. As Figure 5.4 shows, interleaving allows errors in two consecutive bits to be caught in two different code words. If the protection scheme is parity, then this allows one to detect spatial double-bit errors. If the protection scheme is SECDED ECC, then this allows one to detect and correct spatial double-bit errors. In main memory systems composed of multiple DRAM chips, this interleaving is often performed across the multiple chips, with each bit in the DRAM being covered by a different ECC. Then, if an entire DRAM chip experiences a hard fail and stops functioning (called chipkill), the data in that DRAM can be recovered using the other DRAM chips and the ECC. Thus, the same SECDED ECC can not only prevent spatial double-bit errors but also protect the memory system from a complete chip failure.

Scrubbing is a scheme typically used in large main memories, where the probability of a temporal double-bit error can be high. If two different particles flip two different bits in a data word, then SECDED ECC may not be able to correct the error, unless there is an intervening access to the word. This is because typically in computer systems, an ECC correction is invoked only when the word is accessed. If the word is not accessed, the first error will go unnoticed and uncorrected. When the second error occurs, the single-bit error gets converted into a double-bit error, which can only be detected by SECDED but not corrected. Scrubbing tries to avoid such accumulation of errors by accessing the data word, checking the ECC for any potential correctable error, correcting the error, and writing it back. Chipsets normally distinguish between two styles of scrubbing: *demand scrubbing* writes back corrected data back into the memory block from where it was read by the processor and *patrol scrubbing* works in the background looking for and correcting resident errors.

Finally, whether scrubbing is necessary in a particular computer memory system depends on its SER, the target error rate it plans to achieve, and the size of the memory. The next section discusses how to compute the reduction in SER from scrubbing.

5.5   Scrubbing Analysis

This section shows how to compute the reduction in temporal double-bit error rate from scrubbing. First, the SER from temporal double-bit errors in the absence of any scrubbing is computed. Then, the same is computed in the presence of scrubbing. Such analysis is essential for architects trying to decide if they should augment the basic ECC scheme with a scrubber. For both cases—with and without scrubbing—it will be shown how to enumerate the double-FIT rate from first principles [9] and using a compact form proposed by Saleh et al. [18].

5.5.1 DUE FIT from Temporal Double-Bit Error with No Scrubbing

Assume that an 8-bit SECDED ECC protects 64 bits of data, which is referred to as a quadword in this section. A temporal double-bit error occurs when two bits of this 72-bit protected quadword are flipped by two separate alpha particle or neutron strikes. This analysis is only concerned with strikes in the data portion of the cache blocks (and not the tags). To compute the FIT contribution of temporal double-bit errors, the following terms need to be defined:

- Q = number of quadwords in the cache memory. Each quadword consists of 64 bits of data and 8 bits of ECC. Thus, there are a total of 72 bits per quadword

- E = number of random single-bit errors that occur in the population of Q quadwords.

Given E single-bit errors in Q different quadwords, the probability that $(E + 1)^{th}$ error will cause a double-bit error is E/Q. Let Pd[n] be the probability that a sequence of n strikes causes n−1 single-bit errors (but no double-bit errors) followed by a double-bit error on the nth strike. Pd[1] must be 0 because a single strike cannot cause a double-bit error. Pd[2] is the probability that the second strike hits the same quadword as the first strike, or $1/Q$. Pd[3] is the probability that the first two strikes hit different quadwords (i.e., $1 - $ Pd[2]) times the probability that the third strike hits either of the first two quadwords that got struck (i.e., $2/Q$). Following this formulation and using * to represent multiplication, one gets

- Pd[2] = $1/Q$
- Pd[3] = $[(Q−1)/Q]$ * $[2/Q]$
- Pd[4] = $[(Q−1)/Q]$ * $[(Q−2)/Q]$ * $[3/Q]$
- …
- Pd[E] = $[(Q−1)/Q]$ * $[(Q−2)/Q]$ * $[(Q−3)/Q]$ * … * $[(Q−E+2)/Q]$ * $[(E−1)/Q]$.

Then the probability of a double-bit error after a time period $T = Pd[N]$ * $P[N$ strikes in time $T]$ for all N. Using this equation, one can solve for the expected value of T to derive the MTTF to a temporal double-bit error.

There is, however, an easier way to calculate MTTF to a temporal double-bit error. Assume that M is the mean number of single-bit errors needed to get a double-bit error. Then, the MTTF of a temporal double-bit error = M * MTTF of a single-bit error. (Similarly, the FIT rate for a double-bit error = $1/M*$ FIT rate for a single-bit error.) A simple computer program can calculate M very easily as the expected value of Pd[.].

■ EXAMPLE

Compute the DUE rate from temporal double-bit errors in a 32 megabyte cache, assuming a FIT/bit of 1 milliFIT. Use the method described above. Assume $M = 2567$, which can be easily computed using a computer program.

SOLUTION The cache has 2^{22} quadwords. The single-bit FIT rate for the entire cache is $0.001 * 2^{22} * 72 = 3.02 * 10^5$, i.e., the MTTF is $10^9/(3.02 * 10^5) = 3311$ hours. Using a computer program, one can find that $M = 2567$. Then the MTTF to a double-bit error $= 3311 * 2567$ hours $= 970$ years.

Using a Poisson distribution Saleh et al. [18] came up with a compact approximation for double-bit error MTTFs of large memory systems. Derivation of Saleh et al. shows that the MTTF of such temporal double-bit errors is equal to $[1/(72 * f)] * sqrt(pi/2Q)$, where f = FIT rate of a single bit.

■ EXAMPLE

Compute the DUE rate from temporal double-bit errors in a 32-megabyte cache, assuming a FIT/bit of 1 milliFIT. Use the compact equation of Saleh et al.

SOLUTION $f = 0.001$. $Q = 2^{22}$. Then DUE rate $= 0.0085 * 10^9$ hours or 970 years. The answer is the same when computed from first principles or Saleh's compact form.

The above calculation does not factor in the reduced error rates because of the AVF. The single-bit FIT/bit can be appropriately derated using the AVF to compute the more realistic temporal double-bit DUE rate.

It is important to note that the MTTF contribution from temporal double-bit errors for a system with multiple chips cannot be computed in the same way as can be done for single-bit errors. If chip failure rates are independent (and exponentially distributed), then a system composed of two chips, each with an MTTF of 100 years, has an overall MTTF of $100/2 = 50$ years. Unfortunately, double-bit error rates are not independent because the MTTF of a double-bit error is not a linear function of the number of bits. This is also evident in Saleh's compact form, which shows that the rate of such double-bit errors is inversely proportional to the square root of the size of the cache. Thus, quadrupling the cache size halves the MTTF of double-bit errors but does not reduce it by a factor of 4.

5.5.2 DUE Rate from Temporal Double-Bit Error with Fixed-Interval Scrubbing

Fixed-interval scrubbing can significantly improve the MTTF of the cache subsystem. By scrubbing a cache block, one means that for each quadword of the block, one reads it, computes its ECC, and compares the computed code with the existing ECC. For a single-bit error, the error is corrected and the correct ECC is rewritten into the cache. Fixed-interval scrubbing indicates that all cache blocks in the system are scrubbed at a fixed-interval rate, such as every day or every month. Scrubbing can help improve the MTTF because it removes single-bit errors from the cache system (protected with SECDED ECC), thereby reducing the probability of a future temporal double-bit error.

Even in systems without active scrubbing, single-bit errors are effectively scrubbed whenever a quadword's ECC is recalculated and rewritten. This occurs when new data are written to the cache because either the cached location is updated by the processor or the cached block is replaced and overwritten with data from a different memory location. In some systems, a single-bit error detected on a read will also cause ECC to be recalculated and rewritten. The key difference between these passive updates and active scrubbing is that the former provides no upper bound on the interval between ECC updates.

To compute the MTTF with scrubbing, let us define the following terms:

- I = scrubbing interval

- N = number of scrubbing intervals to reach MTTF (with scrubbing active at the end of each interval I)

- pf = probability of a double-bit error from temporally separate alpha or neutron strikes in the interval I.

Then, by definition, MTTF of a temporal double-bit error = $N * I$. Assuming each such scrubbing interval is independent, the probability that one has no double-bit error in the first N intervals followed by a double-bit error in the $N + 1$th interval is $(1 - pf)^N * \text{pf}$. Thus, N is the expected value of a random variable with probability distribution function $(1 - \text{pf})^N * \text{pf}$. So, given an interval I, one computes the number of single-bit errors (say S) that can occur in that interval. pf is equal to the sum of the probabilities of a double-bit error, given 2, 3, 4, ... , S errors. This probability can be computed the same way (as described in the last section) for a system with no scrubbing. Thus, given pf and I, one can easily compute N using a simple computer program.

Figure 5.13 shows how scrubbing once a year, month, and day can improve the MTTF numbers for a system configured with an aggregate 16 gigabytes of on-chip cache and assuming an AVF of 100%. This could, for example, arise from a 64-processor multiprocessor or a cluster, with each processor having 256 megabytes of on-chip cache. Hence, fixed-interval scrubbing can significantly improve the

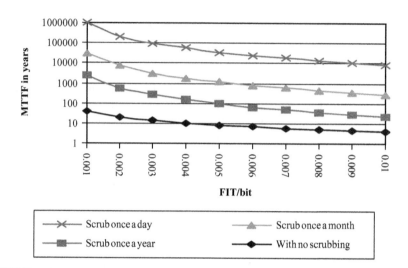

FIGURE 5.13 Impact on fixed-interval scrubbing (once a year, once a week, and once a day) on the MTTF of temporal double-bit errors for a system with 16 gigabytes of on-chip cache. Reprinted with permission from Mukherjee et al. [9]. Copyright © 2004 IEEE.

MTTF of a processor and system's cache subsystem. Thus, for example, for a FIT/bit of 1 milliFIT, the MTTF of temporal double-bit errors are 40 years, 2281 years, 28172 years, and 959267 years, respectively, for a system with no scrubbing and scrubbing once a year, once a month, and once a day.

These numbers come close to compact-form MTTF prediction or temporal double-bit errors with fixed-interval scrubbing of Saleh et al. Closed form of Saleh et al. for this MTTF is $2/[Q * I * (f * 72)2]$. With a FIT/bit (i.e., f) of 0.001, Saleh et al. would predict MTTFs of 2341 years, 28484 years, and 854515 years, respectively, for a system with scrubbing once a year, once a month, and once a day.

Scrubbing reduces the temporal double-bit DUE rate of a chip. In contrast, the next section discusses how to specifically reduce the false DUE rate of a chip.

5.6 Detecting False Errors

This section describes techniques to track false errors and thereby reduce the false DUE AVF. The previous section's discussion on scrubbing analysis, in contrast, reduces the total DUE rate—both true and false—from double-bit errors.

Error detection schemes, such as parity codes, can introduce false errors (see False DUE AVF, p. 86, Chapter 3). An error detection scheme detects not only true errors from faults that will affect the final outcome of the program but also false errors that would not have affected the system's final output in the absence of any error detection (Figure 3.1). For example, a fault in a wrong-path instruction in the instruction queue would not affect any user-visible state. However, the processor is unlikely to know in the issue stage whether or not an instruction is on the correct path and thus may be forced to halt the pipeline on detecting any instruction queue parity error.

Unlike error detection and correction codes that are typically agnostic of the semantics of the underlying architecture, false error detection typically requires this knowledge. Hence, this section illustrates how to reduce the impact of false errors for a specific scenario: in a processor pipeline with structures protected with parity codes. The parity code is written every time a structure is written to. The parity code is read and checked for error every time the corresponding hardware structure is read out. On encountering an error, the processor halts further progress of the pipeline and raises a machine check exception (see Machine Check Architecture later in this chapter).

First, this section discusses the sources of false DUE events in a microprocessor, followed by how the error information can be propagated and how one can eventually distinguish a false error from a true error using this propagated information.

5.6.1 Sources of False DUE Events in a Microprocessor Pipeline

To eliminate false errors from (and thereby reduce the corresponding false DUE rate of) hardware structures in a processor, one must first identify the sources of these errors. Whether a masked fault is flagged as an error is a function of both the architecture and the error detection mechanism. Here five example sources of false errors in a processor pipeline potentially triggered by parity codes are described. The first three are described by Weaver et al. [22], the fourth one by Ergin et al. [4], and the last one by Wang et al. [20].

■ *Instructions whose results the microarchitecture will never commit.* For example, a strike on the bits of a parity-protected hardware structure holding nonop-code bits of a wrong-path instruction may lead to a false error. A wrong-path instruction arises in a processor pipeline that attempts to speculatively execute instructions beyond a control-flow instruction (e.g., branch). If the speculation is incorrect, the pipeline ends up with wrong-path instructions that may flow through different hardware structures in the pipeline but will eventually be squashed at the retire stage. Predicated instruction sets, such as the Itanium® architecture, can have a similar problem if the predicate guarding an instruction evaluates to false. If the pipeline checks for parity errors when the resident bits are read out of the structure, then the pipeline may flag an error and may be forced to halt further progress to avoid any potential SDC. This action is appropriate if it were indeed a true error, but strikes to bits containing nonop-code bits of a wrong-path or a falsely predicated instruction would have been masked in the absence of the parity code.

- *Instruction types that are neutral to errors.* No-ops, prefetches, and branch prediction hint instructions, for example, do not affect correctness. No-ops are instructions that have no effect on program execution but may be necessary for pipeline scheduling or other reasons. Many prefetches often simply move data closer to a processor cache from one farther away, thereby speeding up program execution, but do not affect correctness. Branch prediction hints simply tell the processor pipeline the branch direction to predict, instead of using the underlying hardware prediction. Hence, these instructions are all neutral to errors. Consequently, faults in bits other than the opcode bits of these instructions will not affect a program's final outcome.

- *Dynamically dead instructions.* These instructions generate values that ultimately do not affect the result. One can classify dynamically dead instructions as first level or transitive. First-level dynamically dead (FDD) instructions are those whose results are not read by any other instruction. Transitive dynamically dead (TDD) instructions are those whose results are used only by FDD instructions or other TDD instructions. Depending on whether the instruction writes a register or a memory location, one can further classify the dynamically dead instructions as being tracked via register or memory, respectively. A strike on any bit on a dynamically dead instruction, except the destination register specifier bits, will not change the final outcome of a program. Similarly, a strike on the result (e.g., register or memory value) of a dynamically dead instruction will also not affect the program's outcome. On average, dynamically dead instructions account for 20% of all instructions in the binaries examined by Weaver et al. [22].

- *Narrow values.* Processor pipelines often operate on smaller sized data than allowed by the corresponding data path or registers. For example, instructions could perform addition and subtraction on byte-sized or 8-bit data, whereas the data paths and registers themselves could be 64 bits wide to support operations on 64-bit-wide data. However, if parity is computed for the entire 64 bits, then any error in the 54 bits not holding the byte-sized data would also be flagged as an error.

- *Conditional branches.* Wang et al. [20] identified an additional source of false errors arising from conditional branches. They found that in 40% of the dynamically executed conditional branches in CPU2000 benchmarks, the direction in which the branch went did not matter. This could happen, for example, if a loop is unrolled but also has an exit code that can fully execute the loop with lower performance. If the loop branch is taken, the loop executes fast through the unrolled portion. If the loop branch is not taken, it jumps to the exit code, runs slower, but still produces the correct answer. Thus, when the pipeline detects an error in a branch direction, in the CPU2000 benchmarks, it would flag an error that would have been masked in the absence of the fault detection mechanism 40% of the time.

The next two subsections describe techniques to avoid raising a machine check exception on these classes of false errors. The focus will be on the first four examples because detecting false errors on conditional branches still remains a challenging research problem. This is because it is difficult to determine if the control and data flow after an incorrectly taken conditional branch will converge in the future.

5.6.2 Mechanism to Propagate Error Information

The key challenge in distinguishing false errors from true errors is that the processor may not have enough information to make this distinction at the point it detects the error. For example, when the instruction queue detects an error on an instruction, it may not be able to tell whether the instruction was a wrong-path instruction or not. Consequently, one needs to propagate the error information down the pipeline and raise the error when there is enough information to make this distinction. This section discusses how to propagate this error information for later use. The next section discusses what information one would need to identify false errors.

To propagate error information between different parts of the microprocessor hardware Weaver et al. [22] introduced a new bit called the π bit, which stands for the possibly incorrect bit. A π bit is logically associated with each instruction as it flows down the pipeline from decode to retirement. The π bit is initially cleared to indicate the absence of any error. When the instruction queue receives the instruction, it stores the π bit along with the instruction. On detecting an error (possibly via parity), the instruction queue sets the affected instruction's π bit instead of raising a machine check exception. Subsequently, the instruction issues and flows down the pipeline. When the instruction reaches commit point, one can determine if the instruction was on the wrong path. If so, one can ignore the π bit, avoiding a false DUE event if the bit was set. If not, one has the option to raise the machine check error at the commit point of the instruction. It should be noted that a strike on the π bit itself will result in a false DUE event.

One can easily generalize the π bit mechanism and attach the π bit to different objects flowing through the pipeline, as long as the π bits are propagated correctly from object to object. For example, modern microprocessors typically fetch instructions in multiples, sometimes called chunks. Chunks flow through the front end of the pipeline until they are decoded. One can attach a π bit to each fetch chunk. If the chunk encounters an error, one can set the π bit of the chunk. Subsequently, when the chunk is decoded into multiple instructions, one can copy the π bit value of the chunk to initialize the π bit of each instruction. Thus, one can use the π bit to avoid false DUE events on structures in the front end of the pipeline before individual instructions are decoded. Similarly, the π bit can be propagated from instructions to instructions and registers or from instructions and registers to memory and vice versa. Propagating the π bit between instructions may require the π bit itself to undergo appropriate transformations. For example, on an addition operation, the destination register of the instruction may have to OR the π bits of all the instructions, as well as the operand registers.

In general, a π bit can be attached to any object flowing through the pipeline or to any hardware structure, but the granularity of the π bit depends on the implementation. For example, if a π bit is attached to a 64-bit register value, then a single π bit can only tell that there may have been an error in one of the 64 bits. Alternatively, if there is a π bit per byte, then one could identify the byte among the 64 bits that may have had an error. This may be important to instruction sets that allow byte-level writes. More generally, the granularity of the π bit can be refined to isolate the location of errors in the hardware.

One does not, however, expect all hardware structures in a processor or an entire system to be populated with π bits. For example, an implementation may choose to have π bits in caches but not in main memory. Consequently, when cache blocks are written back from a cache to a main memory, the π bit information would be lost. In such a case, the π bit will go out of scope. When the π bit goes out of scope, an implementation should flag an error if the π bit is set because the system can no longer track the error.

The π bit is also sometimes referred to as the *poison* bit and has been used to track false errors outside the processor cores in some commercial systems. It should be noted that the π bit itself may encounter a fault, so it may need to be protected.

5.6.3 Distinguishing False Errors from True Errors

As discussed earlier, false errors can arise in a processor pipeline from three categories of instructions. This section discusses how one can use the π bit information to avoid false errors on these three instruction categories and narrow values and thereby reduce the false DUE rate of the instruction queue.

False Errors on Uncommitted Instructions

Given the π bit, it is relatively straightforward to avoid false errors on instructions that will never commit their results. As explained earlier, the retire unit can ignore the π bit for the wrong-path and falsely predicated instructions, thereby avoiding false errors on such instructions. The retire unit must, however, examine the π bit of instructions on the correct path and flag an error if the π bit is set. The next two subsections show how to avoid false errors on instructions on the correct path.

False Errors on Neutral Instruction Types

Many instructions, such as the no-ops, prefetches, or branch predict hints, will never affect the final outcome of a program and therefore the hardware need not raise an error on nonopcode bits of such instructions. However, to identify such instructions, the hardware must decode the instruction at every place it wants to avoid a false error. Instead, Weaver et al. [22] proposed using another bit called the anti-π bit, which is associated with every instruction when the instruction is decoded. The anti-π bit is set for neutral instruction types and cleared for others. Then, when the instruction queue gets a parity error on nonopcode bits of an entry, it identifies neutral instructions using the anti-π bit and does not set the π bit on that instruction. In other words, the anti-π bit neutralizes the π bit for those entries. Alternatively, the instruction queue could set the π bit but carry both the anti-π bit and π bit to the retire unit and take the appropriate decision there.

It should be noted that the hardware could also avoid the anti-π bit on every instruction if it decoded the instruction again at the retire unit. Unfortunately, this means that an instruction must be read after it has been issued and completed. This may raise the false DUE AVF by extending the ACE lifetime of the instruction.

The anti-π bit can be generalized to hardware activities that do not affect the correctness of a program. For example, one could attach an anti-π bit to the command and address generated by a hardware data prefetcher. Any soft error on such an activity can be ignored. The anti-π bit provides a concise mechanism to identify such activities.

False Errors on Dynamically Dead Instructions

One can use the π bit to track false errors in dynamically dead instructions. This section illustrates three uses of the π bit. Weaver et al. [22] describe an additional scheme called the postcommit error tracking buffer, which is not covered here.

- π *bit per register*. One can allocate a π bit for every register. An instruction's π bit is propagated to its destination register. An error is signaled when an instruction reads a source register with a π bit set. If no instruction reads the register before it is overwritten, the instruction is FDD, and no error is signalled. This mechanism provides 100% coverage on all FDD instructions. However, when one signals an error, one cannot determine the instruction that originally caused the error. This lack of information may complicate some recovery schemes.

- π *bits on every structure inside the chip, except the memory system*. Although the above two mechanisms avoid false errors on FDD instructions tracked via registers, they do not cover instructions that are transitively dead (TDD) via registers. One easy way to track TDD instructions is to declare the error only when a processor interacts with the memory system or I/O devices. Thus, if there are π bits on every structure in a processor—except caches and main memory—and the same propagation rule for π bits as described earlier is followed, then false errors on TDD instructions can be avoided as well. This would mean signalling errors only when a store instruction or an I/O access is about to commit its data to the caches, memory system, or I/O device. In this case, one can get complete coverage of false errors on

TDD instructions tracked via registers, but like the previous mechanism, one can lose the ability to precisely determine the instruction that originally encountered the error.

■ *π bit on caches and memory*. Finally, if the entire chip and memory system have π bits, then false errors on both FDD and TDD instructions can be tracked via memory as well. In such a case, an error would be raised only when the processor makes an I/O access (e.g., uncached load or store) that has its π bit set. This technique would also allow one to track errors across multiple processors in a shared-memory multiprocessor system.

As the above discussions suggest, the π bit is a powerful mechanism to propagate error information, so that the error can be raised at a later point in time when it can be determined whether the error was actually a false or a true error. Thus, it decouples the detection of an error from the signalling of the error. This allows a microprocessor designer the choice to raise the error either on the use of a value or when the π bit for a value goes out of scope.

False Errors in Narrow Values

As discussed earlier, narrow values embedded in wider data paths can give rise to false errors if the entire data path is protected with parity bits. The key here is to identify the bits that do not matter in the event of an alpha particle or a neutron strike. This can be accomplished by attaching an anti-π bit similar to the one described earlier for neutral instruction types. Thus, if the width of the actual operand (e.g., lower eight bits in a 64-bit register) is known and the anti-π bit is set, then only the parity on the lower operand bits can be computed. This would avoid raising a false error if the nonoperand bits are struck.

Ergin et al. [4] describe a similar mechanism that can identify and ignore the nonoperand bits in a wide register or a data path. Method of Ergin et al. works even in the absence of an error detection mechanism and thereby can reduce the SDC AVF as well. The SDC AVF can arise if sign-extended nonoperand bits are actually used to perform an operation, although the actual operand width is smaller (e.g., if a 64-bit adder is used for an 8-bit addition). In this case, ignoring the nonoperand bits in the registers and sign extending them before the appropriate operation would help reduce the SDC by ignoring any error in the nonoperand bits.

5.7   Hardware Assertions

This section discusses how hardware assertions can be used to detect soft errors. If these assertions are violated during a program's execution, then an error must have occurred. Parity and ECC are a generic form of hardware assertion that does not use architecture-specific knowledge. In contrast, this section discusses examples of architecture-specific hardware assertions. In the previous section, knowledge of the underlying architecture was also necessary to detect false errors. A typical characteristic of these assertions is that they incur lower overhead in area and latency to detect errors compared to error detection and correction codes. The drawback is that the assertion is a customized error detection or correction procedure for each hardware structure and may be hard to generalize like parity or ECC.

One can imagine a variety of such hardware assertions. For example, a *MESI* cache coherence protocol has four states for a cache block participating in the protocol: Modified, Exclusive, Shared, and Invalid. A protocol implementation may require a block to first go from Invalid to Exclusive before it transitions to Modified state. Hence, if a finite-state machine implementing the state transitions detects such a transition from Invalid to Modified state, it can declare an error.

Reddy et al. [17] studied two such hardware assertions in a microprocessor pipeline. The first scheme is related to the instruction issue logic, which the authors call *Timestamp-Based Assertion Checking* (TAC). The second one tracks errors in the register rename mapping, which the authors call *Register Name Authentication* (RNA). Here the TAC mechanism is discussed. The RNA mechanism is more involved, and readers are referred to the paper for the details of how RNA works.

To detect faults, TAC associates timestamps to instructions as they issue to the execution units. Assume instruction A is an addition operation: $R1 = R2 + R3$, which adds registers R2 and R3 and produces the corresponding sum in the destination R1. Also, assume that instruction B writes the destination register R2 before A uses it. Consequently, for many pipelines, the following hardware assertion will hold: Timestamp $(A) \geq$ Timestamp (B). Further, if the latency of instruction B through the execution unit is known to be L, then the following assertion should hold as well: Timestamp $(A) \geq$ Timestamp $(B) + L$. Reddy et al. show how to implement these timestamp counters and check the hardware assertions when instructions retire.

Figure 5.14 shows the result of fault injection into targeted areas in the hardware (e.g., ready bits), which would cause the instruction to issue earlier than the assertions would allow. As seen in the figure, for the nine benchmarks studied with TAC, on average, 80% of the injected faults were detected by TAC (shown as Assert+SDC in the figure). Another 17% of the faults detected by TAC are actually false errors (shown as Assert+Masked). False errors can arise because the final output may still be correct, although an instruction may have issued earlier (e.g., if the previous source value was the same as the new and correct one). The rest are masked and not detected by TAC.

Nakka et al. [10] propose a more generic framework to implement hardware assertions. They propose the use of an offload engine—called the Reliability and Security Engine (RSE)—that will perform the hardware assertions. Communication paths between the main processor and the RSE allow the processor to send the data necessary for the RSE to check the hardware assertion.

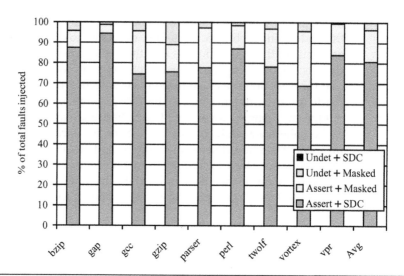

FIGURE 5.14 Breakdown of outcome of TAC fault injection experiments. Assert + SDC = category of faults that would have caused an SDC and detected by TAC. Assert + Masked = false errors detected by TAC. Undet + Masked = not detected by TAC but benign faults. Undet + SDC = not detected by TAC and causes SDC. Reprinted with permission from Reddy et al. [17]. Copyright © 2006 IEEE.

5.8   Machine Check Architecture

The MCA of a processor or a chipset specifies the actions it must take on detecting or correcting a hardware error (e.g., see the Intel manual [7]). Typically, there are three types of actions: informing the OS of the error, recording the information related to the error, and isolating the error to a specific hardware component.

5.8.1 Informing the OS of an Error

On detecting a fault, the hardware would usually raise an uncorrectable machine check interrupt to inform the OS that it cannot guarantee correct operation. This would typically result in a reboot. Nevertheless, in certain cases, the OS may be able to isolate the error to a specific user process. In such a case, the OS can just kill the user process that experienced the error and continue its operation. The overall process-kill DUE rate of the system still remains the same, but the system-kill DUE can be reduced (see Basic Definitions: SDC and DUE, p. 32, Chapter 1).

On correcting an error, the hardware optionally raises a correctable machine check interrupt. This is optional and is typically controlled by a mode bit set by the OS because normal hardware operation can continue correctly. Nevertheless, if the number of correctable interrupts raised is greater than a certain expected threshold, then it may be an indication to the OS that the hardware may not be functioning correctly and an impending uncorrectable error may be expected.

5.8.2 Recording Information about the Error

The hardware can record information about the error in one of two ways: it can provide appropriate information (e.g., error number or type) to the OS when it raises the interrupt. Alternatively, it can log the error in hardware registers. For example, Intel®x86 processors record the error information in hardware register banks [7]. IBM®'s Power architecture records similar information in fault isolation registers [3]. The OS can read these registers when it receives the uncorrectable or correctable machine check interrupt and takes appropriate actions.

5.8.3 Isolating the Error

Isolating an error to a specific hardware chip is typically important for hard errors since the specific unit experiencing an error may need to be replaced. IBM calls such units field-replaceable units (FRUs) [3]. When an error is detected, it can usually be mapped to a specific FRU, unless it happens in the communication channel between two or more FRUs. In this case, the error is typically ascribed to both. This could be a costly procedure since field representatives may have to test and/or replace both units, although only one was probably operating incorrectly. IBM's Power4 processor has enough hardware hooks to minimize such a multiFRU fault isolation.

5.9 Summary

Error coding techniques are used widely in the industry today to protect against transient faults caused by alpha particle and neutron strikes. The coding schemes typically add redundant check bits to a set of data bits in such a way that an error can be either detected or corrected by examining the check bits. Reducing the number of check bits used in an error code is often important to reduce the overhead of error detection and correction.

Parity, SEC, SECDED, DECTED, and CRC codes are common error detection and correction codes used to protect memory structures. Parity codes provide perhaps the simplest form of fault detection. An even parity code needs one check bit that is the XOR of all the data bits. Parity codes can detect single-bit faults and faults in odd numbers of data bits.

SEC codes add a set of check bits to correct single-bit errors. Each check bit in a SEC code is an XOR of a combination of a subset of the data bits. The relationship between the data bits and check bits is typically expressed as a parity check matrix. By multiplying the parity check matrix with a set of incoming data and check bits, one can obtain a set of bits called a syndrome. A zero syndrome indicates no error, whereas a nonzero syndrome can give the exact location of the bit in error if the parity check matrix is set up appropriately to create what is known as a Hamming code.

A SEC code can be extended with an extra bit to create a SECDED code. The extra bit is effectively the parity of the set of bits used by the SEC code. By interpreting the syndrome differently, the SECDED code can correct single-bit errors and detect double-bit errors.

Similarly, a DECTED code can correct double-bit errors and detect triple-bit errors. By interpreting the syndrome differently, one can detect the bit positions of both single-bit and double-bit errors and thereby correct them by inverting the bits.

CRC codes are typically used to detect a burst of errors in signal lines. Although transmission lines do not experience soft errors from particle strikes, CRC codes are still interesting for soft errors because such codes can cover the corresponding send and receive buffers. Polynomial division underlies the principles behind encoding and decoding CRC codes.

Execution units cannot be protected easily with parity, SEC codes, etc. Instead, they can be protected with residue codes or parity prediction circuits. A residue for an integer is the remainder obtained by dividing it by another smaller integer. Residue codes compute residues directly from the result of an operation and separately from the source operands themselves. Then, by comparing the two residues, one can determine if there was an error. Parity prediction circuits work in a similar fashion. Such circuits compute the parity bit directly from the result and separately from the source operands and the corresponding carry chain.

Error codes can come with high overheads, so architects have invented several schemes to reduce their overheads. Reducing the number of 1s in a row of the parity check matrix helps reduce the critical path of encoding and decoding SECDED codes. Other architectures have used out-of-band ECC correction to reduce the overhead. Out-of-band correction does not correct the data in the critical path of the access but does it later.

Scrubbing is another overhead reduction technique that accesses blocks of memory periodically and corrects single-bit errors. This prevents temporal double-bit errors caused by accumulation of single-bit errors. This allows a SECDED code to correct temporal double-bit errors. Alternatively, a single particle strike can affect two or more contiguous bits in some cases causing a spatial multibit error. To prevent such a spatial multibit error, architects often interleave ECC of different data words. A particle strike in such a case will upset separate code words and will be detected and corrected.

Knowledge of the specific architecture can help architects create other fault detection and error correction mechanisms. For example, false errors on wrong-path instructions could be detected and isolated by the hardware since the results of wrong-path instructions will not be committed. Similarly, hardware assertions that assert certain properties of the architecture or instruction flow can be used to detect faults and correct errors.

5.10 Historical Anecdote

Error detection and correction codes have evolved over several decades. In 1948, Claude Shannon's landmark paper, titled "A Mathematical Theory of Communication," perhaps started the formal discipline of coding theory [19]. Working at Bell Labs, Shannon showed that it was possible to encode messages for transmission in such a way that the number of extra bits was minimal. Few years later, Richard Hamming, also in Bell Labs, produced a 3-bit code for four data bits. Hamming invented this code after several failed attempts to punch out a message on a paper using the parity code. Apparently, Hamming expressed his frustration in the following words, "If it can detect the error, why can't it correct it!" Since the early 1950s, coding theory has evolved to cover a variety of fault models and situations. Today, error detection and correction codes are widely used across various forms of computing systems, including the ones sent for space exploration.

References

[1] AMD, "BIOS and Kernel Developer's Guide for AMD Athlon™ 64 and AMD Opteron™ Processors," Publication #26094, Revision 3.14, April 2004. Available at: http://www.amd.com/us-en/assets/content_type/white_papers_and_tech_docs/26094.PDF.

[2] H. Ando, Y. Yoshida, A. Inoue, I. Sugiyama, T. Asakawa, K. Morita, T. Muta, T. Motokurumada, S. Okada, H. Yamashita, Y. Satsukawa, A. Konmoto, R. Yamashita, and H. Sugiyama, "A 1.3 GHz Fifth Generation SPARC64 Microprocessor," in *International Solid-State Circuits Conference*, pp. 1896–1905, 2003.

[3] D. C. Bossen, A. Kitamorn, K. F. Reick, and M. S. Floyd, "Fault-Tolerant Design of the IBM pSeries 690 System Using POWER4 Processor Technology," *IBM Journal of Research and Development*, Vol. 46, No. 1, pp. 77–86, 2002.

[4] O. Ergin, O. Unsal, X. Vera, and A. Gonzalez, "Exploiting Narrow Values for Soft Error Tolerance," *IEEE Computer Architecture Letters*, Vol. 5, pp. 12–12, 2006.

[5] M. Y. Hsiao, "A Class of Optimal Minimum Odd-Weight-Column SEC-DED Codes," *IBM Journal of Research and Development*, Vol. 14, No. 4, pp. 395–401, 1970.

[6] S. Iacobovici, "Residue-Based Error Detection for a Shift Operation," United States Patent Application, filed August 22, 2005.

[7] Intel Corporation, *Intel® 64 and IA-32 Architectures, Software Developer's Manual, Volume 3A: System Programming Guide, Part 1*. Available at: http://www.intel.com.

[8] J.-C. Lo, "Reliable Floating-Point Arithmetic Algorithms for Error-Coded Operands," *IEEE Transactions on Computers*, Vol. 43, No. 4, pp. 400–412, April 1994.

[9] S. S. Mukherjee, J. Emer, T. Fossum, and S. K. Reinhardt, "Cache Scrubbing in Microprocessors: Myth or Necessity?" in *10th IEEE Pacific Rim International Symposium on Dependable Computing (PRDC)*, pp. 37–42, March 3–5, 2004, Papeete, French Polynesia.

[10] N. Nakka, J. Xu, Z. Kalbarczyk, and R. K. Iyer, "An Architectural Framework for Providing Reliability and Security Support," *Dependable Systems and Networks (DSN)*, pp. 585–594, June 2004.

[11] I. A. Noufal and M. Nicolaidis, "A CAD Framework for Generating Self-Checking Multipliers Based on Residue Codes," in *Design, Automation and Test in Europe Conference and Exhibition*, pp. 122–129, 1999.

[12] M. Nicolaidis, "Carry Checking/Parity Prediction Adders and ALUs," *IEEE Transactions on Very Large Scale Integration (VLSI)*, Vol. 11, No. 1, pp. 121–128, February 2003.

[13] M. Nicolaidis and R. O. Duarte, "Fault-Secure Parity Prediction Booth Multipliers," *IEEE Design and Test of Computers*, Vol. 16, No. 3, pp. 90–101, July–September 1999.

[14] M. Nicolaidis, R. O. Duarte, S. Manich, and J. Figueras, "Fault-Secure Parity Prediction Arithmetic Operators," *IEEE Design and Test of Computers*, Vol. 14, No. 2, pp. 60–71, April–June 1997.

[15] W. W. Peterson and E. J. Weldon, Jr., *Error-Correcting Codes*, MIT Press, 1961.

[16] D. K. Pradhan, *Fault-Tolerant Computer System Design*, Prentice-Hall, 2003.

[17] V. K. Reddy, A. S. Al-Zawawi, and E. Rotenberg. "Assertion-Based Microarchitecture Design for Improved Fault Tolerance." in *Proceedings of the 24th IEEE International Conference on Computer Design (ICCD-24)*, pp. 362–369, October 2006.

[18] A. M. Saleh, J. J. Serrano, and J. H. Patel, "Reliability of Scrubbing Recovery Techniques for Memory Systems," *IEEE Transactions on Reliability*, Vol. 39, No. 1, pp. 114–122, April 1990.

[19] C. E. Shannon, "A Mathematical Theory of Communication," *Bell System Technical Journal*, Vol. 27, pp. 379–423, 623–656, July–October, 1948.

[20] N. Wang, M. Fertig, and S. Patel, "Y-Branches: When You Come to a Fork in the Road, Take It," in *12th International Conference on Parallel Architectures and Compilation Techniques (PACT)*, pp. 56–66, 2003.

[21] C. Webb, "z6—The Next-Generation Mainframe Microprocessor," *Hot Chips*, August 19–21, 2007.

[22] C. Weaver, J. Emer, S. S. Mukherjee, and S. K. Reinhardt, "Reducing the Soft Error Rate of a Microprocessor," *IEEE Micro*, Vol. 24, No. 6, pp. 30–37, November–December 2004.

Fault Detection via Redundant Execution

6.1 Overview

Fault detection via redundant execution is a common form of fault detection that has been used for decades. Unlike error coding techniques—described in the previous chapter—that detect faults using redundant information in storage bits or logic units, the techniques described in this chapter detect faults by comparing outputs from redundant streams of instructions. Typically, fault detection via redundant execution can provide greater fault coverage across a processor chip compared to error coding techniques on individual hardware structures. This is because the same redundant execution technique can cover multiple hardware structures, unlike in many error coding implementations where each structure must be protected individually. Further, redundant execution can more easily protect logic and computation blocks that change the data the blocks operate on. In contrast, error coding techniques are typically used for storage and communication that leave the data unchanged over periods of time. Redundant execution techniques can, however, add significant hardware overhead over error coding schemes.

This chapter focuses on two redundant execution schemes commonly used in the industry: Lockstepping and Redundant Multithreading (RMT). In Lockstepping, both redundant copies have exactly the same state in every cycle. Consequently, a fault in either copy may cause the redundant copies to produce different outputs in the same cycle. A fault is detected on such an output mismatch. In contrast, in RMT, only outputs of committed instructions are compared. The internal state of the individual redundant threads in an RMT implementation can be very different.

This chapter discusses the different trade-offs offered by Lockstepping and RMT. To illustrate the trade-offs, this chapter first discusses the concept of the sphere of replication. The sphere of replication determines the logical boundary within which all states are logically or physically replicated. The sphere of replication is critical to understand the outputs in a Lockstepped or an RMT implementation that need comparison.

Then, this chapter describes three Lockstepped implementations—the Stratus ftServer, the Hewlett-Packard NonStop Architecture, and the IBM Z-series processors. These three examples illustrate how varying the size of the sphere of replication can change the trade-offs associated with implementing Lockstepping.

Finally, five RMT implementations are discussed—the Marathon Endurance Server, the Hewlett-Packard NonStop Advanced Architecture (NSAA), simultaneous and redundantly threaded (SRT) processor, the chip-level redundantly threaded (CRT) processor, and dynamic implementation verification architecture (DIVA). Because RMT implementations do not impose the constraint of cycle-by-cycle synchronization, one can implement the redundant threads of RMT in a variety of ways: in a hardware thread, in a processor core, or in a special checker core. These five implementations differ not only in the size of the sphere of replication but also in how the redundant threads are implemented. The first two have been implemented in commercial systems, whereas the last three designs are only paper proposals but have been studied extensively by researchers.

6.2   Sphere of Replication

The concept of the *sphere of replication* [10] makes it easier to understand the mechanics of fault detection schemes based on redundant execution, such as Lockstepping or RMT. In a fault detection scheme with redundant execution, the same program executes as identical and committed instruction streams. For a dual-modular redundancy (DMR) system, there are two identical streams. For a triple modular redundancy (TMR) system, there are three identical streams of execution. Specific outputs are compared from each stream. A fault is flagged when there is a mismatch in the compared outputs.

6.2.1 Components of the Sphere of Replication

The sphere of replication identifies the logical domain protected by the fault detection scheme. That is, any fault that occurs within the sphere of replication and propagates to its boundary will be detected by the fault detection scheme corresponding to the sphere of replication. Figure 6.1 shows an example sphere of replication that includes redundant copies of a microprocessor but excludes memory,

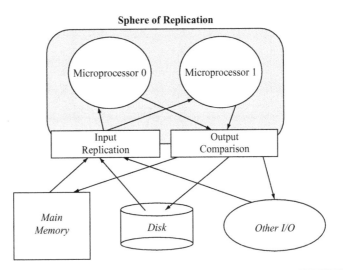

FIGURE 6.1   Sphere of replication (shaded region) that excludes memory, disk, and other I/O.

disks, and other I/O subsystems. There are three questions related to the sphere of replication:

■ For which components will the redundant execution mechanism detect faults? Components outside the sphere of replication are not covered by redundant execution and may need to use alternative techniques, such as parity, ECC, or replication, for fault coverage.

■ Which outputs must be compared? Failure to compare critical values compromises fault coverage. However, needless comparisons increase overhead and complexity without improving coverage.

■ Which inputs must be replicated? Failure to correctly replicate inputs can result in the redundant threads following divergent execution paths.

6.2.2 The Size of Sphere of Replication

The extent of the sphere of replication affects a number of system parameters. A larger sphere typically replicates more state (e.g., memory). However, moving state into the sphere indicates that updates to that state occur independently in each execution copy. Thus, a larger sphere tends to decrease the bandwidth required for output comparison and input replication, potentially simplifying the comparator and replicator circuits.

The size of the sphere of replication also depends on how much control vendors have over the components they use in their machines. For example, vendors, such

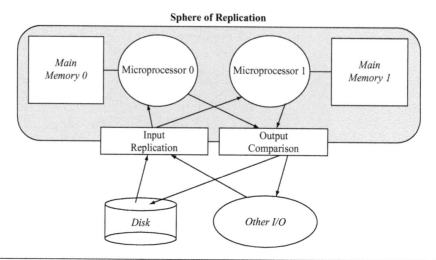

FIGURE 6.2 Sphere of replication (shaded region) that includes memory but excludes disk and other I/O.

as Tandem or Stratus, do not have much control over the microprocessor itself and hence their spheres of replication include two or three microprocessor chips. In contrast, IBM designs its own fault-tolerant microprocessor and hence it can afford to create a sphere of replication that is limited to a single microprocessor, but with redundant pipelines.

In practice, the size of the sphere of replication can vary widely. The sphere of replication in Stratus' ftServer [19] includes redundant copies of both the microprocessor and the main memory (Figure 6.2). In contrast, the sphere of replication of the Hewlett-Packard Himalaya system [23] includes only redundant copies of the microprocessor but not the main memory or I/O system. The sphere of replication in the IBM G5 processor [13], on the other hand, comprises only part of the processor pipeline—fetch, decode, and execution units. Even the architectural register file in the G5 is outside the sphere of replication.

■ E X A M P L E

Compute the DUE FIT rate of the following microprocessor. The microprocessor has two parts: a core and an uncore. The core has 60 SDC FIT and 60 DUE FIT. The uncore has 40 SDC FIT and 40 DUE FIT. Designers decided to add redundant execution to the microprocessor with the sphere of replication covering the entire core. Assume that the false DUE FIT arising from the redundant execution is 10 DUE FIT.

SOLUTION Redundant execution eliminates the SDC FIT for the core since the sphere of replication includes the entire core. The DUE FIT = 10 (falseDUE) + 60 (from core) + 40 (from uncore) = 110 FIT. So, the total SDC FIT reduces from 100 FIT to 40 FIT, but the DUE FIT increases from 100 FIT to 110 FIT.

6.2.3 Output Comparison and Input Replication

Any outputs leaving the sphere of replication must be compared to check for mismatch and corresponding faults. Faults that do not propagate to the output comparator corresponding to the sphere get masked and hence do not require detection or comparison. The output comparator has also been referred to in the literature as the *checker*.

Any inputs into the sphere of replication must be appropriately replicated and delivered to the correct points within the sphere. If the inputs are not replicated correctly, then the redundant copies may still execute correctly but may follow different (but still correct) paths. For example, if the redundant copies read a processor cycle counter but obtain different values for the same cycle counter, then the redundant copies may diverge. Although the individual copies will execute correctly, the output comparison may indicate a mismatch even in the absence of a fault because of divergent execution. Hence, it is critical to ensure that both copies execute the same correct path in the absence of any fault. This can be ensured if the inputs are replicated appropriately.

Broadly, there are two styles of output comparison and input replication. First one is cycle synchronized. That is, each redundant copy within the sphere has exactly the same state in every clock cycle as its redundant copy. The Stratus ftServer, for example, is cycle synchronized. This is typically referred to as *Lockstepping* since the redundant copies within the sphere are exactly in sync in every cycle. The output comparator can compare any set of hardware signals coming from the redundant copies in the sphere. In the absence of any fault, these signals should match exactly in each cycle. It should be noted that the output comparator does not need any semantic information about the signals. As long as the signals match, the comparator can certify that the redundant copies did not encounter a single-bit fault.

The second style is one that compares specific events or instructions, instead of hardware signals, from two executing streams of instructions. For example, the Tandem NSAA [2] has the same sphere of replication as the Stratus ftServer but is not cycle synchronized. Instead, the Tandem NSAA compares I/O events coming from two separate microprocessors. To avoid cycle-by-cycle synchronization, the NSAA takes special care in replicating the inputs, such as interrupts and DMA activities. This is normally referred to as *Loose Lockstepping* or *RMT*. The next two sections describe several styles of Lockstepped and RMT systems.

6.3   Fault Detection via Cycle-by-Cycle Lockstepping

Lockstepping is a well-known fault detection technique and has been used since the 1980s by mainframes and highly reliable computers. Since the 1960s and 1970s, companies, such as Stratus, Tandem, and IBM, have built fault-tolerant computers that failed over to a standby backup machine when the primary failed. However, the fault coverage in these machines was not as evolved as in today's fault-tolerant machines. In the 1980s, more and more business applications, such as online transaction processing, started to run continuously (24 hours a day, 7 days a week). This required the processors themselves to be online 24×7 with very little downtime and high fault coverage and detection capabilities. To meet this market demand, in the 1980s, Stratus and Tandem introduced Lockstepped processors in their systems (besides other enhancements to the memory and I/O subsystem), but IBM relied on extreme levels of error checking throughout its processor and system design. In the 1990s, however, IBM switched to a Lockstepped processor pipeline architecture.

In the cycle-by-cycle Lockstepping, redundant copies of a program are executed through cycle-synchronized redundant hardware. An output comparator compares a set of hardware signals from each redundant copy. On a mismatch, the comparator flags an error. Thus, Lockstepping reduces the SDC rate of a system. Lockstepping by itself, however, does not imply the presence or absence of any accompanying recovery mechanism, such as checkpointing, or the use of TMR that allows systems to make progress even after a fault is detected.

Lockstepping necessitates a cycle-by-cycle synchronization of the redundant hardware. That is, in each cycle, both hardware copies have exactly the same state, execute the same stream of instructions, and produce the same stream of events in exactly the same cycles. In the absence of a cycle-level synchronization, the two redundant copies can diverge rapidly. This effect is even more prominent in dynamically scheduled processors[1], which can speculatively execute instructions that may not be on the correct path. Cycle synchronization is difficult to implement in a logically shared hardware (e.g., within a multithreaded processor core) and hence typically Lockstepping uses physically redundant and identical copies of the hardware. Cycle-by-cycle synchronization also makes input replication easier since the inputs will be correct as long as the replicated inputs are delivered in the same cycle to the redundant hardware.

---

[1]Also known as out-of-order processors.

Lockstepping poses two critical constraints on the system design. First, the redundant copies must be fully deterministic. That is, they must produce the same set of output signals, given the same set of input signals.

Second, the hardware must support deterministic reset. On a reset, a microprocessor's or a chipset's state is set to a specific state. This state must be identical for both redundant copies of the hardware, even for the state that does not affect correct execution of the pipeline. For example, a branch predictor's initial state does not affect the correct execution in the absence of Lockstepping. However, if the redundant microprocessors have different initial state in their branch predictors, then they are highly likely to diverge and cause a Lockstepping mismatch. A branch misprediction could trigger an incorrect path load in one copy, whereas a correct prediction may not trigger the same in the redundant copy causing a Lockstep failure.

The rest of the section discusses the advantages and disadvantages of Lockstepping and describes three commercial implementations of Lockstepped systems. These commercial systems differ in the sizes of the sphere of replication chosen for the Lockstepped implementation.

6.3.1 Advantages of Lockstepping

Lockstepping provides a great degree of fault coverage. It can detect almost all transient faults in the physically redundant copies within the sphere of replication. It cannot detect faults that are masked within the sphere of replication, which are faults one usually does not care about. Lockstepping can also detect most permanent faults in either copy. The only transient and permanent faults Lockstepping cannot detect are the ones that affect the redundant copies in exactly the same way. But the likelihood of two faults affecting the two redundant copies in exactly the same way is extremely low, unless it is caused by a design fault, which could potentially affect both redundant copies in the same way.

Lockstepping can be implemented purely as a hardware layer underneath applications and OS. As long as all inputs—specifically hardware signals feeding the sphere of replication—are correctly replicated to the two redundant copies, all software, including applications and OS, on both the redundant copies of the hardware will execute the same stream of instructions. Then, any mismatch in the stream of instructions due to a fault will be caught by the output comparator. This makes it an attractive solution for system vendors, such as Stratus, who have little control over the microprocessor they may be using or the OS they may have to run.

6.3.2 Disadvantages of Lockstepping

Lockstepping does, however, come with some significant disadvantages. The cost of a Lockstepped system is higher than that of a normal machine since it uses redundant copies of the hardware. At the extreme, the performance-per-unit price of such a system could be nearly half that of a commodity non-fault-tolerant system. In other words, the same dollars could purchase approximately twice as many commodity systems as fault-tolerant systems. Hence, the cost of fault tolerance must be weighed against the customer's penalty incurred from the downtime of the machines.

The validation time for a Lockstepped system can also be higher than that for a normal machine. This is because Lockstepping requires each redundant copy to execute deterministically to produce the same output when fed with the same input. Given up to a billion transistors on today's chips, nondeterminism can arise easily in a microprocessor or a chipset implementation. For example, a floating bit that assumes random values due to circuit marginality may not cause incorrect execution but may easily cause a Lockstep failure. Similarly, clock domain crossing may induce clock skews differently in the redundant copies of the hardware, which may again cause Lockstep failures. The implementation-related nondeterminism must be weeded out for Lockstepping to work correctly.

Validating that a microprocessor will execute deterministically over months or years is a nontrivial job. Typically, testers used in the validation of microprocessors and chipsets can only run for millions of cycles. In a tester, the chip under test is fed with input vectors and compared against a predefined set of output vectors. This process can guarantee fully deterministic operation only for millions of cycles but not for months or years. Hence, a whole range of functional tests are run on the microprocessor to guarantee that the microprocessor will operate correctly over months of operation. These functional tests, however, do not guarantee determinism. Hence, validating that a microprocessor will operate deterministically over months or years is a challenging proposition.

Also, Lockstepping often needs to be accompanied by a recovery mechanism because of its potential to increase the false DUE rate. Recall that the false DUE events arise from benign faults that are detected by the fault detection mechanism (see Silent Data Corruption and Detected Unrecoverable Error, p. 32, Chapter 1). For example, faults in branch predictors may cause one of the Lockstep processors to execute wrong-path instructions that are different from the ones executed by its Lockstep pair (Figure 6.3). This is likely to cause a Lockstep failure and hence is a false DUE event. Similarly, structures that do not have in-line recovery may trigger false DUE events. For example, a parity-protected write-through cache can recover from a strike on the data portion of the cache by refetching the block from a lower level memory. But this refetch operation may cause a timing mismatch with the other Lockstep processor, which may not initiate a refetch in the absence of a fault (Figure 6.4). This would again cause an unnecessary Lockstep failure. Any out-of-band ECC flow would cause a similar problem. One option would be to turn the parity or ECC check off, which may or may not be possible depending on whether the processor had the option to turn it off. Alternatively, this parity or ECC check event could be signalled to the Lockstep output comparator, which could do a fast reset and restart of the whole system. The case studies on Lockstepping in Chapter 3

Cycle	Lockstep Processor A	Lockstep Processor B
...	...	...
n	Correct Path: R1 = R2 + R3	Correct Path: R1 = R2 + R3
n + 1	Wrong Path: R4 = [R1]	Correct Path: R4 = [R1 + 8]
n + 2	Wrong Path: R5 = R4 * 7	Correct Path: R5 = R4 / 7
n + 3	Correct Path: R4 = [R1 + 8]	...
n + 4	Correct Path: R5 = R4 / 7	...
...	...	...

FIGURE 6.3 Lockstep violation due to a strike on a branch predictor of processor A. The notation Rn = [Rm] denotes that register Rn is loaded from the memory location Rm. Each row shows the instructions seen by the execution unit of each processor. The processors A and B are in Lockstep in cycle n. In cycle n + 1, processor A goes down the wrong path due to a strike on its branch predictor. Processor B, however, still remains on the correct path. Processor A returns to the correct path in cycle n + 3, but the two processors are no longer cycle synchronized. Lockstep output comparators that check instruction signals at the memory likely detect this violation immediately. Lockstep output comparators that check signals at the memory or I/O boundary will eventually detect such a timing mismatch.

Cycle	Lockstep Processor A	Lockstep Processor B
...	...	...
n	Correct Path: R1 = R2 + R3	Correct Path: R1 = R2 + R3
n + 1	Correct Path: R4 = [R1 + 8], Parity Error + Refetch	Correct Path: R4 = [R1 + 8], No Error, Cache Hit
n + 2	No commit	Correct Path: R5 = R4 / 7
n + 3	No commit	...
n + 4	Correct Path: R4 = [R1 + 8], No Error, Cache Hit	...
...	...	...

FIGURE 6.4 Lockstep violation due to a parity check followed by refetch of cache line in processor A. Each row shows the instruction committed in the specific cycle in each processor. In cycle n + 1, processor A's cache gets a parity error and must refetch the cache line. Processor B does not get a parity error and proceeds without any hiccup. Processor A eventually commits the offending instruction in cycle n + 4, but by that time, both processors are already out of Lockstep.

(see Case Study: False DUE from Lockstepped Checkers, p. 87) show that false DUE can contribute significantly to the total DUE rate of the Lockstepped system.

False DUE events can also arise in Lockstep processors from faults in un-ACE instructions that may not cause a timing mismatch. For example, an alpha particle or a neutron strike on the result of a dynamically dead instruction is un-ACE since there is no further consumer of the result value. But a fault detection resulting from a mismatch in the result of the corresponding dynamically dead instructions in a Lockstep processor pair will be a false DUE event.

Chapter 7 discusses recovery mechanisms to reduce both true and false DUE rates arising in a Lockstep system. The next three subsections describe three different commercial Lockstep systems, each with a different sphere of replication.

■ EXAMPLE

Assume that a processor write-back cache has a DUE FIT of 500 when protected with parity and DUE FIT of 0 when protected with SECDED ECC. When the ECC is invoked, it requires an extra cycle to correct the error. Also, assume that the processor branch predictor has an intrinsic SDC rate of 100 FIT. The branch predictor's AVF is zero for a single processor, so the total SDC rate of the branch predictor is zero in non-Lockstep mode. Two such processors are now Lockstepped. What is the false DUE rate for this Lockstepped pair? Assume false DUE arises only from the cache and branch predictor.

SOLUTION   From the cache, the false DUE rate for Lockstepping is 500 × 2 = 1000 FIT (the factor 2 arises because of two processors in a pair). From the branch predictor, the SDC FIT rate is 100 × 2 × 0.1 = 20 FIT, if one assumes a Lockstep AVF of 10%. The total false DUE FIT arising from Lockstepping is 1020 FIT.

6.3.3 Lockstepping in the Stratus ftServer

The Stratus ftServer (Figure 6.5) provides a very high level of fault tolerance using redundant Lockstepped processors and redundant I/O components [19]. Stratus Technologies has a long history of building custom fault-tolerant computers since the early 1980s [12]. Like Stratus' previous machines, the ftServer is targeted for use in mission-critical applications (see SDC and DUE Budgets, p. 34, Chapter 1), which can tolerate extremely low levels of SDC and DUE rates. Unlike Stratus' prior fault-tolerant computers, however, the ftServer uses commodity off-the-shelf components, such as Intel x86 processors and Windows Server 2003 and Linux OS. The processor nodes themselves can either be multicore or symmetric multiprocessors (SMP).

The ftServer comes in two configurations: DMR or TMR. In the DMR configuration, the sphere of replication comprises dual-redundant Lockstepped processors, dual copies of main memory, and dual copies of the chipset. The "fault detection and isolation" component in Figure 6.5 consists of the output comparator and the input replicator. Output comparison is done at the I/O boundary before traffic goes

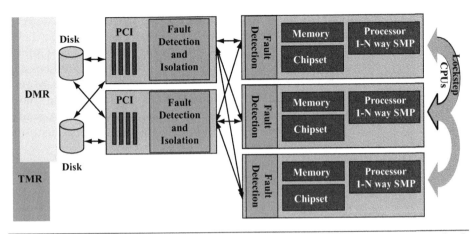

**FIGURE 6.5   Stratus ftServer. Reprinted with permission from Somers [19].
Copyright © 2002 Stratus Technologies.**

into the PCI[2] bus. Input replication is done at the I/O boundary as well. The output
comparator, input replicator, and the I/O components themselves (e.g., PCI bus,
disks) are mirrored for added fault tolerance. The Ethernet network adapter is not
mirrored but has three backup adapters. During transmission, all four adapters are
used, but packets can be received only on a single adapter. The ftServer has fault
detection mechanisms on other mechanical components, such as the power supply
and the fans.

In the DMR system, if one output is in error, then the node that generated that
output is shut down, but the other node continues execution. In the event that a
miscompare occurs between two outputs, the DMR comparator waits for a corresponding error signal to propagate from one of the two Lockstepped nodes. The
error signal could arise from a number of sources, including internal parity errors,
power errors, and thermal errors in one of the Lockstepped pairs. Then, the comparator knows not only that there is an error but also which of the two systems is
faulty. Based on this information, it removes the offending node (processor, memory, and chipset) from service and initiates a service call with Stratus. In the absence
of the internal error signal, the ftServer removes a node from service based on a
software algorithm that takes various heuristics, such as age of the components,
into account. In all cases, the system continues to operate without interruption.

In the TMR mode, however, the ftServer does not need to make use of the error
signal. This is because on a single fault, two of the three systems must be correct.
Once it identifies the faulty system, it removes the offending node from service

---

[2]PCI stands for Peripheral Component Interconnect. It is a standard bus that allows I/O
devices to connect to the rest of the computer system.

and continues in the reduced DMR mode. It also initiates a service call to Stratus to repair the degraded system.

Interestingly, the ftServer can run a commodity OS without any significant change because the underlying hardware is completely Lockstepped. Each individual OS image in the ftServer has no knowledge that there are other redundant images of the same OS working in Lockstep. In contrast, the Hewlett-Packard NSAA (see RMT in the Hewlett-Packard NonStop® Advanced Architecture, p. 225) needs to modify its OS because it does not implement strict cycle-by-cycle Lockstepping.

The use of the Windows OS does, however, require the ftServer to take some additional steps[3]. Errant device drivers are acknowledged as the root cause of many of the Windows crashes. Microsoft itself has estimated that device drivers are responsible for 30% of Windows NT® reboots. Hence, the ftServer takes additional steps to ensure that device drivers do not cause Lockstep failures or crashes. For example, the ftServer only installs device drivers that pass the Microsoft Windows 2003 WHQL (Windows Hardware Quality Labs) tests. Further, Stratus has hardened the device drivers through extensive testing and additional error checks either by licensing the device driver from the vendor or by working with the vendor. The ftServer also isolates the PCI adapters from the rest of the system when it detects device driver problems.

6.4   Lockstepping in the Hewlett-Packard NonStop Himalaya Architecture

Like Stratus Technologies, Tandem Computer Systems has been building fault-tolerant computer systems for mission-critical applications since the late 1970s [12]. In the late 1980s, Tandem introduced custom-built Lockstepped processors into its fault-tolerant computers to respond to market demands for highly available fault-tolerant computers. Tandem started using commodity MIPS®-based RISC processors in its Lockstepped machines from the early 1990s. However, unlike Stratus' recent move to use a commodity OS, Tandem continues to use its own in-house fault-tolerant OS called the NonStop kernel. In 2001, Hewlett-Packard acquired Compaq Computer Corporation, which in turn had acquired Tandem Computer Systems in 1997. Tandem continues to design and sell fault-tolerant computers under the Hewlett-Packard NonStop brand name.

Figure 6.6 shows a block diagram of the Hewlett-Packard NonStop Himalaya system [23]. It uses Lockstepped off-the-shelf MIPS microprocessors. Like the Stratus ftServer, it also is targeted to tolerate extremely low levels of SDC and DUE. The sphere of replication comprises the microprocessor (and its associated off-chip caches) and the interface ASICs responsible for output comparison and input

---

[3]See http://www.stratus.com for white papers on this subject.

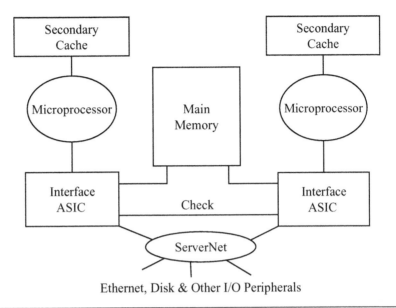

FIGURE 6.6 Hewlett-Packard NonStop Architecture with Lockstepped microprocessors and ServerNet as the interconnect fabric.

replication. Unlike the Stratus ftServer system, main memory in Hewlett-Packard's NonStop Architecture is outside the sphere of replication. Consequently, the NonStop Architecture does not have to replicate the entire main memory. Instead, it protects main memory using ECC.

The I/O subsystem is also outside the sphere of replication. The I/O system is connected to the processors using an in-house network called ServerNet, which forms the backbone of the current NonStop Architecture. ServerNet is a fast, scalable, reliable, and point-to-point system area network that can connect a large number of processors and I/O peripherals. ServerNet provides independent redundant paths for all traffic via independent X and Y subnetworks.

To provide recovery from faults, the NonStop Himalaya servers use *process pairs* implemented by the NonStop kernel. One of the processes in a process pair is designed as the primary and the other one as the backup. The primary process runs on a pair of Lockstepped processors and regularly sends checkpointing information to the backup process, which runs on a pair of Lockstepped processors as well. When the primary process experiences a Lockstep error, the backup process takes over and continues to run. Hewlett-Packard Corporation also offers a TMR solution in its NonStop Integrity line of servers, which do not require the use of process pairs.

Recently, Hewlett-Packard moved away from Lockstepped systems for Redundantly Multithreaded systems. Hewlett-Packard calls this its NSAA. This implementation is discussed later in this chapter (see RMT in the Hewlett-Packard NonStop® Advanced Architecture, p. 225).

6.5   Lockstepping in the IBM Z-series Processors

IBM has been designing fault-tolerant machines since the late 1960s but waited to introduce its Lockstepped processor architecture as late as the mid-1990s. In the early 1990s, IBM moved the processors used in its high-end fault-tolerant machines from emitter-coupled logic (ECL) to CMOS technology. Successive generations of these CMOS processors were codenamed G1–G6, and z900 and more recently z990 and z9 EC. In the ECL processors and in G1 and G2 CMOS processors, IBM used extensive per-structure error checking logic in its pipeline. IBM estimates the total area overhead from such error checking logic to be between 20% and 30% of single nonchecked pipeline [13].

With the move to CMOS technology, IBM designers found it increasingly difficult to increase the processor frequency with extra levels of error checking logic in the critical path of the pipeline. Hence, with the introduction of the G3 microprocessor, IBM moved to a Lockstepped pipeline implementation in which the error checking logic was removed from the critical path. Instead, the instruction fetch and execution units were replicated and checked for errors at the end of the pipeline. Spainhower and Gregg [20] estimate the area overhead from this Lockstepped implementation to be 35%. IBM used this dual fetch and execution unit through its z990 architecture. In the more recent z6 architecture, however, IBM has reverted to a non-Lockstepped implementation that has as many as 20 000 error checkers sprinkled across the microprocessor [22].

Figure 6.7 shows the IBM G5's Lockstepped processor architecture. The G5 consists of four units: the buffer control element (BCE), the I unit, the E unit, and the R unit. The BCE consists of the L1 cache and TLBs. The I unit is responsible for the instruction fetch, decode, address generation, and issue queue that issues instructions to the E unit. The E unit consists of the execution units and local copies of the register files. The R unit is responsible for fault detection and recovery. It holds checkpointed copy of the entire microarchitectural state of the processor (including the architecture register file), timing facility, and other miscellaneous state information.[4]

The sphere of replication in the G5 consists of replicated I and E units. Figure 6.7 shows signals from replicated units (dashed arrows). Any updates to the architecture register file (in the R unit) or caches (in the BCE) must be first checked for faults (output comparison). Similarly, any inputs into the I or E unit must be replicated appropriately (input replication). Unlike Stratus's ftServer or Tandem's NonStop Himalaya Architecture, the sphere of replication in the IBM machine is much smaller. The G5's sphere of replication excludes not only main memory and I/O components but also the caches and architected register file. Consequently, all

---

[4]Historical Anecdote on p. 248 describes the origin of the name R unit.

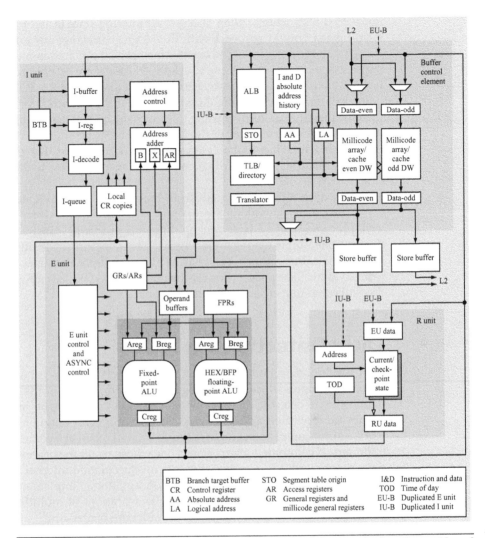

FIGURE 6.7 IBM G5 microprocessor. Reprinted with permission from Slegel et al. [13]. Copyright © 1999 IEEE.

arrays in the BCE and R unit must be protected to allow complete fault coverage. For example, the L1 cache is write-through, so it uses parity codes to recover from faults. The store buffer and the architecture register file are protected with ECC. Although the architecture register file lives outside the sphere of replication in the R unit, the G5 maintains shadow copies of the register file in the E units to speed up execution.

When the R unit or L1 cache detects a fault, the pipeline commits pend-ing updates from retired instructions and discards any state corresponding to nonretired instructions. Then, it resets all internal state in the I and E units. The R unit is read out in sequence with ECC logic, correcting any errors in the registers. The shadow copies of the register file are also updated in parallel. The R unit registers are read a second time to ensure that there are no hard correctable errors. Finally, the E unit restarts instruction fetching and execution. If the Lockstep violation was caused by a transient fault, such as an alpha particle or a neutron strike, then the pipeline can recover using this mechanism. However, if the second read of the R unit indicates an error or the E unit cannot successfully retire an instruction, then the machine has probably encountered a hard error. The pipeline is halted with a different sequence of diagnosis and recovery invoked to deal with the latter case.

Forwarding replicated signals from the I and E units to R unit and L1 cache can cause extra delays along the pipeline. To avoid slowing down the pipeline, the z990 processor [14]—a later version of the G5 processor—skews the signals from the replicated units by one cycle. Signals from the primary I and E units arrive at the output comparators a cycle earlier than the replicated ones. This allows the primary units to be packed close to the nonreplicated units, such as the caches, and the replicated units to be placed further apart.

6.6   Fault Detection via RMT

RMT is a fault detection mechanism, which like cycle-by-cycle Lockstepping, runs redundant copies of the same program and compares outputs from the redundant copies to detect faults. Recall that in Lockstepping outputs are compared and inputs are replicated at a hardware clock or at a cycle boundary. In contrast, in RMT, output comparison and input replication happen at a committed instruction boundary. A committed instruction is one whose result the processor commits and does not discard, for example, due to a misspeculation internal to the processor. Like "multithreading," RMT itself is a concept, not an implementation. RMT can be implemented on most, if not all, implementations of multithreading, such as simultaneous multithreading (SMT) or multicore processors.

Comparing outputs and replicating inputs at a committed instruction boundary enables RMT to relax many of the constraints imposed by cycle-by-cycle Lockstepping. RMT does not require the hardware running the redundant copies of a program to have exactly the same microarchitectural state in every cycle. Examples of structures that hold microarchitectural state are the branch predictor, instruction queue, caches, etc. Instead, RMT only requires that the redundant copies have the same architectural state (e.g., architectural register file, memory) at a committed instruction boundary. Unlike microarchitectural state, architectural state of a machine corresponding to program is visible to a user or a programmer.

Relaxing the constraint of cycle synchronization allows great flexibility in RMT implementations. As is shown later, RMT can be implemented across whole systems, within a single system, across two similar and different processor cores, and within a single-processor core. Chapter 8 discusses how RMT can be implemented purely in software as well. For pure hardware RMT implementations, relaxing the constraint of cycle synchronization makes functional validation of the RMT implementation much easier since much of the validation can be done in a presilicon design phase.

Relaxing the constraint of cycle synchronization, however, makes input replication more complicated in RMT than in Lockstepping. This section shows how five different RMT implementations deal with input replication in different ways. How the sphere of replication differs in each of these implementations is also explained.

6.7 RMT in the Marathon Endurance Server

Marathon Technologies Corporation was perhaps the first to implement an RMT-based fault-tolerant machine using redundant processors in a commercially available server called Endurance^{TM} 4000. The company was founded in 1993 on the premise of this technology. The key to Endurance's success is its use of commodity OSs, such as Windows OS, and commodity microprocessors, such as the Intel Pentium processors. Another key feature of the Endurance machine is its ability to tolerate disasters (e.g., terrorist attack) by separating the redundant processors up to 1.5 km apart. Because of such loose coupling between the redundant processors, Marathon had to adopt the RMT model instead of the cycle-by-cycle Lockstepping.

Recently, Marathon Technologies has moved away from the Endurance servers, which required some custom hardware. Today, Marathon's EverRun servers provide fault tolerance using a pure software implementation based on virtualization technology (see Chapter 8).

Figure 6.8 shows a block diagram of the Endurance machine [3]. It consists of two compute elements (CEs) and two I/O processor boxes. All four processors—one in each CE and one in each I/O—are commodity Intel Pentium microprocessors that run the Windows NT OS. The CEs execute the redundant threads as an RMT pair. The sphere of replication consists of the two CEs, which include the microprocessor, the main memory, and a proprietary Marathon InterConnect (MIC) card. The MIC connects the two CEs and the two I/O processors. The MIC links can be up to 1.5 km in length, thereby allowing a CE and an I/O processor slice to be located physically apart from its pair.

The I/O processors do the output replication. The CEs forward their I/O requests to both I/O processors, each of which compares both requests for any mismatch. The storage system is mirrored across both I/O processors. The network adapter is also replicated. Network packets can be received in either I/O processor, but packets can only be sent out through one of the two I/O processors. If one of the sending adapters fails, the other one takes over. In an implementation with reduced coverage, it is also possible to have only one I/O processor.

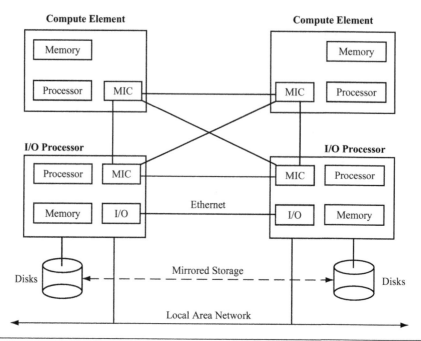

FIGURE 6.8 Block diagram of the Marathon Technology Endurance™ Series Machine. MIC = Marathon's Proprietary Interconnect ASIC.

The input replication is a little more involved. To ensure that both the redundant threads execute the same committed instruction stream, the inputs—including asynchronous interrupts and I/O responses—must be replicated to both threads at exactly the same instruction boundary. When the two redundant processors run independently of each other, this condition is hard to guarantee. The Hewlett-Packard NSAA, described in the next section, is able to do this with OS support. But the Marathon Endurance machine runs a commodity OS and hence does not have this advantage.

Instead, the Endurance machine divides up the instruction stream into successive quanta. At the beginning of each quantum, the performance counter is set to trigger a callback after a specific number of instructions have been committed. All other interrupts are disabled. When a quantum expires, each redundant processor enters into a special mode in which each tries to ensure that it has reached the same instruction in the execution stream with the same state. This special mode consists of a complex combination of instruction execution, single stepping, and breakpointing. The synchronization requests during this interval are sent to the I/O processors, which try to determine if both processors have reached the same point of execution. At the end of this synchronization phase, both processors reach the same instruction boundary. At this point, all inputs are replicated via the MIC to the redundant processors. This includes also the logical time delivered to both the processors by the I/O processor, which acts as a time server.

The Endurance machine prevents any SDC since I/O writes are checked before written to disk. On an RMT mismatch, the I/O processors try to determine the CE that failed. If the diagnosis is successful, then the failed CE will be restarted or replaced and restarted as the case may be. When the failed CE comes back up, it automatically synchronizes with the running CE, which continues running in a nonredundant mode throughout the recovery period.

If the I/O processors cannot, however, determine the CE that had the data corruption, then there are two options available to the system administrator for configuring the system response to indeterminate CE data corruption. If the system was configured for data availability, the CE that is removed is the slave CE (recipient of the last synchronization). If the system was configured for data integrity, then the entire system is rebooted and restarted.

I/O processor errors are handled similarly. For data availability, the I/O processor with the poorest selection of devices will be removed (I/O boot drive, CE boot drive, CE Data drives, Ethernet ports, memory, etc). If the I/O processors are indistinguishable in capabilities, then the I/O processor that did not boot the master CE (source of the last synchronization) is removed. For data integrity, the entire system must be rebooted.

6.8   RMT in the Hewlett-Packard NonStop® Advanced Architecture

Recently, Hewlett-Packard moved its NonStop servers from a cycle-by-cycle Lockstepped architecture (see Lockstepping in the Hewlett-Packard NonStop Himalaya Architecture, p. 218) to what they call "Loose Lockstepping" [2], which is a form of RMT. Bernick et al. [2] believe that cycle-by-cycle Lockstepping will become significantly harder in the future because of five reasons. First, minor nondeterministic behavior, such as arbitration of asynchronous events, will continue to exist in future microprocessors. These are not easy to deal with in a Lockstepped processor pair. Second, power management techniques with variable frequencies—critical to current and future processors—may cause Lockstep failures and may need to be turned off for Lockstepping to work. Third, multiple clocks and clock domain crossings in a microprocessor create asynchronous interfaces that are very difficult to deal with in Lockstepping. Fourth, low-level fix-up routines (e.g., in microcode or millicode) triggered in a microprocessor to correct soft errors can complicate Lockstep operation. Finally, Lockstepping microprocessor pairs today implies Lockstepping multiple cores. A problem in one of the cores can bring down the whole multicore chip, which can be an undesirable property.

Figure 6.9 shows a picture of Hewlett-Packard's NSAA. A processing element is a processor core. A slice consists of several PEs and main memory. One or more

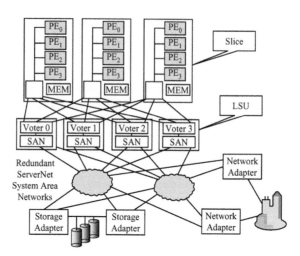

FIGURE 6.9 Hewlett-Packard NonStop Architecture. Reprinted with permission from Bernick et al. [2]. Copyright © 2005 IEEE.

PEs make up a processor socket, which is currently from Intel's Itanium® processor line. A logical processor consists of one PE per slice. Two or three PEs—one from each slice—make up a logical processor that runs redundant RMT code. In a DMR configuration, two PEs make up a logical processor. In a TMR configuration, the number of PEs in a logical processor is three. Each logical processor is associated with at least one logical synchronization unit (LSU), which acts as the output comparator and input replicator. Each LSU is completely self-checking internally. One LSU per logical processor is sufficient, but for increased availability, the NSAA may allow a machine to configure two LSUs per logical processor in the future. Thus, a TMR machine using three 4-way SMPs will have three slices, four logical processors, and four LSUs. The logical processors are, however, not allowed to communicate with other logical processors within the SMP system using shared memory.

The sphere of replication in the NSAA is somewhat similar to that of the Endurance machine. The sphere consists of either duplicate (for DMR) or triplicate (for TMR) copies of a slice. A PE in each slice forms one of the two (for DMR) or three (for TMR) elements of a system. The LSUs check outgoing stores to main memory and I/O requests. For a DMR system, the LSUs check for mismatch between the two redundant copies. In a TMR system, the LSUs check outputs from three slices and vote on which one is the correct one. Like the Marathon Endurance machine, the I/O subsystem in NSAA is mirrored to allow greater level of fault tolerance. It also has end-to-end checksums for disk accesses to allow for better data integrity.

As discussed earlier, processor and I/O completion interrupts in an RMT implementation must be delivered to redundant instruction streams at exactly the same instruction boundary, which is referred to as a "rendezvous" point by the NSAA. Normally, such interrupts would arrive at each PE at slightly different times and may not be delivered at the same instruction boundary in each thread. Hence, on receiving an interrupt, each PE writes its proposal for a future rendezvous point in LSU hardware registers. The proposal consists of a voluntary rendezvous opportunity (VRO) sequence number along with the delivered interrupt. Once an LSU collects all the rendezvous proposals, it writes them back into a special block of memory in each slice. Each PE compares its own rendezvous proposal with the ones posted by other PEs and selects the highest VRO sequence number. When each PE reaches that VRO, it can symmetrically schedule the interrupt handler for execution.

The VRO is defined by a small set of instructions that must be embedded throughout the OS and user applications because the VRO code constitutes the only rendezvous points in the NSAA. If a process executes for a long time without entering any VRO code, then the NSAA uses a combination of fast-forwarding and state copying to bring the two threads at the same instruction boundary. Readers are referred to Bernick et al. [2] for a detailed description of these algorithms. In an alternate implementation, Hewlett-Packard's Integrity S2 systems use performance counters to define a rendezvous point without the need for explicit embedding of VRO code. The NSAA also disables some Itanium-specific data and control speculation that may cause the redundant streams to execute different sequences of instructions.

Mechanisms to recover from a fault are different in the DMR and TMR configurations. In the DMR configuration, an output comparison error followed by a self-identifying fault, such as a bus error, allows the NSAA to precisely identify the faulty slice and isolate it. Operations can continue on the good slice. Alternatively, if the output comparator cannot determine the faulty slice, then the application could fail over to a different logical processor and continue execution. In the TMR configuration, an output comparison error allows the voting logic to identify the faulty slice (since only one of the three will usually experience a fault) and isolate it, and allows the nonfaulty ones to continue execution.

When a faulty slice resumes execution after a reboot (if it is a soft error) or replacement (if it is a hard error), its state must be made consistent with the state of the other slice or slices with which it synchronizes its redundant execution. The NSAA refers to this as a "reintegration" operation. To facilitate reintegration, the NSAA provides a high-bandwidth unidirectional ring network that connects the slices together. The entire memory state of a nonfaulty slice is copied to the slice being reintegrated. The copy operation also intercepts and copies any in-flight store to memory issued by the nonfaulty slices during the reintegration process.

6.9 RMT Within a Single-Processor Core

This section and the next two sections describe RMT implementations that have a smaller sphere of replication than the Marathon Endurance server or the Hewlett-Packard NSAA. Specifically, this section examines RMT implementations whose spheres of replication are limited to a single-processor core, similar to that used in the IBM Z-series processors (see Lockstepping in the IBM Z-series Processors, p. 220).

Two such implementations with different spheres of replication are described: one that includes the architecture register file but not the caches or main memory (SRT-Memory) and another that excludes the architecture register file and caches and main memory (SRT-Register). Both SRT implementations rely on an underlying processor architecture called SMT. To the best of the author's knowledge, no current commercial machine implements RMT within a single core. Numerous researchers are, however, investigating techniques to improve the RMT design. Because this is an active area of research, this section covers the RMT nuances within a single core in greater detail than the other RMT implementations covered earlier. First, this section describes an example SMT processor. Then, it will discuss how to extend the SMT implementation to incorporate SRT enhancements.

6.9.1 A Simultaneous Multithreaded Processor

SMT is a technique that allows fine-grained resource sharing among multiple threads in a dynamically scheduled superscalar processor [18]. An SMT processor extends a standard superscalar pipeline to execute instructions from multiple threads, possibly in the same cycle. To facilitate the discussion in this section, a specific SMT implementation is used (Figure 6.10). Mukherjee et al. describe an alternate implementation of SMT in a commercial microprocessor design that was eventually canceled [6]. In the SMT implementation in Figure 6.10, the fetch stage feeds instructions from multiple threads (one thread per cycle) to a fetch/decode queue. The decode stage picks instructions from this queue, decodes them, locates their source operands, and places them into the register update unit (RUU). The RUU serves as a combination of global reservation station pool, rename register file, and reorder buffer. Loads and stores are broken into an address and a memory reference. The address generation portion is placed in the RUU, while the memory reference portion is placed into a similar structure, the load/store queue (LSQ) (not shown in Figure 6.10).

Figure 6.10 shows instructions from two threads sharing the RUU. Multiple IPCs are issued from the RUU to the execution units and written back to the RUU without considering thread identity. The processor provides precise exceptions by committing results for each thread from the RUU to the register files in program order. Tullsen et al. [17] showed that optimizing the fetch policy—the policy that determines the thread from which the instructions are fetched in each cycle—can improve the performance of an SMT processor. The best-performing policy Tullsen, et al. examined was named ICount. The ICount policy counts the number of instructions from active threads that are currently in the instruction buffers and fetches instructions from the thread that has the fewest instructions. The assumption is that the thread with the fewest instructions moves

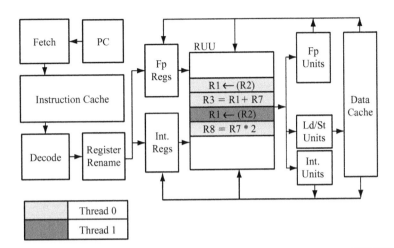

FIGURE 6.10 Sharing of RUU between two threads in an SMT processor. Reprinted with permission from Reinhardt and Mukherjee [10]. Copyright © 2000 IEEE.

instructions through the processor quickly and hence makes the most efficient use of the pipeline.

6.9.2 Design Space for SMT in a Single Core

One can modify an SMT processor to detect faults by executing two redundant copies of each thread in separate thread contexts. Unlike true SMT threads, each redundant thread pair appears to the OS as a single thread. All replication and checking are performed transparently in hardware. This class of single-core RMT implementations within an SMT processor is referred to as an SRT processor [10]. One of the two redundant threads in an SRT processor will be designed to run ahead of the other. Hence, the two redundant threads will be referred to as the leading thread and the trailing thread.

This section focuses on how to design an SRT processor with two thread contexts to support a single-visible-thread SRT device. One can easily extend such designs to support two OS-visible threads on an SMT machine with four thread contexts.

Unlike the RMT implementations in the Marathon Endurance server or the Hewlett-Packard NonStop Architecture, the sphere of replication in an SRT processor is physically less distinct because replication occurs through both time redundancy and space redundancy. For example, the corresponding instructions from redundant threads may occupy the same RUU slot in different cycles (time redundancy), different RUU slots in the same cycle (space redundancy), or different slots in different cycles (both). Like other RMT systems, SRT processors detect faults by comparing outputs of redundantly executing instruction streams.

6.9.3 Output Comparison in an SRT Processor

The sphere of replication determines the values that need to be compared. Output comparison techniques for the three different single-core RMT implementations are described.

Output Comparison in SRT-Memory

When the register file lies inside the sphere (SRT-Memory, Figure 6.11a), there are three types of values that exit the sphere:

- **Stores.** The output comparator must verify the address and data of every committed store before it forwards it outside the sphere of replication. One can use an ordered, noncoalescing store buffer shared between the redundant threads to synchronize and verify store values as they retire in program order from the RUU/LSQ. Each thread has an independent tail pointer into the buffer. If a thread finds its tail entry uninitialized, it writes the address and data value of its store into the entry. The second thread will find this entry initialized, so it will compare its address and data with the existing values. On a match, the entry is marked as verified and issued to the data cache. A mismatch indicates a fault. In this implementation, misspeculated stores never send their data outside the sphere of replication, so they do not need checking. To provide each thread with a consistent view of memory, the store

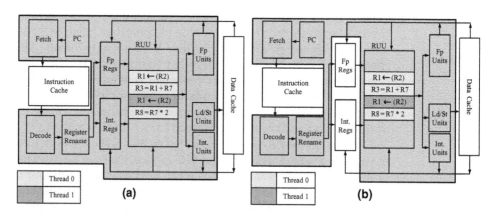

FIGURE 6.11 Two spheres of replication for an SRT processor. The shaded box in (a) shows a sphere of replication that includes the entire SMT pipeline shown in Figure 6.10, except the first-level data and instruction caches (SRT-Memory). The shaded box in (b) shows a sphere of replication that excludes the architectural register file, the first-level data cache, and the first-level instruction cache (SRT-Register). Reprinted with permission from Reinhardt and Mukherjee [10]. Copyright © 2000 IEEE.

buffer forwards data to subsequent loads only if the store has retired in the thread issuing the load.

■ **Cached load addresses.** Although cached data and instruction fetch addresses leave the sphere of execution, they do not affect the architectural state of the machine, so they do not require checking. If a faulty address leads to an incorrect load value or instruction, any resulting error will be detected via other output comparison checks before affecting architectural state outside the sphere. It will be seen later that allowing one thread to issue cache fetches early (and without output comparison), effectively prefetching for the other thread, is critical to the efficiency in SRT processors.

■ **Uncached load addresses.** Unlike cached loads, uncached loads typically have side effects in I/O devices outside the sphere of replication, so these addresses must be checked. However, unlike stores, uncached load addresses must be compared before the load commits. Fortunately, in most processors, uncached loads issue nonspeculatively and only after all prior loads and stores commit. Also, no load or store after the uncached load in program order can issue until the uncached load commits. Thus, an uncached load can simply stall in the execute stage until the corresponding instruction for the other thread arrives, at which point the addresses can be compared. An undetected fault could occur if an uncached load address was erroneously classified as cached and allowed to proceed without being checked. Adequate precautions must be taken to prevent this specific case, such as additional physical address cacheability checks.

Output Comparison in SRT-Register

The second sphere of replication (Figure 6.11b) does not contain the register file (SRT-Register), so it requires output comparison on values sent to the register file—i.e., on register write-backs of committed instructions. As with stores, both the address (register index) and value must be verified. Register write-back comparison could be done as instructions retire from the RUU. However, forcing every instruction to wait for its equivalent from the other thread significantly increases RUU occupancy. Since the RUU is a precious resource, one could instead use a register check buffer, similar to the store buffer, to hold results from retired but unmatched instructions. The first instance of an instruction records its result in the buffer. When the corresponding instruction from the other thread leaves the RUU, the index and value are compared and, if they match, the register file is updated.

As with the store buffer, results in the register check buffer must be forwarded to the subsequent instructions in the same thread to provide a consistent view of the register file. The design can avoid complex forwarding logic by using the separate per-thread register files of the SMT architecture as "future files" [15]. That is, as each instruction retires from the RUU, it updates the appropriate per-thread register file, as in a standard SMT processor. This register file then reflects the up-to-date but unverified register contents for that redundant thread. As register updates are verified and removed from the register check buffer, they are sent to a third register file, which holds the protected, verified architectural state for the user-visible thread.

Having a protected copy of the architectural register file outside the sphere of replication simplifies fault recovery on an output mismatch, as the program can be restarted from the known good contents of the register file (as in the IBM Z-series microprocessors [13]). However, this benefit requires the additional costs of verifying register updates and protecting the register file with ECC or similar coverage. Although the register check buffer is conceptually similar to the store buffer, it must be significantly larger and must sustain higher bandwidth in updates per cycle to avoid degrading performance.

6.9.4 Input Replication in an SRT Processor

Inputs to the sphere of replication must be handled carefully to guarantee that both execution copies follow precisely the same path. Specifically, corresponding operations that input data from outside the sphere must return the same data values in both redundant threads. Otherwise, the threads may follow divergent execution paths, leading to differing outputs that will be detected and handled as if a hardware fault occurred. As with output comparison, the sphere of replication identifies values that must be considered for input replication: those that cross the boundary into the sphere.

Input Replication in SRT-Memory

For the first sphere of replication (SRT-Memory, Figure 6.11a), four kinds of inputs enter the sphere:

- **Instructions.** If the contents of the instruction space do not vary with time, then unsynchronized accesses from redundant threads to the same instruction address will return the same instruction without additional mechanisms. Updates to the instruction space require thread synchronization, but these updates already involve system-specific operations to maintain instruction-cache consistency in current processors. These operations can be extended to enforce a consistent view of the instruction space across redundant threads.

   The instruction replication itself can be implemented in a couple of ways. One possibility would to replicate instructions directly from the instruction cache but to allow unsynchronized access to it from both threads [10]. Another possibility would be to forward retired instructions from the leading thread to the trailing thread's fetch unit [6]. The latter is very precise because only the committed instruction stream is forwarded to the trailing thread, thereby avoiding any branch misprediction in the trailing thread. In fact, a branch misprediction in the trailing thread in this case would be flagged as an error.

In effect, the branch direction and address computation logic for the trailing thread acts as an output comparator.

- **Cached load data.** Corresponding cached loads from replicated threads must return the same value to each thread. Unlike instructions, data values may be updated by other processors or by DMA I/O devices between load accesses. An out-of-order SRT processor may also issue corresponding loads from different threads in a different order and in different cycles. Because of speculation, the threads may even issue different numbers of loads. Later in this section, two mechanisms for input replication for cached load data—active load address buffer (ALAB) and load value queue (LVQ)—are described.

- **Uncached load data.** As with cached load data, corresponding loads must return the same value to both threads. Because corresponding uncached loads must synchronize to compare addresses before being issued outside the sphere of replication, it is straightforward to maintain synchronization until the load data return and then replicate that value for both threads. Other instructions that access nonreplicated, time-varying state, such as the Alpha rpcc instruction that reads the cycle counter, are handled similarly.

- **External interrupts.** Interrupts must be delivered to both threads at precisely the same point in their execution. Three solutions are possible. The first solution forces the threads to the same execution point by stalling the leading thread until the trailing thread catches up and then delivers the interrupt synchronously to both threads. The second solution delivers the interrupt to the leading thread, records the execution point at which it is delivered (e.g., in committed instructions since the last context switch), and then delivers the interrupt to the trailing thread when it reaches the identical execution point. The third solution rolls both threads back to the point of the last committed register write. Rolling back may, however, be difficult if memory state is committed and exposed outside the sphere of replication.

Input Replication in SRT-Register

As with output comparison, moving the register file outside the sphere means that additional values cross the sphere boundary. In the case of input replication, it is the register read values that require further consideration. However, each thread's register read values are produced by its own register writes, so corresponding instructions will receive the same source register values in both threads in the absence of faults (and assuming that all other inputs are replicated correctly). In fact, many source register values are obtained not from the register file but by forwarding the results of earlier uncommitted instructions from the RUU (or from the "future file" as discussed in the previous section). Hence, input replication of register values requires no special mechanisms even when the register file is outside the sphere of replication.

6.9.5 Input Replication of Cached Load Data

Input replication of cached load data is problematic for both SRT-Memory and SRT-Register implementations because data values can be modified from outside the processor. For example, consider a program waiting in a spin loop on a cached synchronization flag to be updated by another processor. The program may count the number of loop iterations in order to profile wait times to adaptively switch synchronization algorithms. To prevent redundant threads from diverging, both threads must spin for an identical number of iterations. That is, the update of the flag must appear to occur in the same loop iteration in each thread, even if these corresponding iterations are widely separated in time. Simply invalidating or updating the cache may cause the leading thread to execute more loop iterations than the trailing thread. Hence, special attention needs to be given to input replication of cached data. Here two mechanisms for input replication of cached load data, the ALAB and the LVQ, are described.

Active Load Address Buffer

The ALAB provides correct input replication of cached load data by guaranteeing that corresponding loads from redundant threads will return the same value from the data cache. To provide this guarantee, the ALAB delays a cache block's replacement or invalidation after the execution of a load in the leading thread until the retirement of the corresponding load in the trailing thread.

The ALAB itself comprises a collection of identical entries, each containing an address tag, a counter, and a pending-invalidate bit. When a leading thread's load executes, the ALAB is searched for an entry whose tag matches the load's effective address; if none is found, a new entry is allocated. Finally, the entry's counter is incremented to indicate an outstanding load to the block. When a trailing thread's load retires, the ALAB is again searched and the matching entry's counter is decremented.

When a cache block is about to be replaced or invalidated, the ALAB is searched for an entry matching the block's address. If a matching entry with a nonzero count is found, the block cannot be replaced or invalidated until all the trailing thread's outstanding accesses to the block have completed. At this point, the counter will be zero and the block can be released. To guarantee that the counter eventually reaches zero, the cache sets the ALAB entry's pending-invalidate bit to indicate that it is waiting; leading-thread loads that attempt to increment an entry with a set pending-invalidate bit are stalled. Because the trailing thread can always make forward progress, the outstanding loads will eventually complete. Although the block for which the cache receives the invalidation request cannot be immediately invalidated, the cache can still send an acknowledgement for the invalidation request. However, depending on the memory consistency model, the processor may have to stall other memory requests until the corresponding ALAB counter reaches zero.

The ALAB must reliably track every outstanding load and therefore cannot evict a nonzero ALAB entry. A leading-thread load must stall if it cannot allocate a new entry due to ALAB mapping conflicts or if it would increment the entry's counter beyond the maximum value. Leading-thread stalls may lead to a deadlock. The processor can detect this deadlock condition by recognizing when (1) the trailing thread has caught up to the leading thread and (2) the leading thread's oldest instruction is a load stalled because it cannot perform its ALAB increment. In this case, the processor must flush the leading thread from the pipeline (as on a misspeculation) and decrement the ALAB counters corresponding to executed but squashed loads. The stalled load can then reissue; because the trailing thread has committed all outstanding loads, it is guaranteed not to stall on the ALAB increment. Evaluation of an SRT processor by Reinhardt and Mukherjee [10] showed that a 64-entry, fully associative ALAB with 7-bit counters avoids this deadlock entirely across all their benchmarks.

Load Value Queue

The LVQ is an alternate mechanism to implement input replication of cached load values. Compared to the ALAB, the LVQ is much simpler to implement. In this mechanism, one of the redundant threads is designated as a leading thread and the other one as a trailing thread. Only the leading thread issues cached loads to the data cache outside the sphere of replication. When a cached load commits, the leading thread sends the load's address and value to the trailing thread via the LVQ, which is a first-in first-out buffer.

The LVQ can be read out both in program order and out-of-order. In the in-order implementation, the trailing thread performs loads in program order and nonspeculatively. The loads in the trailing thread proceed through the regular processor pipeline. However, instead of probing the data cache, the trailing thread waits for its corresponding load address and value from the leading thread to show up at the head of the LVQ. Input replication is guaranteed because the trailing thread receives the same value for the load that the leading thread used.

Alternatively, loads from the trailing thread can issue out of order. In this case, the leading thread must attach tags to the loads it inserts in the LVQ and must transmit these tags to the trailing thread. One way to achieve this is to forward the instruction tags of the leading thread to the replicated instructions of the trailing thread. This is easy to do in the second instruction replication mechanism outlined earlier (see Input Replication in an SRT Processor, p. 232).

Figure 6.12 shows the lookup operation performed in either an in-order or an out-of-order LVQ. Typically, a load would look up both a data cache and a store queue simultaneously to see the structure that has the most recent value required by the load instruction. This is the path followed by loads from a leading thread. However, loads from the trailing thread simply look up the LVQ, bypass the data cache and store queue, and forward their data back to the registers waiting for the loaded values.

The LVQ provides a couple of additional advantages to the SRT implementation. The LVQ reduces the pressure on the data cache ports because, unlike the

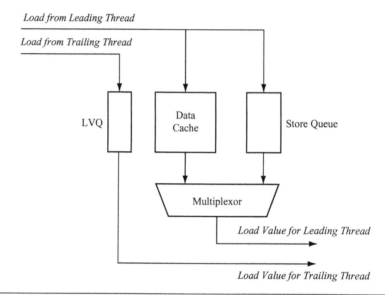

FIGURE 6.12 Lookup operation of an LVQ in an SRT processor. The figure shows only the case when a leading thread's load hits in the data cache or the store queue.

ALAB design, only one thread probes the data cache. Additionally, the LVQ can accelerate fault detection of faulty addresses by comparing the effective address of the leading thread from the LVQ with the effective address of the trailing thread.

6.9.6 Two Techniques to Enhance Performance of an SRT Processor

An SRT processor can improve its performance using one thread to improve cache and branch prediction behavior for the other thread. Two such mechanisms—slack fetch and branch outcome queue—that achieve this performance improvement are described. During execution, these mechanisms allow the trailing thread to use fewer pipeline resources, such as compute bandwidth and pipeline structure space, which improves the overall performance of an SRT processor. Hence, typically an SRT processor—running a single logical program replicated as redundant threads—would complete the logical program faster than the underlying SMT pipeline would when running two independent copies of the same program.

Slack Fetch

The slack fetch mechanism tries to maintain a constant slack of instructions between the leading and the trailing threads. Ideally, branches from the trailing thread should probe the branch predictor after the corresponding branch from the leading thread has executed and updated the predictor. Similarly, accesses from the trailing thread should probe the instruction and data caches after the corresponding accesses from the leading thread have incurred any misses and brought in the cache blocks. Slack fetch achieves this by forcing instructions from the trailing thread to execute later in time than the corresponding instructions from the leading thread.

In one implementation, slack fetch can be implemented using a signed counter that is decremented when the (predesignated) leading thread commits an instruction and can be incremented when the (predesignated) trailing thread commits an instruction. The counter is initialized at reset to the target slack. By adding the counter's value in each cycle to the trailing thread's instruction count, the ICount fetch policy (see A Simultaneous Multithreaded Processor, p. 228) favors the leading thread until the target slack is reached, automatically guiding the fetch stage to maintain the desired slack.

Alternatively, slack fetch can also be implemented using the second instruction replication mechanism described earlier (see Input Replication in an SRT Processor, p. 232). By forwarding committed instructions from the leading thread to the fetch unit of the trailing thread, an inherent slack is introduced, such that cache misses incurred by the leading thread's instructions are already resolved before the corresponding trailing-thread loads probe the cache or LVQ, as the case may be. In this implementation, the trailing thread's instructions can be prioritized over that of the leading thread to help reduce pressure pipeline structures, such as the store queue.

Branch Outcome Queue

The branch outcome queue reduces misspeculation more directly and effectively than the slack fetch mechanism. This technique uses a hardware queue to deliver the leading thread's committed branch outcomes (branch PCs and target addresses) to the trailing thread. In the fetch stage, the trailing thread uses the head of the queue much like a branch target buffer, reliably directing the thread down the same path as the leading thread. Consequently, in the absence of faults, the trailing thread's branches never misfetch or mispredict and the trailing thread never misspeculates. To keep the threads in sync, the leading thread stalls in the commit stage if it cannot retire a branch because the queue is full. The trailing thread stalls in the fetch stage if the queue becomes empty.

Alternatively, if the committed instructions from the leading thread are forwarded to the fetch stage of the trailing thread, then one does not need a branch outcome queue. This is because the trailing thread receives the precise committed path of the leading thread, which achieves the same effect as the branch outcome queue.

6.9.7 Performance Evaluation of an SRT Processor

Mukherjee et al. evaluated the performance of an SRT implementation on a commercial SMT processor using all the 18 SPEC95 benchmarks [6]. They showed that on average SRT degrades performance over running just the single thread (without any redundant copies) by 32%. SRT techniques improve performance over running two redundant copies of the same program (without any input replication or output comparison) by 11%. This improvement is due to the positive effects of the LVQ and instruction replication in the SRT processor. The LVQ reduces data cache misses in two ways: the trailing thread cannot miss because it never directly accesses the cache and the leading thread thrashes less in "hot" cache sets because it does not compete with the trailing thread. They found that their SRT processor, on average, has 68% fewer data cache misses than the base processor running redundant copies of two threads.

Mukherjee et al. also found that output comparator for stores is one of the key bottlenecks in the SRT design. The store comparator increases the lifetime of a leading thread's stores, which must now wait for the corresponding stores from the trailing thread to show up before they can retire. On average, for one logical thread, the store comparator increased the lifetime of a leading thread's store by 39 cycles. Eighteen of these cycles represent the minimum latency for the trailing-thread store to fetch and execute; the extra 21 cycles came from queuing delays in the instruction replication mechanism and processor pipeline.

Consequently, increasing the size of the store queue has significant impact on performance because this allows other stores from the leading thread to make progress. A per-thread store queue (with 64 entries per thread) instead of a combined store queue for both leading and trailing threads improved performance by 4%, bringing the degradation to only roughly 30%. Completely eliminating the impact of the store comparator perhaps with an even bigger store queue would improve performance by another 5% and reduce the performance degradation to 26%.

In a multicore processor with SMT threads, turning on SRT also reduces the overall throughput of the processor itself since half the threads are running redundantly instead of contributing to the overall performance. By turning on SRT, the benefits of a second SMT thread running on the same core are lost. SRT recovers some of this performance loss because of reduced pressure on the memory system (since only one thread accesses memory) and branch predictor (since only one thread gets its predictions from the branch predictor). However, a better underlying SMT implementation will usually cause less penalty in latency (i.e., increase in execution time of a single program) from SRT. But a better SMT implementation would also cause higher degradation in throughput from a multicore SRT processor.

6.9.8 Alternate Single-Core RMT Implementation

Rotenberg's single-core RMT implementation (AR-SMT) design [11] is an alternate implementation of RMT within a single-core SMT processor. AR-SMT incorporates two redundant threads: the "active," or A-thread, and the "redundant," or R-thread. Committed register write-backs and load values from the A-thread are placed in a *delay buffer*, where they serve as the alternate execution stream against which R-thread results are checked and predictions to eliminate speculation on the R-thread. Thus, the delay buffer combines SRT's register check buffer and branch outcome queue. In addition, the R-thread uses the delay buffer as a source of value predictions to speculate past data dependencies.

AR-SMT is one point in the SRT design space; its sphere of replication is the SRT-Register's sphere in which the register file resides outside. In AR-SMT, the R-thread register file serves as the architectural file: register write-back values are verified before updating the R-thread registers, and the R-thread file is considered to be a valid checkpoint for fault recovery. As with the register files in SRT-Register, the A-thread register file serves only to bypass uncommitted register updates still in the delay buffer. Thus, replication does not provide fault coverage for the R-thread register file, so this register file must be augmented with an alternate coverage technique, such as ECC. Otherwise, a fault in an R-thread register value would lead to a mismatch in A-thread and R-thread results. AR-SMT would correctly detect this fault but may improperly recover by restarting from the corrupted R-thread register file contents.

Fundamentally, for fault detection using redundant computation, one needs two redundant computation units or threads and an output comparator. Typically, these are three distinct components of a fault detecting system. AR-SMT, however, combines the R-thread with the output comparator, which potentially saves hardware but results in reduced fault coverage compared to an SRT-style design.

AR-SMT also varies significantly from the SRT designs described above since in AR-SMT entire main memory is inside the sphere of replication. This scheme provides better memory fault detection than ECC. Nevertheless, doubling the physical memory of a system can be very expensive. Because the R-stream has a separate memory image distinct from that of the A-stream, AR-SMT requires modifications in the OS to manage the additional address mappings needed to replicate the address space. To make this replication simpler, the design disables redundant threading on OS calls, leaving kernel code vulnerable to transient hardware faults.

6.10 RMT in a Multicore Architecture

This section discusses how RMT can be implemented in a chip multiprocessor (CMP), more popularly known today as multicore processors. This is referred to as chip-level redundant threading or CRT. CRT achieves core Lockstepping's permanent fault coverage while maintaining SRT's low-overhead output comparison and efficiency optimizations. The basic idea of CRT is to generate logically redundant threads, as in SRT, but to run the leading and trailing threads on separate processor cores, as shown in Figure 6.13 [6,7].

The trailing threads' LVQs and branch outcome queues now receive inputs from leading threads on the other processor core. Similarly, the store comparator, which compares store instructions from redundant threads, receives retired stores from the leading thread on one processor core and trailing thread on another processor core. To forward inputs to the LVQ, the branch outcome queue, and the store comparator, moderately wide datapaths between the processors are needed. It may be possible to lay out the processor cores on the die such that such datapaths do not traverse long distances. These datapaths will be outside the sphere of replication and may require protection with some form of information redundancy, such as parity.

CRT processors provide two advantages over Lockstepped microprocessors. First, in Lockstepped processors, all processor output signals must be compared for mismatch, including miss requests from the data and instruction caches. This

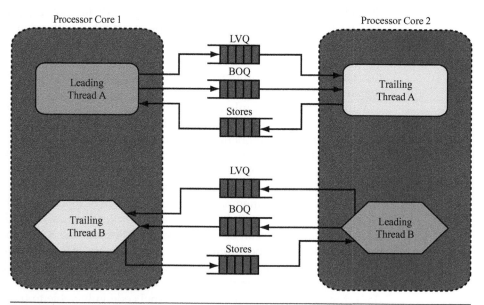

FIGURE 6.13   Block diagram of a CRT implementation using two cross-coupled processor cores. BOQ = branch outcome queue.

comparison is in the critical path of the cache miss and often adversely affects performance. More generally, the output comparator must interpose on every logical signal from the two processor cores, check for mismatch, and then forward the signal outside the sphere of replication. Of course, a CRT processor incurs latency to forward data to the LVQ, the branch outcome queue, or the store comparator, but these queues serve to decouple the execution of the redundant threads and are not generally in the critical path of data accesses.

Second, CRT processors can run multiple independent threads more efficiently than Lockstepped processors. By pairing leading and trailing threads of different programs on the same processor, the overall throughput can be maximized. A trailing thread rarely or never misspeculates, freeing resources for the other application's leading thread. Additionally, in our implementation, trailing threads do not use the data cache or the load queue, freeing up additional resources for leading threads. Evaluation of Mukherjee et al. shows that such a CRT processor performs similarly to core Lockstepping for single-program runs (in non-SMT mode) but can outperform Lockstepping by 13% on average (with a maximum improvement of 22%) for multithreaded program runs.

6.11 DIVA: RMT Using Specialized Checker Processor

As discussed earlier, in both the SRT and CRT implementations, the trailing thread consumes significantly less pipeline bandwidth than the leading thread. Austin took this idea further by designing a custom lightweight checker core that is paired up with a normal processor core. Austin calls this the *Dynamic Implementation Verification Architecture (DIVA)* [1].

To understand how DIVA works, let us look at the following sequence of two dynamic instructions: (Inst1) R1 = R2 + R3 and (Inst2) R5 = R1 + R4. The instruction Inst1—executed first—reads source registers R2 and R3, computes the sum of the values in the two registers, and then writes them back into destination register R1. The second instruction Inst2 does the same—reads values in source registers R1 (just produced) and R4, computes the sum of these values, and then writes them back into destination register R4. To verify that this sequence of instructions executes correctly, two properties must be ensured:

- Given the source register values, each add operation computes the result value correctly. For example, given the source register values for Inst2, it must be verified that the add operation actually computes the value in R5 correctly.

- The source register values flow correctly and without errors to every instruction. For example, it must be verified that Inst2 receives the correct values for R1 and R4. In a pipelined implementation with pipeline bypasses, the value in R1 may flow directly from Inst1 to Inst2 without going through the

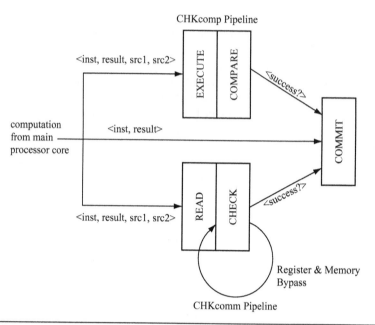

FIGURE 6.14 DIVA checker architecture.

architecture storage register R1. In this case, it must be verified that this bypass flow is correct.

Figure 6.14 shows how DIVA implements these two verification steps using separate checker pipelines: the CHKcomp pipeline that verifies the first property and the CHKcomm pipeline that verifies the second. Because the two pipelines are simpler than the main processor core, Austin argues that they can run at a much slower frequency, allowing a much simpler pipeline implementation than the main processor. Using a performance simulator of the DIVA architecture, Austin has also shown that on a set of selected SPEC benchmarks, the DIVA checker slows down the main processor core on average by only 3% (with a maximum degradation of 15%).

The CHKcomp Pipeline

The CHKcomp pipeline verifies that an instruction has been executed correctly by the main processor. For a given instruction executed in the main processor core, the CHKcomp pipeline receives the instruction, the source operand values, and the result operation. In the EXECUTE stage, the CHKcomp pipeline recomputes the result based on the source operand values it receives. Then, in the COMPARE stage, it compares the result value just generated with the result value it receives from the main processor. If the compared values match, then the DIVA checker can certify that the values have been computed correctly. If the compared values do not match due to a particle strike in either the main processor or the DIVA checker, it will be flagged as an error.

The CHKcomp pipeline does not have any dependences. It simply computes the result value based on the incoming source operand values. Consequently, as many CHKcomp pipelines can be instantiated as may be necessary to match the instruction throughput of the main processor. For example, if the main processor can commit four instructions in a DIVA checker cycle, one can potentially instantiate four CHKcomp pipelines to avoid slowing down the main processor. This allows the DIVA CHKcomp pipeline to match the main processor's bandwidth even with a much slower clock.

The CHKcomm Pipeline

The CHKcomm pipeline verifies that operand values flow correctly from one instruction to another. Like the CHKcomp pipeline, it also receives instructions with their source operands. In the READ stage, it reads the values of these source operands from its own architectural register file. The CHECK stage verifies that the source operands received and source operands read from the register file match. For example, for Inst1, the CHKcomm pipeline will read R2 and R3 from its own register file and verify that the incoming source operand values match these. Then, in the cycle after the READ stage, the CHKcomm pipeline will write back the result value into the destination register. For Inst1, the pipeline will write back R1's value it received from the main processor into the register file.

Inst2, however, poses a little complication since it must receive the value of R1 from Inst1. Hence, the pipeline needs to implement a bypass path, so that Inst1's result value is forwarded as a source operand for Inst2. Thus, the CHECK stage can receive the source operands either from the register file or from the bypass path. In either case, however, it simply compares the source operands while the write-back of the result operand proceeds in parallel.

Unlike the CHKcomp pipeline, replicating the CHKcomm pipeline poses some difficulties because of the bypass operations. To match the instruction bandwidth of the main processor, the CHKcomm pipeline can be replicated multiple times. But each of these pipelines must implement a complicated bypass network that allows values to be forwarded arbitrarily across any of these pipelines. For example, if the main processor can commit four instructions per DIVA checker cycle, then four CHKcomm pipelines can be implemented to avoid slowing down the main processor. However, if Inst1 is executed on CHKcomm pipeline 1 and Inst2 is executed on CHKcomm 2, then Inst1's result value must be forwarded from pipeline 1 to pipeline 2. In the general case, Inst1's result may need to be forwarded to all of the other three pipelines. Such bypass paths may make it challenging to scale the CHKcomm pipeline to match the main processor bandwidth.

Trade-offs in the DIVA Processor

The DIVA architecture poses two significant advantages beyond detecting transient faults. First, it can detect both permanent faults and design errors in the main processor. In contrast, in an SRT processor, permanent faults on structures shared between the redundant threads may go undetected. A CRT processor would detect such permanent faults, but since the redundant cores would typically be replicated, it will not be able catch design errors. Also, DIVA does not require an underlying SMT architecture like an SRT processor.

DIVA does, however, pose a few other challenges in its design. First, in the presence of I/O devices or multiple processors, the DIVA architecture can detect false transient faults due to external memory writes. On detecting a fault, the DIVA checker proposes to execute the instruction entirely on the DIVA checker itself. This property can potentially lead to a livelock if the external agent (I/O device or processor) continues to provide the same external memory write but with different values. It should be possible to augment DIVA with an ALAB or LVQ to avoid this problem.

Second, to guarantee forward progress in the presence of permanent faults in the main processor, DIVA assumes that the checker is always correct and proceeds using the checker's result in the case of a mismatch. Also, for uncached loads and stores that cannot be executed twice, DIVA relies on the checker to execute them nonredundantly. In these cases, faults in the checker itself—including transient faults—must be detected or avoided through alternative techniques. For example, Austin suggests that design errors could be avoided by formally verifying the checker. Alternatively, the DIVA checkers can be replicated and run in Lockstep to detect faults in the checker itself.

6.12   RMT Enhancements

Enhancements in the RMT design are an active area of research today. The primary goal of most RMT optimizations is to reduce the performance degradation of a single program caused by the redundant thread. Such enhancements may offer reduced fault coverage in some cases. That may still be worthwhile if the SER is within a product's SDC and DUE budgets.

An RMT implementation's performance degradation can be reduced using three broad techniques: relaxing constraints on input replication, relaxing constraints on output comparison, and partial RMT techniques that avoid executing some number of instructions in one of the redundant threads.

6.12.1   Relaxed Input Replication

Input replication, such as of that of load values, may sometimes prevent the trailing thread from making progress. This could be because the communication latency between the leading and the trailing threads is high. Alternatively, this could also be because the trailing thread starts to march ahead of the leading thread, which could happen in some RMT designs. Load value predictors—explored in microarchitecture research in the past decade—can help the trailing thread make progress without waiting for its inputs from the leading thread's LVQ, for example.

6.12.2 Relaxed Output Comparison

Output comparison necessitates committed instructions or selected instructions of both the threads to be available for comparison. This can cause performance degradation if this comparison causes one of the threads to be held up. The SRT and CRT implementations already relax this constraint somewhat by requiring output comparison only on selected instructions, such as stores, without increasing a processor's SDC rate.

Even in an SRT implementation, the leading thread can be held up by the trailing thread. This may disallow the leading thread to make further progress possibly because structures, such as the store queue, may have filled up. To allow the leading thread to make progress, an SRT implementation could choose to not compare stores when the leading thread cannot make progress. This would incur a higher SER, which could be computed through an AVF analysis.

Vijaykumar et al. [21] proposed and Gomaa et al. [4] refined a scheme called dependence-based checking elision (DBCE) to reduce the number of instructions that needs to be checked for faults. As instructions execute, a transient fault propagates through the instruction chains via either its control flow (e.g., branches) or its dataflow (e.g., registers). As instructions execute, the DBCE scheme builds short sequences of instructions to create these dependence chains. Given such a DBCE chain, only the last instruction in the chain needs to be checked for faults. Instructions that mask faults, such as OR instruction, cannot be part of such a chain unless they are the last instruction of a chain. Otherwise, they can mask faults, whose effect may not be captured by the last instruction in a DBCE chain. Alternatively, masking instructions can be allowed to be part of a chain but that will reduce the fault coverage of the RMT implementation.

Alternatively, a transaction-based implementation can offer significantly greater flexibility in fault detection but without sacrificing fault coverage. In a transaction-based system, a program is constructed as a series of transactions, where each transaction may consist of one or more instructions. The result of a transaction is committed to the global state atomically at the end of a transaction. By running transactions redundantly and comparing their outputs before the transaction commit point, one can detect faults in such transactions. This requires output comparison only when a transaction commits, thereby reducing the need for both redundant threads to be synchronized at every instruction.

6.12.3 Partial RMT

Another way to reduce performance degradation from an RMT implementation is to avoid executing selected regions of one of the redundant threads. By not executing instructions from one of the threads, for example, an SRT implementation can free up execution resources for the other thread. There have been a number of proposals that exploit this:

- **Instruction reuse.** Sodani and Sohi proposed instruction reuse to expedite the execution of a single program [16]. Sodani and Sohi created an instruction reuse buffer that tracks one or more instructions' input and output values. If an instruction or a sequence of instructions is executed again and can be matched against the instructions present in the reuse buffer, then the pipeline can simply obtain the output of the instructions without executing them. In an SRT implementation, the contents of the reuse buffer can be updated after the output comparator certifies that an instruction is fault free. In subsequent executions, the trailing thread can use values being passed to it by the leading thread (e.g., via an RVQ) to probe the reuse buffer, obtain the result values in case of a hit, and thereby avoid executing the instruction itself. Parashar et al. [8] and Gomaa and Vijaykumar [5] have explored variants of this scheme.

- **Avoid execution of dynamically dead instructions.** Parashar et al. [9] have proposed a scheme called SlicK that tracks backward slices of instructions in the leading thread to determine, and thereby eliminate, any dynamically dead instruction from the trailing thread. Recall that a dynamically dead instruction is one whose result will never be used by any instruction in future. For example, if register R1 is written by an instruction and subsequently overwritten by another instruction without any intervening writes to R1, then the first instruction that wrote R1 is dynamically dead. SlicK would potentially avoid executing any such dynamically dead instruction tracked through the architectural registers.

- **Store value prediction.** In the basic SlicK implementation, all stores are compared to ensure full fault coverage. Only the dynamically dead instructions are not executed. Parashar et al. [9] enhanced this scheme further by using store value prediction. The enhanced SlicK implementation avoided executing any instruction leading up to a store, whose value could be predicted by a store predictor.

- **Turn trailing thread off in high IPC regions.** Another way to reduce the performance degradation of RMT is to completely turn off the trailing thread in certain regions of a program. These regions could be regions with high IPC, low AVF, or high power dissipation. Gomaa and Vijaykumar [5] studied a scheme in which they turned off the trailing thread (thereby reducing fault coverage and increasing AVF) during regions of a program they found to have high IPC. The key challenge of such a scheme is to recreate the correct state of the trailing thread from which it can be restarted. Gomaa and Vijaykumar recreated this state by forwarding architectural state updates from the leading to the trailing thread. Alternatively, one could use an RVQ or an LVQ to recover the trailing thread's state as well since both these structures contain the updates to architectural state.

6.13 **Summary**

In fault detection via redundant execution, identical copies of the same program are executed redundantly. The outputs of the redundantly executing streams are compared for mismatch. A mismatch signals a fault and can initiate a number of possible recovery actions.

The concept of the sphere of replication helps explain how such redundant execution streams work. The sphere of replication identifies the logical domain protected by the fault detection scheme. Any component within the sphere must be logically or physically replicated. Any output leaving the sphere of replication must be compared to check for mismatch and corresponding faults. The output comparator has also been referred to in the literature as the *checker*. Any inputs into the sphere of replication must be appropriately replicated and delivered to the correct points within the sphere.

There are two ways of performing such redundant execution: cycle-by-cycle Lockstepping or RMT. Lockstepping has been used for decades in mainframe systems, whereas RMT has been introduced in the past decade. RMT has also been referred to as loose Lockstepping. In Lockstepping, redundant streams are usually run on two separate, but identical, processor cores. The processor cores must have the exact same state in each cycle. Consequently, output comparison involves comparing the values of signals coming from each processor core in the sphere of replication. Lockstepping does not require any semantic information from these signals. Because both processor cores have the same state in each cycle, any pair of hardware signals from the redundant processor cores must have the exact same value in the absence of a fault. Inputs into the sphere must be replicated and delivered to both processor cores at the same cycle boundary.

The Stratus ftServer, Hewlett-Packard Himalaya, and the IBM Z-series machines are all Lockstepped. Nevertheless, the designs differ significantly in their spheres of replication, which affects the hardware overhead incurred by each design. The ftServer includes main memory in its sphere of replication, which implies that main memory is replicated for each redundant execution stream. Himalaya's sphere of replication does not include main memory. Instead, Himalaya compares outputs arising from the processors themselves. The Z-series processors replicate processor pipelines on the processor chip itself. The processor pipelines constitute the sphere of replication in the Z-series processors.

In contrast, in RMT, only committed instructions from redundant streams are compared for mismatch. The underlying hardware contexts running the redundant threads can be either processor cores or hardware threads in a multithreaded processor. The hardware contexts running the redundant threads may not have the same state in the same cycle. What outputs must be compared depends on the size of the sphere. If the sphere of replication includes processor cores and memory, then only outputs to I/O devices must be compared for mismatch. Inputs must be replicated carefully in an RMT system since the underlying redundant contexts have different states in the same cycle. Inputs are typically delivered at the same committed instruction boundary, which ensures that the redundant execution streams do not diverge.

The Marathon Endurance Server and the Hewlett-Packard NSAA implement RMT with similar spheres of replication. Both include memory and processors in their spheres of replication and compare I/O outputs for mismatches. The machines, however, differ significantly in the way they do input replication. The Marathon Endurance Server raises an interrupt after a predetermined number of committed instructions. The two instruction streams coordinate and determine a point to replicate inputs. In NSAA, however, the instruction streams explicitly poll the machine for input replication requests and post requests to obtain the replicated inputs. This requires modifications to the OS, which is possible because the NSAA runs its own custom OS, unlike the Endurance Server that runs Windows, which is a commodity OS.

Other proposed RMT designs, such as SRT, CRT, and DIVA, implement RMT within a single chip. In SRT, the redundant threads are implemented within a single multithreaded processor core, whereas in CRT the redundant threads are executed on different processor cores on the same chip. In contrast, DIVA creates a specialized checker core to act as one of the redundant threads. Each of these designs offers different trade-offs and hardware overhead.

6.14 Historical Anecdote

Contributed by Dr. Phil Emma of IBM Corporation:

In a bold move in the early 1990s, IBM transitioned mainframe design over to CMOS. For the first time, IBM was going to build its large S390 Servers (now called "series") using a new custom CMOS microprocessor. The internal name for this system was "Alliance," signifying the joint "alliance" formed between research engineers and development engineers to create this processor.

I was given the role of leading the specification and design of the "R unit," which was an entirely new unit that took a new approach toward RAS within the new context of CMOS. S390 historically featured "bulletproof" RAS required by many mainframe customers and particularly those in financial businesses. Our goal was to make each server generation more reliable than the previous generation. We wanted to catch as many faults as possible and to recover from as many errors as possible.

Historically, in our bipolar designs, checking circuitry was integral to the computation circuitry. Most combinational logic had integral parity prediction, and most state logic was encoded using fail-safe techniques and state checking. There were many registers put in for retaining state (to recover to), and all state was scannable. In those machines, the timing was not dominated by wire paths, and the powerful ECL logic was not very sensitive to capacitive loading, so these well-evolved techniques made sense.

CMOS, however, is extremely sensitive to loading, and in the new custom design, wire lengths were critical. Therefore, in a radical departure from what we had done in the past, we decided to merely replicate the processor (but not the cache). The pair of processors were operated as a single logical processor. A new "R unit" verified—on a per-instruction basis—that both processors produced the same results. The R unit also maintained a "golden" (ECC-protected) checkpoint of all architected state. The R unit could effect recovery actions and log errors when they were found. It also contained lots of other odds and ends, time-of-day clocks, timers, service interfaces, trace arrays, etc.

The R unit was designed as a fully replicated and fully cross-checked "dual pipeline." That is, in addition to doing bulletproof checking of the rest of the machine, it also did a bulletproof checking of itself. We estimated that the chip area overhead for having a redundant processor and an R unit was about 30%. By not putting parity prediction circuits into the processors, we kept their areas (hence wire-bound paths) and capacitive loadings small. This gave us a significant performance advantage for nearly the same area penalty, and an arguably smaller design effort.

People have conjectured why this was called the "R unit." We do not know for sure. Charles Webb, the chief architect of the machine, started calling it the "R unit" from day 1. "R" never stood for anything. It was just a letter (like the I and E units). Many years after the fact, I have heard sales and marketing people say that it was the "Register" unit (because of all the checkpointing), the "Recovery" unit, or the "Reliability" unit. In reality, it was just the "R" unit, and "R" did not stand for anything.

Vijay Lund was a rising-star executive in charge of the project. Initially, it was clear that he was a little uncomfortable with having research engineers embedded in the development design team and more than a little uncomfortable (in those days) with having a researcher like me leading part of the design. Given our cultural differences, I fully understood this, and it was not an unreasonable point of view. But what was especially perplexing to all executives—and to Vijay—was that although we had two processors in there, we were using them like they were a single processor. Most executives had a very hard time accepting this as being the right approach.

Every time I bumped into Vijay (and most other executives), I would be asked how I knew that we are going to have errors at all in the silicon. This was a new technology. We had no evidence that we would have errors at all. Would it not be better to just forget about the R unit and run the chip as a dual-core chip? I would always tell them that they should think about RAS exactly like they thought about term life insurance. If they are extremely lucky, then it is a total waste of their money. Nonetheless, every time I would see Vijay, he would ask the same question. I think that he enjoyed teasing me to keep me on my toes.

One day I came into Vijay's conference room to give my weekly status report to the management team. At the end of my presentation, Vijay said that he had finally figured out what "R" stood for. So I asked "What does 'R' stand for?" He said "It means Removable." As I said before, Vijay knew how to keep a team on its toes.

References

[1] T. M. Austin, "DIVA: A Reliable Substrate for Deep Submicron Microarchitecture Design," in *32nd Annual International Symposium on Microarchitecture (MICRO)*, pp. 196–207, 1999.

[2] D. Bernick, B. Bruckert, P. D. Vigna, D. Garcia, R. Jardine, J. Klecka, and J. Smullen, "NonStop® Advanced Architecture," in *Proceedings. International Conference on Dependable Systems and Networks (DSN)*, pp. 12–21, Yakohama, Japan, June/July 2005.

[3] T. D. Bissett, P. A. Leveille, E. Muench, G. A. Tremblay, "Loosely-Coupled, Synchronized Execution," United States Patent 5,896,523, issued April 20, 1999.

[4] M. A. Gomaa, C. Scarbrough, T. N. Vijaykumar, and I. Pomeranz, "Transient Fault-Recovery for Chip Multiprocessors," in *Proceedings of 30th Annual International Symposium on Computer Architecture (ISCA)*, pp. 98–109, June 2003.

[5] M. A. Gomaa and T. N. Vijaykumar, "Opportunistic Fault Detection," in *32nd Annual International Symposium on Computer Architecture (ISCA)*, pp. 172–183, Madison, Wisconsin, USA, June 2005.

[6] S. S. Mukherjee, M. Kontz, and S. K. Reinhardt, "Detailed Design and Evaluation of Redundant Multithreading Alternatives," in *Proceedings of the 29th Annual International Symposium on Computer Architecture (ISCA)*, pp. 99–110, Anchorage, Alaska, USA, May 2002.

[7] R. Nair and J. E. Smith, "Method and Apparatus for Fault-Tolerance Via Dual Thread Crosschecking," United States Patent Application, publication date September 19, 2002.

[8] A. Parashar, S. Gurumurthi, and A. Sivasubramaniam, "A Complexity-Effective Approach to ALU Bandwidth Enhancement for Instruction-Level Temporal Redundancy," in *31st Annual International Symposium on Computer Architecture (ISCA)*, pp. 376–386, June 2004.

[9] A. Parashar, S. Gurumurthi, and A. Sivasubramaniam, "SlicK: Slice-Based Locality Exploitation for Efficient Redundant Multithreading," in *12th Annual International Conference on Architectural Support for Programming Languages and Operating Systems (ASPLOS)*, pp. 95–105, October 2006.

[10] S. K. Reinhardt and S. S. Mukherjee, "Transient Fault Detection via Simultaneous Multithreading," in *27th Annual International Symposium on Computer Architecture (ISCA)*, pp. 25–36, Vancouver, British Columbia, Canada, USA, June 2000.

[11] E. Rotenberg, "AR-SMT: A Microarchitectural Approach to Fault Tolerance in Microprocessors," in *29th Annual Fault-Tolerant Computing Systems (FTCS)*, p. 84, Madison, Wisconsin, USA, June 1999.

[12] D. P. Sieiorek and R. S. Swarz, *Reliable Computer Systems: Design and Evaluation*, A. K. Peters, 1998.

[13] T. J. Slegel, R. M. Averill III, M. A. Check, B. C. Giamei, B. W. Krumm, C. A. Krygowski, W. H. Li, J. S. Liptay, J. D. MacDougall, T. J. McPherson, J. A. Navarro, E. M. Schwarz, K. Shum, and C. F. Webb, "IBM's S/390 G5 Microprocessor Design," *IEEE Micro*, pp. 12–23, March/April 1999.

[14] T. J. Slegel, E Pfeffer, and J. A. Magee, "The IBM eServer z990 Microprocessor," *IBM Journal of Research and Development*, Vol. 48 No. 3/4, pp. 295–309, May/July 2004.

[15] J. E. Smith and A. R. Pleszkun, "Implementing Precise Interrupts in Pipelined Processors," *IEEE Transactions on Computers*, Vol. 37, No. 5, pp. 562–573, May 1988.

[16] A. Sodani and G. S. Sohi, "Dynamic Instruction Reuse," in *24th Annual International Symposium on Computer Architecture (ISCA)*, pp. 194–205, Denver, Colorado, USA, June 1997.

[17] D. M. Tullsen, S. J. Eggers, J. S. Emer, H. M. Levy, J. L. Lo, and R. L. Stamm, "Exploiting Choice: Instruction Fetch and Issue on an Implementable Simultaneous Multithreading Processor," in *23rd Annual International Symposium on Computer Architecture (ISCA)*, pp. 191–202, May 1999.

[18] D. M. Tullsen, S. J. Eggers, and H. M. Levy, "Simultaneous Multithreading: Maximizing On-Chip Parallelism," in *22nd Annual International Symposium on Computer Architecture (ISCA)*, pp. 392–403, Italy, June 1995.

[19] J. Somers, "Stratus ftServer—Intel Fault Tolerant Platform," Intel Developer Forum, Fall 2002.

[20] L. Spainhower and T. A. Gregg, "IBM S/390 Parallel Enterprise Server G5 Fault Tolerance: A Historical Perspective," *IBM Journal of Research and Development*, Vol. 43, No. 5/6, pp. 863–873, September/November 1999.

[21] T. N. Vijaykumar, I. Pomeranz, and K. Cheng, "Transient Fault Recovery using Simultaneous Multithreading," in *Proceedings of the 29th Annual International Symposium on Computer Architecture (ISCA)*, May 2002.

[22] C. Webb, "z6—The Next-Generation Mainframe Microprocessor," *Hot Chips*, August 2007.

[23] A. Wood, R. Jardine, and W. Bartlett, "Data Integrity in HP NonStop Servers," in *2nd IEEE Workshop on Silicon Errors in Logic and System Effects (SELSE)*, Urbana-Champaign, April 2006.

7

Hardware Error Recovery

7.1 Overview

As discussed in Chapter 1, soft errors can be classified into two categories: SDC and DUE. Some of the error coding techniques, such as parity, discussed in Chapter 5 and the fault detection techniques outlined in Chapter 6 convert SDC to DUE. In some cases, these techniques may result in additional DUE rates arising from the mechanics of the fault detection technique. This chapter discusses additional hardware recovery techniques that will reduce both SDC and DUE rates arising from transient faults. The same recovery techniques may not be as useful for permanent hardware faults because fixing the permanent fault typically would involve a part replacement, which is not discussed in this chapter.

This chapter uses the term *error recovery* instead of fault recovery, even though the process of detecting a malfunction was referred to as fault detection. This is because a hardware fault is a physical phenomenon involving a hardware malfunction. The effect of the fault will typically propagate to a boundary of a domain where the fault will be detected by a fault detection mechanism. Once a fault is detected, it becomes an error. Thus, a recovery mechanism will typically help a system recover from an error (and not a fault).

The reader should also note that an error recovery mechanism reduces the SDC and DUE rates of the *domain* it is associated with. For example, a domain could simply include the microprocessor chip. Alternatively, a domain could include an entire system. (For a detailed discussion on how faults and errors relate to the domain of fault detection and error recovery, the reader is referred to the section Faults, p. 6, Chapter 1.) To characterize the SDC or DUE rate of a domain, one can compute the domain's MTBF as the sum of its MTTF and MTTR. At the system level, one often talks about availability, which is the ratio of system uptime (MTTF) divided by total system time (MTBF). Availability is often expressed in *number of 9s*. For example, five 9s would mean that the system availability is 99.999%. This would mean a system downtime of 5.26 minutes per year. Similarly, six 9s would mean a system downtime of 31.56 seconds.

Recovery schemes can be broadly categorized into forward and backward recovery schemes. In forward error recovery, the system continues fault free execution from its current state even after it detects a fault. This is possible because forward recovery schemes maintain concurrent and replicated state information, which allows them to execute forward from a fault free state. In contrast, in a backward error recovery, usually the state of the machine is rewound backward to a known good state from where the machine begins execution again.

This chapter discusses various forms of forward and backward error recovery schemes. Forward recovery schemes discussed in this chapter include fail-over systems that fail over to a standby spare, DMR systems that run two copies of the same program to detect faults, TMR systems that run triplicate versions of the same program, and pair-and-spare systems that run a pair of DMR systems with one being the primary and the other secondary standby.

The design of a backward error recovery scheme depends on where the fault is detected. If a fault is detected before an instruction's result register values are committed, for example, then the existing branch misprediction recovery mechanism can be used to recover from an error. This chapter examines backward recovery schemes in various systems: systems that detect faults before register values are committed, systems that detect faults before memory values are committed, and systems that detect faults before I/O outputs are committed.

7.2 Classification of Hardware Error Recovery Schemes

Fundamental to any error recovery mechanism is the "state" to which the system is taken when the error recovery mechanism is triggered. For example, Figure 7.1 shows state transition of a system consisting of two bits. The initial state of the system is 00. It goes through two intermediate states—01 and 10. Eventually, it reaches the state 11 when it gets a particle strike. This changes the state incorrectly to 01. If this incorrect state transition can be detected, then the error can be flagged. Also, it is assumed that the identity of the bit that was struck is unknown, so the state cannot be reconstructed. Then, to recover from the error, there are several choices that are described below.

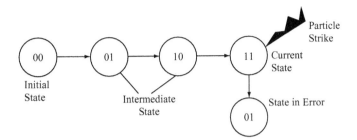

FIGURE 7.1 State to recover to on a particle strike. The system can revert to the initial state (reboot) or the intermediate state (backward error recovery) or continue forward from its current state (forward error recovery).

7.2.1 Reboot

The system can be reverted to its initial state, and execution can be restarted. This may require rebooting a system. In Figure 7.1, this would correspond to reverting to the initial state 00. This is a valid recovery mechanism for transient faults, if the latency of reexecution is not critical. For hard errors, however, this mechanism may not work as well since on reexecution the system may get the same error again.

7.2.2 Forward Error Recovery

The second option would be to continue the system from its current state after a fault is detected. In Figure 7.1, this would correspond to the current state 11 when the fault is encountered. This style of error recovery is usually known as *forward error recovery*. The key in forward error recovery is to maintain redundant information that allows one to reconstruct an up-to-date error-free state. ECC is a form of forward error recovery in which the state of the bits protected by the ECC code is reconstructed, thereby allowing the system to make forward progress.

Forward error recovery of full computing systems, including logic and ALU components, however, can be much more hardware intensive since this may use redundant copies of the entire computing state. An output comparator would identify the faulty component among the redundant copies. The correct state can be copied from one of the correct components to the faulty one. Then the entire system can be restarted.

This could be possible, for example, if there are triply redundant copies of the execution system and state. Before committing any state outside the redundant copies, the copies can vote on the state. If one of them disagrees, then the other two are potentially correct. The faulty copy can be disabled, and the correct ones could proceed. For transient faults, the faulty copy can be scrubbed of any faults and reintegrated into the triply redundant system. For permanent faults, the faulty component may have to be replaced with a fault free component.

7.2.3 Backward Error Recovery

Backward error recovery is an alternate recovery scheme in which the system can be restored to and restarted from an intermediate state. For example, in Figure 7.1, the system could revert to the state 01 or 10. Backward recovery often requires less hardware than forward error recovery mechanisms. Backward error recovery does, however, require saving an intermediate state to which one can take the system back to when the fault is detected. This intermediate saved state is usually called a *checkpoint*. How checkpoint creation relates to fault detection, when outputs can be committed, when external inputs can be consumed, and how the granularity of fault detection relates to recovery are discussed below.

How Fault Detection Relates to Checkpoint Content

Broadly, there are three kinds of states in a computer system: architectural register files, memory, and I/O state. A checkpoint can comprise one or more of these states. Typically, in a computer system and particularly in a microprocessor, a register file is small and frequently written to. Memory is significantly bigger and committed to either when a store instruction executes and commits its value to memory or an I/O device transfers data to a pre-allocated portion of memory. Finally, the I/O state (e.g., disks) is usually the biggest state in a computer system and usually committed to less frequently than register files or main memory.

What constitutes a checkpoint in a system also depends on where the fault detection point is with respect to when these states are committed. This chapter describes four styles of backward error recovery depending on where the fault detection occurs. These include system where

- Fault detection before any state—register, memory, or I/O state—is committed.

- Fault detection after register state is committed but before memory or I/O state is committed.

- Fault detection after register and memory states are committed but before I/O state is committed.

- Fault detection after I/O state is committed.

Output and Input Commit Problems

A system with backward error recovery must also be careful about the output and input commit problems. A recovery scheme can only recover the state of a system within a certain boundary or a sphere of recovery. For example, assume that a sphere of recovery consists only of a processor chip where the fault detection occurs before memory or I/O is committed to. The checkpoint in such a recovery scheme may consist only of processor registers. On detecting an error, the processor reloads its entire state from the checkpoint and resumes execution.

The output commit problem arises if a system allows any output, which it cannot recover from, to exit the sphere of recovery. In the example just described, the processor cannot recover from memory writes it propagates to main memory or I/O operations it propagates to disks because the checkpoint it maintains does not comprehend memory or I/O state. Consequently, to avoid the output commit problem, this system cannot allow any corrupted store or I/O operation to exit the sphere of recovery until it has certified that all prior operations leading to the store or I/O operation are fault free.

Similarly, the input commit problem arises if a system restarts execution from a previous checkpoint but cannot replay inputs that had arrived from outside the sphere of recovery. During the course of normal operation in the example that was just described, a processor will receive inputs, such as load values from the memory system and external I/O interrupts. When the processor rolls back to a previous checkpoint and restarts execution, it must replay all these events since these events may not arrive again. How the backward recovery schemes solve the output and input commit problems is described in this chapter.

Granularity of Fault Detection

To only detect faults in a given domain and reduce the SDC rate, it is often sufficient to do output comparison at a coarser granularity. For example, to prevent corrupted data to exit a processor, one can compare selected instructions, such as stores, that update memory or I/O instructions, such as uncached loads and stores. Once one detects such a fault, one can halt the system and prevent any SDC from propagating to memory or disks. The same granularity of comparison may work for forward error recovery. In a triply redundant system, the component that produces the faulty store output can be identified by the output comparator, isolated, and then restarted using correct state from one of the other two components.

In contrast, the same granularity of fault detection may not work for backward error recovery. If a store output mismatch from two redundant copies is detected, it is not known which one is the faulty copy. Further, checkpoint may not be fault free since the checkpoint may consist of the architectural registers in a processor. A faulty instruction can update the architectural registers, but its effect may show up much later through a store at the output comparator. If the processor is rolled back to a previous architectural checkpoint, it cannot be guaranteed that the checkpoint itself is fault free (Figure 7.2). Hence, backward error recovery to reduce DUE necessitates fault detection at a finer granularity than what is simply needed to prevent SDC.

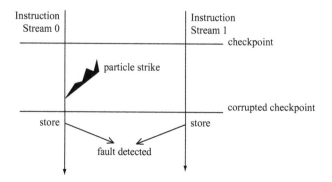

FIGURE 7.2 Granularity of fault detection in a backward error recovery executing redundant instruction streams. If only store instructions are compared, then the fault can be eventually detected. But the checkpoint prior to where the fault is detected may already have been corrupted. If the instruction stream reverts to the last checkpoint after the fault is detected, then it may not be able to correct the error.

This chapter discusses several examples of backward error recovery schemes that will highlight different aspects of the relationship between fault detection and checkpoint creation, output and input commit problems, and granularity of fault detection.

7.3 Forward Error Recovery

Forward error recovery schemes allow a system to proceed from its current state once the system detects a fault. Four styles of forward error recovery schemes are discussed in this section: fail-over systems, DMR systems, TMR systems, and pair-and-spare systems.

7.3.1 Fail-Over Systems

Fail-over systems typically consist of a mirrored pair of computing slices: a primary slice and a standby slice. The primary executes applications until it detects a fault. If the system can determine that the fault has not corrupted any architectural state, then it can copy the state of the primary slice to the standby. The standby then takes over and continues execution. This is often possible if the effect of a fault is limited to a single process. The entire system can continue execution from the point the fault is detected, but the failed process may need to restart from the beginning. The early fault-tolerant computing systems built in the 1960s and 1970s by IBM, Stratus, and Tandem were primarily fail-over systems. Marathon's recent Endurance server used fail-over principles to recover from hardware errors (see RMT in the Marathon Endurance Server, p. 223, Chapter 6).

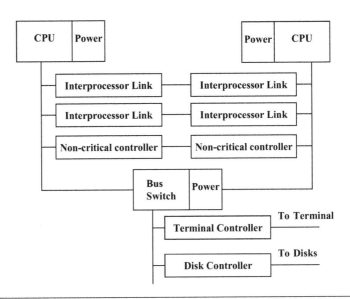

FIGURE 7.3 An example of an early fail-over, fault-tolerant computer system from Tandem [2].

Figure 7.3 shows an example of an early fail-over system from Tandem Computer systems. Such a configuration had replicated processors or CPUs, power supplies, interprocessor links, etc. in each slice of the fail-over system. A bus switch monitored the primary slice for faults. If it found a fault, then it would copy the state to the standby and resume execution.

Fail-over systems are often good at recovering from software bugs or system hangs but may have trouble recovering from a transient fault that corrupts the architectural state of the machine. In such cases, the standby can take over but may not be able to guarantee forward error recovery because the state of the primary slice has already been corrupted. Hence, the system will have to be rebooted, and all applications will be restarted. The next three mechanisms described in forward error recovery try to address this problem and provide better forward error recovery guarantees.

7.3.2 DMR with Recovery

DMR systems can be designed to recover from transient faults. Typically, DMR systems can detect faults using Lockstepping or RMT (see Chapter 6). An output comparator will compare the outputs of two redundantly executing instruction streams. An output mismatch will indicate the presence of a fault (Figure 7.4a). In some cases, an output mismatch can also occur due to the two slices taking correct but divergent paths. Stratus ftServer and Marathon Endurance machines both support a DMR mode with recovery.

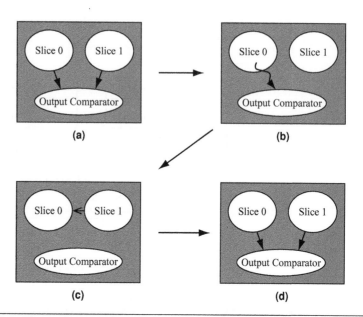

FIGURE 7.4 Flow of operations in a DMR machine. (a) Normal operation with dual systems whose outputs are checked by an output comparator. (b) An internal error checker in System 0 fires, and the signal propagates to the output comparator. (c) The entire system is frozen. The entire system state is copied from System 1 to System 0. (d) Normal operation is resumed.

Once an output mismatch is caught, the output comparators can wait for the arrival of an internal error signal (Figure 7.4b). For example, when a bit in Slice 0 flips in a structure protected with parity, Slice 0 can propagate the parity error signal to the output comparator. Without this signal, the output comparator cannot determine the slice is in error, although it detects the existence of a fault. After determining the slice is in error, the output comparator can initiate a copy of the internal machine state from Slice 1 to Slice 0 (Figure 7.4c). This ensures that both slices have the same state. Then it can resume execution (Figure 7.4d).

Such a scheme cannot recover from all errors. How much this scheme reduces the DUE depends on how much internal error checking one has in each slice. Also, in Lockstepped DMR systems, the false DUE can be particularly high because of timing mismatches. In such a case, a recovery mechanism, such as this one, can be beneficial.

7.3.3 Triple Modular Redundancy

A TMR system provides much lower levels of DUE than a DMR system. As the name suggests, a TMR system runs three copies of the same program (Figure 7.5) and compares outputs from these programs. The fault detection technique itself

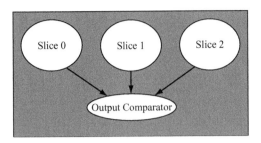

FIGURE 7.5 A TMR system. The output comparator compares outputs from three copies of the same program.

could be either cycle-by-cycle Lockstepping or RMT (see Chapter 6). If the output comparison logic (often called a *voter* for a TMR system) determines that outputs from the three slices executing the same program redundantly are the same, then there is no fault and the output comparison succeeds.

If the output comparison logic finds that only two of the three outputs match, then it will signal that the slice producing the mismatched output has experienced a fault. Typically, the TMR system will disable the slice in error but let the other two slices continue execution in a degraded DMR mode. The TMR system now has a couple of choices to bring online the slice in error. One possibility would be for the TMR system to log a service call for a technician to arrive at the site to debug the problem and restart the system.

Alternatively, it can try to fix the error and bring the slice in error into the TMR domain again. For transient faults, this would be the ideal solution. To achieve this, the degraded DMR system will be halted at an appropriate point, and the entire state of the correct slices would be copied to the one in error. Then, the TMR system can resume execution. The system will come back up automatically, but during this state copy, the entire TMR system may be unavailable to the user. This would add to the downtime of a TMR system and may reduce its availability.

The time to copy state from a correct slice of the TMR machine to the incorrect slice can be prohibitive, particularly if the TMR system copies main memory, as in Stratus' ftServer or Hewlett-Packard's NSAA (see Chapter 6). To reduce this time, the NSAA provides two special hooks for this reintegration. First, it provides a direct ring-based link between the memories used in the three slices of the TMR system. This allows fast copying during reintegration. Figure 7.6 shows a picture of this direct connection. The LSUs (described in Chapter 6) serve as the output comparators and input replicators for the TMR system. Memory in any slice can receive updates to it either from its local processor or from another slice coming from reintegration link.

Second, the NSAA allows execution to proceed while the underlying system copies memory state from the correct to the faulty slice, thereby reducing downtime. The degraded DMR system—consisting of the correct pair of working slices—can continue to generate writes to main memory while this copy is in progress.

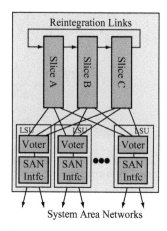

FIGURE 7.6 Reintegration in the Hewlett-Packard NSAA design. Reprinted with permission from Bernick et al. [4]. Copyright © 2005 IEEE.

Because the reintegration links connect to the path between the chipset and main memory, all writes to the main memory from a slice are forwarded on the reintegration links and copied to the slice being reintegrated. Eventually, the processors in the DMR system will flush their caches. Write-backs from the caches will again be forwarded to the reintegration links and copied over. Finally, the processors will be paused, and their internal state copied over to a predefined region in memory. The reintegrated processor will receive these updates and copy the state from memory to the appropriate internal registers. Thereafter, the TMR system can resume execution.

7.3.4 Pair-and-Spare

Like TMR, pair-and-spare is a classical fault-tolerance technique, which maintains a primary pair of slices and a spare pair as standby (Figure 7.7). The pairs themselves are used for fault detection with a technique such as Lockstepping or RMT. The spare pair receives continuous updates from the primary pair to ensure that the spare pair can resume execution from the point before the primary failed. This is termed *forward error recovery* because the spare pair does not roll back to any previous state. This saves downtime by avoiding any processor freeze to facilitate reintegration of the faulty slice. But it requires more hardware than TMR.

In the Tandem (and now Hewlett-Packard) systems, a software abstraction in its OS—called the NonStop kernel—called *process pair* facilitated the implementation of pair-and-spare systems [3]. A process pair refers to a pair of logically communicating processes. Each process in turn runs on a pair of redundantly executing Lockstepped processors. The process pair abstraction allows the two communicating logical processes of the pair to collectively represent a named resource, such as

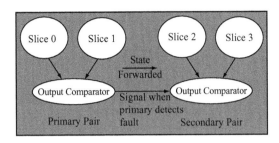

FIGURE 7.7 System with pair-and-spare.

a processor. One process functions as the primary unit at any point in time and sends necessary state to the other logical process. If the primary process detects a fault, the spare process would take over. The NonStop kernel would transparently redirect requests to the spare without the application seeing an error. The NonStop kernel takes special care to ensure that the named resource table that provides this redirection transparency is fault free.

7.4 Backward Error Recovery with Fault Detection before Register Commit

Backward error recovery often requires less hardware than forward error recovery schemes described in the previous section. This is because in a backward error recovery, as each instruction executes, the system does not have to maintain an up-to-date fault free state. Maintaining this state—beyond what is required for fault detection only—can be quite hardware intensive (e.g., as in TMR). Instead, in a backward error recovery, the system keeps checkpoints to which it can roll back in the event of an error. A checkpoint can be more compact and less hardware intensive than maintaining a full-blown slice of the machine (e.g., as done in TMR).

This section discusses recovery and checkpointing techniques when the fault detection happens before register values are committed by a processor. Subsequent chapters examine how the sphere of recovery can be expanded by doing the fault detection before memory values or I/O values are committed. If fault detection occurs prior to committing register values, then a backward error recovery scheme can make use of the state that a modern processor already keeps to recover from a misspeculation, such as branch misprediction.

Broadly, such backward error recovery techniques can be classified into two categories: one that requires precise fault detection and one that detects faults probabilistically. Four techniques that require precise detection are discussed: Fujitsu SPARC64 V's parity with retry, the IBM Z-series Lockstepping with retry, simultaneous and redundantly threaded processor with recovery (SRTR), and chip-level redundantly threaded processor with recovery (CRTR). For probabilistic techniques, two proposals are discussed: exposure reduction with pipeline squash and fault screening with pipeline squash and reexecution. SRTR, CRTR, and the probabilistic mechanisms are still under investigation and have not been implemented in any commercial system.

The output commit problem is solved relatively easily in these systems, but the input commit problem requires careful attention. The processor does not commit any value before the values are certified to be fault free. Hence, there is no danger of exposing a fault free state that cannot be recovered from. This solves the output commit problem. The input commit problem can arise from four sources: architectural register values, memory values, I/O devices accessed through uncached accesses (e.g., uncached loads), and external interrupts. Architectural registers are idempotent and cannot be directly changed by another processor or I/O device, so they can be reread a second time after the processor recovers. Memory values can be changed by an I/O device or by another processor, but they can be reread without loss of correctness (even if the value changes). Uncached loads must be handled carefully because an uncached load may return a value that cannot be replayed a second time. Hence, uncached loads are handled carefully by these systems. The addresses are typically compared and certified fault free before the uncached load is issued into the system. Finally, interrupts are only delivered at a committed instruction boundary, so that interrupts do not have to be replayed.

7.4.1 Fujitsu SPARC64 V: Parity with Retry

Perhaps the simplest way to implement backward error recovery is to reuse the checkpoint information that a processor already maintains. Typically, to keep a modern processor pipeline full, the processor will predict the direction a branch instruction will take before the branch is executed. For example, in Figure 7.8, the SPARC64 V processor [1] will predict the direction of the branch in the Fetch stage of the pipeline, but the actual outcome of the branch will not be known until the Execute stage.

Because these branch predictions can sometimes be incorrect, these processors have the ability to discard instructions executed speculatively (in the shadow of the mispredicted branch) and restart the execution of the pipeline right after the last instruction that retired from the pipeline. To allow restart, a pipeline would checkpoint the architectural state of the pipeline every time a branch is predicted. Then, when a branch misprediction is detected, the transient microarchitectural state of the pipeline can be discarded and execution restarted from the checkpoint associated with the branch that was mispredicted. More aggressively, speculative microprocessors, such as the SPARC64 V, checkpoint the state every cycle, thereby having the ability to recover from any instruction—whether it is a branch or not.

Every time a SPARC64 V instruction is executed speculatively, its result is written to the reorder buffer for use in other instructions. The reorder buffer acts as a buffer that maintains the program order in which instructions retire, as well as the values generated speculatively by each instruction. Once an instruction retires, it is removed from the reorder buffer, and its register value is committed to the register

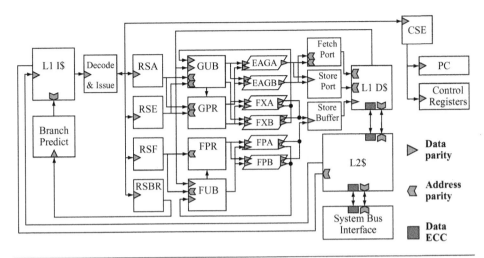

FIGURE 7.8 Error checkers in the Fujitsu SPARC64 V pipeline. Instructions are fetched from the L1 instruction cache (I$) with the help of the branch predictor. Then they are sent to the I-Buf. Thereafter, they are decoded and sent to the reservations stations (RSA, RSE, RSF, and RSBR) for dispatch. Registers GUB, GPR, FPR, and FUB are read in the Reg-Read stage. Instructions are executed in the Execute stage. EAGA, EAGB, FXA, FXB, FPA, and FPB represent different execution units. The memory stage has a number of usual components, such as store buffer, L1 data cache, L2 cache. Reprinted with permission from Ando et al. [1]. Copyright © 2003 IEEE.

files GPR (for integer operations) and FPR (for floating-point operations). Thus, the architecture register file automatically contains checkpointed states up to the last committed instruction. This way when a misspeculation is detected, the pipeline flushes intermediate states in the reorder buffers (as well as other more obscure states) and returns to correct state in one cycle.

The same mechanism can be used to recover from transient faults. The Fujitsu SPARC64 V processor (Figure 7.8) uses such a mechanism. It protects microarchitectural structures, address paths, and datapaths with parity codes, and ALUs and shifter with parity prediction circuits. Parity codes and parity prediction circuits can detect faults but not correct errors. When a parity error is detected, this pipeline prevents the offending instruction from retiring, throws away any transient pipeline state, and restarts execution from the instruction after the last correctly retired instruction.

7.4.2 IBM Z-Series: Lockstepping with Retry

Compared to the Fujitsu SPARC64 V, the IBM Z-series processors are more aggressive in their DUE reduction through the use of Lockstepping with retry. The Z-series processors (until the z990) implement Lockstepped pipelines that detect transient faults before any instruction's result is committed to the architectural register file (see Lockstepping in the IBM Z-series Processors, p. 220 in Chapter 6).

The Z-series uses three copies of the register file. One copy each is associated with each of the Lockstepped pipelines. As instructions retire, speculative register state is written into these register files, so that subsequent instructions can read their source operands and make progress. The third copy is the architectural register file to which results are committed if and only if the Lockstep comparison indicates no error. This third copy is protected with ECC because its state is used to recover the pipeline when a fault is detected.

If the Lockstepped output comparator indicates an error, then the pipeline state is flushed and architectural register file state is loaded back in both the pipelines. Execution restarts from the instruction after the last correctly retired instruction. The Z-series also has other mechanisms to diagnose errors and decide if the error was a permanent or a soft error. The Z-series processors are used in IBM's S/390 systems that have both system-level error recovery and process migration capabilities.

7.4.3 Simultaneous and Redundantly Threaded Processor with Recovery

An SRTR processor [21] is an enhancement of an SRT processor that was discussed in Chapter 6 (see RMT Within a Single-Processor Core, p. 227). Instead of using Lockstepping as its fault detection mechanism, like the IBM Z-series machines, an SRTR processor uses an RMT implementation called SRT as its fault detection mechanism. An SRT processor was designed to prevent SDC but not recover from errors and reduce DUE. Hence, an SRT processor need not compare the output of every instruction. An SRT processor only needed to do its output comparison for selected instructions, such as stores, uncached loads and stores, and I/O instructions, to prevent any data corruption to propagate to memory or disks.

The SRTR mechanism enhances an SRT processor with the ability to recover from detected errors using the checkpointing mechanism that already exists in a modern processor today. This requires two key changes to an SRT processor. First, to have an error-free checkpoint, the granularity of fault checking must reduce to *every* instruction, unlike the SRT version that checks for faults on selected instructions (see Figure 7.2). Second, the fault check must be performed *before* an instruction commits its destination register value to the architectural register file.

Using the above two changes, an SRTR processor—like an IBM Z-series processor that uses Lockstepping for fault detection—can recover from errors. However, detecting the faults before register commit makes it difficult for an SRTR processor to obtain the performance advantage offered by an SRT processor. Recall that an SRT processor's performance advantage came from skewing the two redundant threads by some number of committed instructions. Because the leading thread would run several tens to hundreds of instructions ahead of the trailing thread, it would resolve branch mispredictions and cache misses for the trailing thread. Consequently, when the trailing thread probes the branch predictor, it would get the correct branch direction. Similarly, the trailing thread would rarely get cache misses because the leading thread would have already resolved them.

Achieving the same effect in an SRTR processor is difficult because of two reasons. First, in an SRT processor, the leading thread's execution may not slow down significantly, even with the skewed execution of the redundant threads. This is because the leading thread is allowed to commit instructions, thereby freeing up internal pipeline resources. In an SRTR processor, however, instructions from the leading thread cannot be committed until they are checked for faults. This makes it difficult to prevent performance degradation in an SRTR processor.

Second, in an SRT processor, the leading thread forwards only committed load or register values to the trailing thread. In contrast, an SRTR processor must forward uncommitted results to the trailing thread. This may not, however, skew the redundant threads far enough to cover the branch misprediction and cache miss latencies. Hence, Vijaykumar et al. [21] suggest forwarding results of instructions speculatively from the leading to the trailing thread before the leading thread knows if its instructions are on the wrong path or not.

Forwarding results speculatively leads to additional complexity of having to roll back the state of the thread in the event of a misspeculation. To manage this increased complexity, the SRTR implementation forces the trailing thread to follow the same path—whether correct path or incorrectly speculated path—as the leading thread. In contrast, in an SRT processor, the trailing thread only follows the same correct path as the leading thread but is not required to follow the same misspeculated path.

The rest of this section describes how an SRTR processor augments an SRT processor to enable transparent hardware recovery. Specifically, the key structures necessary are a prediction queue, an active list (AL), a shadow active list (SAL), a modified LVQ, a register value queue (RVQ), and the commit vectors (CV). These are described below. Figure 7.9 shows where these structures can be added to an SMT pipeline. Vijaykumar et al. have shown through simulations that SRTR

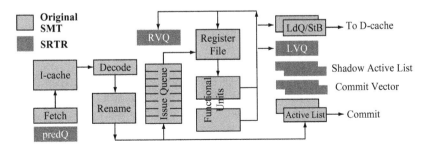

FIGURE 7.9 The SRTR pipeline. Reprinted with permission from Vijaykumar et al. [21]. Copyright © 2002 IEEE.

degrades the performance of an SRT pipeline by only 1% for SPEC 1995 integer benchmarks and by 7% for SPEC 1995 floating benchmarks.

Prediction Queue (predQ)

To force both threads to follow the same path, SRTR uses a structure called the prediction queue (predQ) to forward predicted PCs—correct or incorrect—from the leading to the trailing thread (Figure 7.9). The trailing thread uses these predicted PCs, instead of its branch predictor, to guide its instruction fetch. The leading thread must clean up predQ on a misprediction and roll back both its state and the trailing thread's state. After instructions are fetched, they proceed to the issue queue and the AL.

Active List and Shadow Active List

The AL is a per-thread structure that holds instructions in predicted order. When an instruction is issued and removed from the issue queue, the instruction stays in its AL to allow fault checking. Because SRTR forces leading and trailing threads to follow the same path, corresponding instructions from the redundant threads occupy the same positions in their respective ALs. When the leading thread detects a misprediction, it removes wrong-path instructions from its AL and sends a signal to the trailing thread to do the same. Eventually, instructions commit from the AL in program order.

The SRTR pipeline maintains an additional structure called the SAL. The SAL entries correspond to the ones in the AL. The SAL holds pointers to the LVQ and RVQ entries.

Load Value Queue

The SRTR pipeline maintains the LVQ in the same way the SRT does, except it introduces a level of indirection through the SAL. The SAL maintains pointers to both LVQ and RVQ entries, thereby allowing ease of checking. Further, to facilitate the rollback of the LVQ, branches place the LVQ tail pointer in the SAL at the time they enter the AL. Because the LVQ is in (speculative) program order, the LVQ tail pointer points to the LVQ entry to which the LVQ needs to be rolled back, if the branch mispredicts. A mispredicted branch's AL pointer locates the LVQ tail pointer in the SAL, and the LVQ is rolled back to the pointer.

Register Value Queue

The SRTR pipeline uses the RVQ to check for faults in nonmemory instructions. After the leading instruction writes its result back into its own copy of the register file, it enters the *fault-check* stage. In the fault-check stage, a leading thread's instruction puts its result value in the RVQ using the pointer from the SAL. The instruction then waits in the AL to commit or squash due to faults or mispredictions. Because the fault-check stage is after write-back, the stage does not affect branch misprediction penalty or the number of bypass paths.

The trailing thread's instructions also use the SAL to obtain their RVQ pointers and find their leading counterparts' values. Because either the leading or trailing thread—in theory—can reach the RVQ first, the SRTR uses a full/empty bit to indicate the thread that stored its value first in an RVQ entry. When the corresponding instruction from the redundant thread reaches the fault-check stage, it is compared for mismatch with the existing entry. An RVQ entry is relinquished in queue order after the checking is done. To avoid the bandwidth pressure to compare every instruction, Vijaykumar et al. introduced a scheme called DBCE, which was discussed in Chapter 6 (see RMT Enhancements, p. 244).

Recovery in SRTR

Register values, store addresses and values, and load addresses are checked in the RVQ (store buffer) and LVQ, respectively. To facilitate the check, SRTR introduced a structure called the CV. Each entry in the CV corresponds to an entry in the AL. As instructions are placed in the AL, their CV entries are set to a *not-checked-yet* state. As instructions retire, they are stalled at commit until they are checked. If the check succeeds, the CV entries corresponding to the leading and trailing instructions are set to the *checked-ok* state. Corresponding instructions from the leading and trailing threads commit only if its CV entry and its trailing counterpart's CV entry are in the checked-ok state.

If a check fails, the CV entries of the leading and trailing instructions are set to the *failed-check* state. When a failed-check entry reaches the head of the leading AL, all later instructions are squashed. The leading thread waits until the trailing thread's corresponding entry reaches the head of the trailing AL before restarting both threads at the offending instruction. Because there is a time gap between the setting and the reading of the CV and between the committing of leading and trailing counterparts, the CV is protected by ECC to prevent faults from corrupting it in the time gap.

There are errors from which SRTR cannot recover: if a fault corrupts a register after the register value is written back (committed), then the fact that leading and trailing instructions use different physical registers allows SRTR to detect the fault on the next use of the register value. However, SRTR cannot recover from this fault. To avoid this loss of recovery, one solution is to provide ECC on the register file.

7.4.4 Chip-Level Redundantly Threaded Processor with Recovery (CRTR)

Gomaa et al. [9] extended the SRTR concept to a CMP or what is more popularly known as multicore processors today. Instead of running the redundant RMT threads within a single core, CRTR runs the redundant threads on two separate cores in a multicore processor, similar to what CRT does for fault detection (see RMT in a Multicore Architecture, p. 240, Chapter 6).

CRTR differs from SRTR in one fundamental way. A leading thread's instructions are allowed to commit their values to the leading thread's own register file before the instructions are checked for faults. Before a trailing thread's instruction commits, it compares its output with the corresponding instruction from the leading thread. This scheme allows the leading thread to march ahead and introduce the slack needed to resolve the cache misses and branch mispredictions before the corresponding instruction from the trailing thread catches up. This is necessary in CRTR because the latency of communication between cores is longer than that observed by SRTR.

Recovering the leading thread is, however, more complex because the leading thread can commit the corrupted state to its own register file. The trailing thread flushes its speculative state when it detects a fault, and it copies its state to the leading thread, thereby recovering the leading thread as well. Then it can restart the pipeline from the offending instruction. Also, a CRTR processor does not need an AL or an SAL since instructions are compared for faults in a program's commit order.

7.4.5 Exposure Reduction via Pipeline Squash

In this section on backward error recovery with fault detection before register commit, four techniques were described, all of which precisely detect the presence of a fault prior to triggering an error recovery operation. This subsection and the next one examine two techniques that do not precisely detect the presence of a fault and may trigger the error recovery speculatively.

This subsection describes an eager scheme that will squash the pipeline state on specific pipeline events, such as a long-latency cache miss. The next subsection reviews a lazier mechanism. The basic idea in the eager scheme is to remove pipeline objects from vulnerable storage, thereby reducing their exposure to radiation. Because these pipeline events are usually more common than soft errors, this scheme squashes the speculative pipeline state only when the pipeline may be stalled, thereby minimizing the performance degradation from this scheme.

For example, microprocessors often aggressively fetch instructions from protected memory, such as the main memory protected with ECC or a read-only instruction cache protected with parity (but recoverable because instructions can be refetched from the main memory on a parity error). However, these instructions may stall in the instruction queue due to pipeline hazards, such as lack of functional units or cache misses. The longer such instructions reside in the instruction queue, the higher the likelihood that they will get struck by an alpha particle or a neutron.

In such cases, one could squash (or remove) instructions from the instruction queue and bring them back when the pipeline resumes execution. No architectural state is committed to the register file because the state squashed is purely in flight and speculative. This reduces an instruction's exposure to radiation, thereby lowering the instruction queue's SDC and/or DUE rate. Weaver et al. [23] introduced this scheme and evaluated it for an instruction queue. Gomaa and Vijaykumar [8] later evaluated such squashing for a full pipeline.

Triggers and Actions

Mechanisms to reduce exposure to radiation can be characterized in two dimensions: triggers and actions. A *trigger* is an event that initiates an *action* to reduce exposure. The goal is to avoid having instructions sit needlessly in processor structures, such as the instruction queue, for long periods of time. Hence, the trigger must be an event that indicates that queued instructions will face a long delay. Cache misses provide such a trigger. Instructions following a load that misses in the cache may not make progress while the miss is outstanding, particularly in an in-order machine. The situation is similar, though not as pronounced, for out-of-order machines in which instructions dependent on a load miss cannot make progress until the load returns data. Hence, it is fair to expect that removing instructions from the pipeline during the miss interval should not degrade performance significantly.

Once the processor incurs a cache miss, one possible action could be to remove existing instructions from the processor pipeline. These instructions can be refetched later when the cache miss returns the data. Such instruction squashing attempts to keep instructions from sitting needlessly in the pipeline for extended periods. To avoid removing instructions that could be executed before the miss completes, the pipeline should squash only those instructions that are younger than the load that missed in the cache. Fetch throttling can also be another action. Fetch throttling prevents new instructions from being added to the pipeline by stalling the front end of the machine.

This section illustrates how to reduce the AVF by squashing instructions in a pipeline (the action) based on a load miss in the processor caches (the trigger). The AVF reduction techniques are illustrated using an instruction queue—a structure that holds instructions before they are issued to the execution units in a dynamically scheduled processor pipeline.

Analyzing Impact on Performance

Traditionally, the terms $MTBF$ and $MTTF$ have been used to reason about error rates in processors and systems (see Metrics, p. 9, Chapter 1). Although MTTF provides a metric for error rates, it does not allow one to reason about the trade-off between error rates and the performance of a processor. Weaver et al. [23] introduced the concept of MITF as one approach to reason about this trade-off. MITF tells us how many instructions a processor will commit, on average, between two errors. MITF is related to MTTF as follows:

$$\text{MITF} = \frac{\text{number of committed instructions}}{\text{number of errors encountered}}$$

$$= \frac{\text{number of committed instructions}}{\frac{\text{total execution time in cycles}}{\text{frequency} \times \text{MTTF}}}$$

$$= \text{IPC} \times \text{frequency} \times \text{MTTF}$$

As with SDC and DUE MTTFs, one has corresponding SDC and DUE MITFs. Hence, for example, a processor running at 2 GHz with an average IPC of 2 and DUE MTTF of 10 years would have a DUE MITF of 1.3×10^{18} instructions.

A higher MITF implies a greater amount of work done between errors. Assuming that, within certain bounds, increasing MITF is desirable, then one can use MITF to reason about the trade-off between performance and reliability. Since MITF = 1/(raw error rate × AVF), one has

$$\text{MITF} = \frac{\text{IPC} \times \text{frequency}}{\text{raw error rate} \times \text{AVF}} = \frac{\text{frequency}}{\text{raw error rate}} \times \frac{\text{IPC}}{\text{AVF}}$$

Thus, at a fixed frequency and raw error rate, MITF is proportional to the ratio of IPC to AVF. More specifically, SDC MITF is proportional to IPC/(SDC AVF), and DUE MITF is proportional to IPC/(DUE AVF). It can be argued that mechanisms that reduce both the AVF and the IPC, such as the one proposed in the previous section, may be worthwhile only if they increase the MITF, that is, if they increase the IPC-to-AVF ratio by reducing AVF relative to the base case to a greater degree than reducing IPC.

Although one can use MITF to reason about performance versus AVF for incremental changes, one needs to be cautious not to misuse it. For example, it could be argued that doubling processor performance while reducing the MTTF by 50% is a reasonable trade-off as the MITF would remain constant. However, this explanation may be inadequate for customers who see their equipment fail twice as often.

Benefits of Pipeline Squash for an Instruction Queue

Table 7-1 shows how the average IPC and average AVFs change when all instructions in the instruction queue are squashed after a load miss in the L1 and the L0 caches [23]. The simulated machine configuration is the same as used in ACE Analysis Using the Point-of-Strike Fault Model, p. 106, Chapter 3. The L0 cache is the smallest data cache closest to the processor pipeline. The L1 cache is larger than L0 but is accessed only on an L0 miss. The SDC arises when the instruction queue is not protected, whereas the DUE arises if the instruction queue is protected with parity.

TABLE 7-1 ■ **Impact of Squashing on IPC and an Instruction Queue's SDC and DUE AVFs**

Design Point	IPC	SDC AVF	DUE AVF	IPC/SDC AVF	IPC/DUE AVF
No squashing	1.21	29%	62%	4.1	2.0
Squash on L1 load misses	1.19	22%	51%	5.6	2.3
Squash on L0 load misses	1.09	19%	48%	5.7	2.3

In this machine, when instructions are squashed on load misses in the L1 cache, the IPC decreases only by 1.7% (from 1.21 to 1.19) for a reduction in SDC and DUE AVFs by 26% (from 29% to 22%) and 18% (from 29% to 1.09%), respectively. However, when instructions are squashed on L0 misses, the IPC decreases by 10% for a reduction in SDC and DUE AVFs of only 35% and 23%, respectively. Squashing based on L0 misses provides a greater reduction in AVF for a correspondingly higher reduction in performance. Nevertheless, squashing on L1 misses appears more profitable because the SDC MITF (proportional to IPC/SDC AVF) and DUE MITF (proportional to IPC/DUE AVF) go up 37% and 15%, respectively.

Gomaa and Vijaykumar [8] studied squashing instructions on an L1 miss for an out-of-order processor. They found that the SDC AVF of the instruction queue of their simulated machine decreases by 21%, which is close to what Weaver et al. [23] reported for an in-order pipeline, shown in Table 7-1. However, Gomaa et al. [9] found that the IPC decreases by 3.5%, which is almost two times higher than that reported by Weaver et al. [23]. This is probably because an out-of-order pipeline can more effectively hide some of the cache miss latency by continuing to issue instructions following the load that missed in the cache. Squashing these instructions that could be issued out of order in the shadow of the load miss would degrade performance. In contrast, an in-order pipeline may not be able to issue instructions in the shadow of a load miss. Therefore, squashing such instructions would not cause significant performance degradation.

7.4.6 Fault Screening with Pipeline Squash and Re-execution

Fault screening is a mechanism that—like exposure reduction via pipeline squash—speculatively squashes pipeline state to recover from errors. Unlike exposure reduction that eagerly squashes pipeline state on long-latency pipeline events, fault screening is a lazier mechanism that tries to predict the presence of a fault in the system based on the current state of the microarchitecture and a program's behavior. Then they will trigger a recovery operation through pipeline squash.

Basic Idea

Fault screening is a mechanism that classifies program, architectural, or microarchitectural state into fault free state and faulty state. This is much like screening cancerous cells as benign or malignant. Because fault screening is much like diagnosing diseases from a patient's symptoms, the mechanism has also been referred to as *symptomatic* fault detection.

Fault screeners differ from a traditional fault detection mechanism, such as a parity code, in three ways. First, typically traditional fault detectors can precisely detect faults a detector is designed to catch (e.g., single-bit faults). In contrast, fault screeners can only probabilistically identify whether a given state is faulty or fault free at a given point of time.

Second, often a fault detection mechanism is assumed to be fail-stop—that is, on detecting a fault, the detection mechanism works with the processor or OS to halt further progress of a program to avoid any SDC. The probabilistic nature of fault screeners makes it difficult for fault screeners to be fail-stop.

Third, a fault screener can identify propagated faults, whereas often a fault detection mechanism cannot do the same. For example, if a fault occurs in an unprotected structure and then propagates to a parity-protected structure, the parity code cannot detect the fault. This is because the parity is computed on the already faulty bits. A fault screener screens faults based on program behavior and does not rely on computing a code based on incoming values. Hence, fault screeners are adept at identifying faults that propagate from structure to structure.

A fault screener identifies faulty state by detecting departure of program state from expected or established behavior. Racunas et al. [16] refer to such a departure as a *perturbation*. A perturbation may be *natural*, resulting from variations in program input or current phase of the application. A perturbation may also be *induced* by a fault. One can consider a static instruction in an algorithm that generates a result value between 0 and 16 the first thousand times it is executed. Its execution history suggests that the next value it generates will also be within this range. If the next instance of the instruction instead generates a value of 50, this value is a deviation from the established behavior and can be considered a perturbation. The new value of 50 could be a natural perturbation resulting from the program having moved to process a new set of data or it could be an induced perturbation resulting from a fault.

Interestingly, perturbations induced by a fault resulting from a neutron or an alpha particle strike far exceed the natural perturbation in a program. For example, a strike to a higher order bit of an instruction address is highly likely to crash a program, causing an induced perturbation. In the absence of software bugs, such extreme situations are unlikely during the normal execution of a program. The next subsection illustrates this phenomenon.

Natural versus Induced Perturbations

To illustrate how induced perturbations may far exceed natural perturbations, Racunas et al. [16] used departure from a static instruction's established result value space as an operational definition of a perturbation. A static instruction is an instruction that appears in a program binary at a fixed address when loaded into program memory but can have numerous dynamic instances as it is executed throughout a program. For example, incrementing the loop counter in a program loop can be a static instruction, but when it is executed multiple times for each iteration of the loop, it becomes a dynamic instantiation of the instruction. A static instruction's result value space is the set of values generated by the dynamic instances of that static instruction. Racunas et al. [16] classify this value space into three classes. The first class consists of static instructions each of whose working set of results is fewer than 256 unique values 99.9% of the time. For a static instruction falling into this class, a perturbation is defined as any new result that cannot be found in this array of 256 recently generated unique values.

The second class consists of static instructions whose result values have an identifiable stride between consecutive instances of the instruction 99.9% of the time. Racunas et al. [16] maintain an array of 256 most recent unique strides for each static instruction. Each stride is computed by subtracting the most recent result value generated by the instruction from its current result value. For static instructions falling into this class, a perturbation is defined as any new result that produces a stride that cannot be found in the array of 256 unique strides.

The third class consists of any static instruction not falling into either of the first two classes. For these instructions, the first result value they generate is recorded. Each time a new result value is generated, it is compared to this first result, and any bits in the new result that differ from the first result are identified. A bitmask is maintained that represents the set of all bits that have never changed from the first result value through the entire course of program execution. For this class of static instructions, a perturbation is defined as any new result value that differs from the first result value in a bit location that has previously been invariant. After each perturbation occurs, the bitmask is changed to mark the new bit as a variant. Hence, a static instruction with a 32-bit result can be responsible for a maximum of 32 perturbations in the course of the entire benchmark run. The method of detecting invariant bits in this class is similar to the method of software bug detection proposed by Hangal and Lam [11].

Figure 7.10 compares natural and induced perturbations caught by the fault screening mechanism described. For this graph, Racunas et al. [16] ran each benchmark 10 000 times. On each program run, they picked a random segment of dynamic

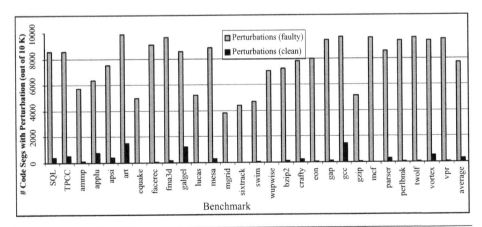

FIGURE 7.10   Perturbation in program segments. "Perturbations (clean)" are natural perturbations in a program. "Perturbations (faulty)" are induced perturbations. Reprinted with permission from Racunas et al. [16]. Copyright © 2007 IEEE.

was detected. In such a case, overhead to create the checkpoint is minimal, but the time to recover depends on how fast the error log can be traversed. This section describes a backward error recovery scheme that uses a *history buffer* as the log for incremental checkpointing.

In the periodic checkpointing, the processor periodically takes a snapshot of its state, which in this case is the processor register state. Before committing its state to a checkpoint, the processor or the system will typically ensure that the snapshot is fault free. Then, on detecting a fault, the processor reverts to the previous checkpoint and restarts execution. Periodically checkpointing the processor state reduces the need to do fault detection on every instruction execution (to guarantee that the checkpoint is fault free). However, it creates two new challenges. First, a large amount of state may need to be copied to create the checkpoint. Second, a large amount of state may be necessary to be compared to ensure that the newly created checkpoint is fault free. This section describes a scheme called *fingerprinting* that reduces the amount of state necessary to be compared to ensure a fault free checkpoint.

Because neither system—incremental nor periodic checkpointing—described in this section commits memory or I/O values outside the sphere of recovery, the output and input commit problems are solved relatively easily. No store or I/O operation is allowed until the corresponding operation is certified fault free, which solves the output commit problem. This also makes it easier to solve the input commit problem. Because stores are not committed till they pass their fault detection point and loads are idempotent, these systems can allow loads to be reissued by the processor after they roll back and recover from a previous checkpoint. Any uncached load must, however, be delayed until the load address is certified fault free.

Although incremental checkpointing and periodic checkpointing are discussed in the context of backward error recovery before memory commit, these techniques can also be extended to backward error recovery schemes that allow memory or I/O to commit before doing the fault check. Neither of these techniques, to the best of my knowledge, has been implemented in any commercial system.

7.5.1 Incremental Checkpointing Using a History Buffer

The key requirement to move the fault detection point of an instruction beyond its register commit point is to preserve the checkpointed state corresponding to an instruction. In SRTR, the processor frees up this state as soon as an instruction retires. Instead, an instruction could be retired, but its state must be preserved until the instruction passes its fault detection point. This state can be preserved via the use of a history buffer [12]. Smith and Pleszkun [18] first proposed a history buffer to support precise interrupts, but this section shows how it can be used for backward error recovery as well.

Structure of the History Buffer

The history buffer consists of a table with a number of entries. Each entry contains information pertinent to a retired instruction, such as the instruction pointer, the old destination register value, and the physical register to which the destination architecture register was mapped. The instruction pointer can be either the program counter or an implementation-dependent instruction number. The old destination value records the value of the register before this instruction rewrote it with its destination register value. Finally, the register mapping is also needed to identify the physical register that was updated. The number of entries in the history buffer is dependent on the implementation and must be chosen carefully to avoid stalls in the pipeline. The history buffer mechanism can work with many fault detection mechanisms, such as Lockstepping and SRT, which detect faults after register values have been committed.

Adding Entries to the History Buffer

When an instruction retires, but before it writes its destination register to the register file, an entry is created in the history buffer, and the entry's contents are updated with the instruction pointer, old register value, and register map. Reading the old register value may require an additional read port in the register file. An additional read port may increase the size of the register file and/or increase its access time. Nevertheless, there are known techniques to avoid this performance loss (e.g., by replicating the register file). In effect, by updating the history buffer when an instruction retires, an incremental checkpoint of the processor state is created.

Freeing up Entries in the History Buffer

When a retired instruction passes its fault detection point, the corresponding entry in the history buffer is freed up. This is because the checkpoint corresponding to an instruction that has been certified to be fault free is no longer needed.

Recovery Using the History Buffer

If the fault detection mechanism detects a fault in a retired instruction, then the recovery mechanism kicks in. The following are the steps involved in recovering a processor that uses SRT as its fault detection mechanism:

- For both redundant threads, flush all speculative instructions that have not retired.

- For the trailing thread, flush its architectural state, including the architectural register file.

- Reconstruct the correct state of the architectural register file of the leading thread using the existing contents of the register file and the history buffer. The architectural register file contains values up to the last retired instruction, which can be older than the instruction experiencing the fault. Hence, the values have to be rolled back to the state prior to the instruction with the fault. This can be accomplished by finding the oldest update to that register from the history buffer. This procedure is repeated for every register value that exists in the history buffer.

- Then the history buffer is flushed.

- The contents of the leading thread's register file are loaded into the trailing thread's register file.

- Both threads are restarted from the instruction that experienced the fault.

7.5.2 Periodic Checkpointing with Fingerprinting

Unlike incremental checkpointing that incrementally builds a checkpoint, periodic checkpointing takes a snapshot of the processor state periodically. Although this reduces the necessity to do a fault detection on every instruction to ensure a fault free checkpoint, it does require copying a large amount of state to create the checkpoint and to ensure that the checkpoint is fault free.

Smolens et al. [19] introduced the concept of *fingerprinting* to reduce the amount of state that may be necessary to be compared for fault detection prior to checkpoint creation. In some ways, fingerprinting is a data compression mechanism. As and when each value is generated, it merges the generated value into a global running value or the fingerprint. Then, instead of comparing each and every value, the fingerprint can be compared prior to generating a checkpoint to ensure that the checkpoint is fault free.

Fingerprinting may be appropriate for a system that uses Lockstepping or RMT as its fault detection mechanism and where the communication bandwidth between the redundant cores, threads, or nodes is severely constrained. For example, if there are two processor chips or sockets forming the redundant pair, then the bandwidth between the chips may be limited. Copying state to create a checkpoint within a processor chip, however, may not incur a significant performance penalty since the communication for this copy operation may be limited to within a processor chip. But the state comparison requires off-chip communication, and off-chip bandwidth is often severely limited. Fingerprinting can help reduce the performance degradation from comparing large amounts of off-chip state. The rest of this subsection discusses fingerprinting and how it can help reduce the bandwidth requirements in the context of *chip-external* detection.

Fingerprint Mechanism

A fingerprint provides a concise view of the past and present program states. It contains a summary of the outputs of any new register values created by each executing instruction, the new memory values (for stores), and the effective addresses (for both loads and stores). By capturing all updates to architectural state, a fingerprint can ensure that it can help create a fault free checkpoint. It should be noted that this implementation of fingerprinting allows not only register values to be committed to the register file but also memory values to be committed to the processor caches. Nevertheless, the caches are not allowed to write back their modified data to main memory till the fault check has been done.

For fault detection with RMT processors, the fingerprint implementation must monitor only committed register values. However, for fault detection with Lockstepped processors, the fingerprint implementation can monitor both speculative and committed updates to the physical register file since both Lockstepped processors are cycle synchronized.

To create the fingerprint, Smolens et al. [19] propose hashing the generated values into a cyclic code. There are two key requirements for the code. First, the code must have a low probability of undetected faults. Second, the code should be small for both easy computation and low bandwidth comparison. For a p-bit CRC code, the probability of an undetected error is at most 2^{-p}. Smolens et al. [19] used a 16-bit ($p = 16$) CRC code for their evaluation. A 16-bit CRC code has a 0.000015 probability that an error will go undetected. The reader is referred to Cyclic Redundancy Check, p. 178, Chapter 5, for a discussion on CRC codes.

Evaluation Methodology

To evaluate chip-external fault detection using fingerprinting for full-state comparison, Smolens et al. [19] simulated the execution of all 26 SPEC CPU 2000 benchmarks using SimpleScalar sim-cache [6] and two commercial workloads—TPC-C and SPECWeb—using Virtutech Simics. The simulated processor executed one IPC at a clock frequency of 1 GHz. The only microarchitecture parameter relevant to this evaluation was the level 2 (L2) cache configuration. The simulated processor has an inclusive 1-megabyte four-way set-associative cache with 64-byte lines.

The benchmarks used for evaluation can be classified into three categories: SPEC integer, SPEC floating point, and commercial workloads. Using the prescribed procedure from SimPoint [14], the authors simulated up to eight predetermined 100 million instruction regions from each SPEC benchmark's complete execution trace. Using Simics, the authors ran two commercial workloads on Solaris 8: a TPC-C-like online transaction processing (OLTP) workload with IBM DB2 and a SPECWeb. The 100-client OLTP workload consisted of a 40-warehouse database striped across five raw disks and one dedicated log disk. The SPECWeb workload serviced 100 connections with Apache 2.0. The authors warmed both commercial workloads until the CPU utilization reached 100% and the transaction rate reached steady state. Once warmed, the commercial workloads executed for 500 million instructions.

Bandwidth Required for Full-State Comparison Using Lockstepped Processors

Smolens et al. [19] evaluated fault detection mechanisms based on their state-comparison bandwidth demand between mirrored processors. The bandwidth

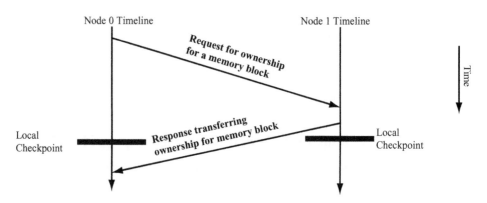

FIGURE 7.13 Demonstration of how local uncoordinated checkpoint can lead to inconsistent global state.

a memory block. Node 1 sends a transfer of ownership response back to node 0. At this point, both nodes 0 and node 1 take their local checkpoints. Then, node 0 receives the transfer of ownership response. The two local checkpoints shown in the figure are, however, inconsistent and cannot portray a single consistent global state. This is because node 1 thinks node 0 is the owner of the block, and node 0 thinks node 1 is the owner. Hence, checkpoint generation must be coordinated to generate a single global consistent state.

ReVive generates a single consistent global state by stopping all processing nodes and coordinating the generation of their individual checkpoints. In contrast, SafetyNet generates local checkpoints within a node but has enough intelligence built into the system to roll back to the appropriate local checkpoints that together make up a single global state.

The three systems solve the output and input commit problems slightly differently. For LVQ-based recovery, memory is outside the sphere of recovery. It only allows fault free stores to propagate to memory. However, because it allows multiple stores to commit to main memory since it created the checkpoint, it must record and replay all load values issued so far to ensure that on reexecution all the stores already committed are seen again. In contrast, for ReVive and SafetyNet memory is inside the sphere of recovery but I/O is not. They solve the output commit problem by not allowing I/O operations to commit until they can take a checkpoint. They solve the input commit problem by recording all external events, such as I/O interrupts, and by replaying them during reexecution after a rollback and recovery.

7.6.1 LVQ-Based Recovery in an SRT Processor

Log-based recovery is a well-established technique in the fault-tolerant systems area [7]. The concept applies to any system that is piecewise deterministic, i.e., in

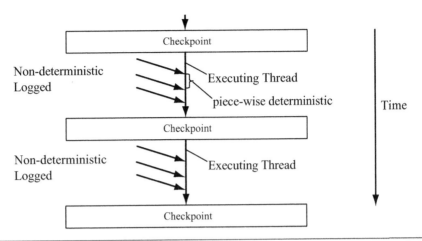

FIGURE 7.14 Log-based transparent recovery in a piecewise deterministic system.

which execution consists of deterministic intervals separated by nondeterministic events (Figure 7.14). In this context, an event is considered nondeterministic if its execution is not determined entirely by the state inside the system. The state of a piecewise deterministic system at any point can be recreated, given an initial condition and a complete description of all nondeterministic events up to the desired point.

In the context of a processor's transparent hardware recovery, the initial condition is the checkpointed architectural state, which includes any architecturally visible state, such as the architectural register file. The nondeterministic events are the following:

- Load values, since memory contents may be modified external to the sphere of replication by other processors or I/O devices

- Asynchronous interrupts

- Timing-dependent operations, such as ones that read on-chip cycle counters.

Interestingly, these events are identical to those required to maintain determinism between the leading and trailing threads in an SRT processor. As long as execution is deterministic, one can rely on both threads to follow the same execution path. Thus, it is exactly these same nondeterministic events that must be captured to keep the two threads consistent. This is the key insight behind log-based recovery in an SRT processor [17].

A designer may want to keep the checkpoint and the log fault free at creation and protected against transient faults (e.g., by ECC) to increase the fault coverage. After the completion and fault free validation of a checkpoint, both the previous checkpoint and all log entries prior to the new checkpoint can be discarded. The minimum checkpoint frequency is determined by the maximum available log storage and the maximum tolerable recovery latency.

The rest of this subsection describes the relationship between fault detection, checkpointing, and logging in SRT; how it can handle faults during regular execution (i.e., nonrecovery mode), during checkpoint creation, and in recovery mode; and how it can log load values and asynchronous interrupts.

Fault Detection, Checkpointing, and Logging

As discussed earlier (RMT within a Single-Processor Core, p. 227, Chapter 6), an SRT processor's performance advantage over Lockstepping comes from doing the fault detection in the larger SRT-Memory sphere. Nevertheless, when checkpoints are created, it may also be necessary to check them for faults. That is, all new register values from the two redundant threads must be compared for mismatch. In the absence of a mismatch, they can be sent to the ECC-protected checkpoint storage. Because the checkpoint includes the state of the architectural register file, fault detection during checkpoint creation is similar to the checking required for the smaller SRT-Register's sphere used in SRTR. However, the logging mechanism may avoid the performance problems suffered by SRTR due to limited slack. The log of external events, such as loads and asynchronous interrupts, ensures that the internal state of the machine at the point of a detected fault can be reproduced starting from the last checkpoint. This capability decouples the fault detection in SRT-Memory (used to verify data sent out externally) from the fault detection in SRT-Register (used to validate checkpoints).

This decoupling has two effects. First, checkpoint validation is not in the critical path of the program execution and thus can be done at leisure and in the background. Second, the log-based method does not have to create checkpoints corresponding to every output point. Rather, checkpoints can be created at any desired frequency. As discussed above, decreases in checkpoint frequency come at the cost of increased log storage and increased recovery latency.

Handling Faults in Nonrecovery Mode

In such a design, during normal execution (i.e., in a nonrecovery mode), an output mismatch for data exiting the SRT-Memory sphere (e.g., in a store address or in a value) indicates the presence of a fault. Then, both the leading- and the trailing-thread contexts are flushed because one cannot tell the thread that contains the fault. Both thread contexts are reloaded from the latest complete checkpoint. Then execution is resumed in recovery mode in which both the threads are driven by the saved event log.

This recovery mode is indistinguishable from the normal operation for the trailing thread but is distinct for the leading thread. Thus, for example, leading thread loads will be satisfied from the log rather than from the data cache, as in a normal operation. These recovery operations continue until all outputs generated since the checkpoint have been regenerated. This point is detected by maintaining a counter that tracks the number of outputs generated since the previous checkpoint, decrementing this counter for each regenerated output and exiting recovery mode when the counter reaches zero. The regenerated outputs, such as stores, are discarded because they have already been exposed outside the SRT-Memory.

Handling Faults during Checkpoint Creation

Checkpoint creation may occur either continuously in the background or at periodic intervals. Register values from both threads are compared for mismatch. In the absence of any mismatch, the outputs are committed to the external checkpoint storage. However, on a mismatch, both threads are flushed and restored to the previous complete checkpoint. Then the same procedure is followed as outlined earlier. It should be noted that external checkpoint storage must be capable of storing a full valid checkpoint until a subsequent checkpoint can be fully validated, potentially requiring twice the capacity of the architectural register file.

Handling Faults during Recovery Mode

Faults could occur during the recovery procedure itself. To detect these faults, one can run both the leading and trailing threads and compare both threads' outputs that leave the SRT-Memory sphere. A detected fault during recovery mode could imply the presence of a permanent fault, when one could flag this to the OS to let it decide future action. Alternatively, if system designers believe that transient faults could be likely in such a situation, then they could simply restart both the redundant threads from the last checkpoint and run them off the saved event log.

Logging Loads Using the LVQ

As discussed earlier, the log required for recovery is a simple extension of the input replication mechanism already needed by an SRT processor to provide fault detection. Hence, one can leverage the input replication mechanisms, which recreate nondeterministic events from the leading thread in the trailing thread, to handle these same events for recovery. RMT replicates load values from the leading to the trailing thread using an LVQ. The LVQ is a queue of load values as seen by the leading thread. Loads from the trailing thread obtain their values from the LVQ instead of the data cache.

Extending the LVQ for recovery consists of maintaining LVQ entries after the trailing thread has consumed them until a checkpoint is completed. If checkpoints are frequent, entries may be kept in the same physical LVQ structure used for input replication. Otherwise, a larger log may be required, when LVQ entries may be copied from the primary LVQ to a separate recovery-only LVQ structure. Because the performance of recovery may not be critical, the recovery LVQ may be larger, slower, and farther from the core than the replication LVQ. Even if a single physical structure is used, it is useful to logically distinguish replication LVQ entries (which have not been consumed by the trailing thread) from recovery LVQ entries (which have been consumed by the trailing thread but are not yet incorporated in the latest checkpoint).

Logging Asynchronous Interrupts

Asynchronous interrupts are replicated in an SRT processor by logging the point in the dynamic instruction stream at which the interrupt is taken in the leading thread and by generating the interrupt at the same point in the trailing thread, or by forcing the thread copies to synchronize to the same execution point and delivering the interrupt simultaneously. In the former scenario, asynchronous interrupts could be logged and replayed in the same manner as loads. The mechanism for generating interrupts from the log can leverage the logic used to generate interrupts in the trailing thread in a normal operation. In the latter scenario, interrupts are not stored and transferred between threads, so there is no mechanism to leverage. In this case, a more direct approach would be to force a checkpoint after the threads synchronize, eliminating the need to log interrupt events. Other nondeterministic events should be addressed on a case-by-case basis, though this framework of leveraging the replication mechanism for logging (or forcing checkpoints) should apply in general.

7.6.2 ReVive: Backward Error Recovery Using Global Checkpoints

ReVive is a backward error recovery technique targeted for cache-coherent shared-memory multiprocessors [15]. Such a shared-memory multiprocessor is composed of multiple nodes. Each node consists of a processor, caches, and its local main memory. The caches are kept coherent using a cache coherence protocol. Unlike the LVQ-based recovery in SRT, ReVive maintains system-wide global checkpoints that can help the system recover from errors in either the memory system or the coherence protocol logic.

ReVive has three basic components: distributed parity to detect faults in memory, an undo log that tracks the first writes to memory locations after a consistent checkpoint is established, and a method to create a globally consistent checkpoint. In the LVQ-based recovery just described, the SRT processor periodically copies its architectural state to a separate checkpoint memory. This maybe feasible since the architectural state is typically small, consisting of architectural registers and processor state. Then it logs nondeterministic events and input load values to replay them later during recovery.

In contrast, ReVive uses the entire memory itself as its checkpoint since copying it may be infeasible. First, ReVive ensures that the entire memory across the system is in a consistent state. This may require flushing the processor caches, so that individual processors do not have a more recent up-to-date copy of modified blocks. ReVive's log keeps track of the first memory write to any memory location since the consistent global state was created. On detecting a fault, ReVive uses the combination of the current memory state and the undo log to roll back the entire memory state of the system to a globally consistent state.

Distributed Parity

ReVive implements distributed parity to both detect faults and recover from errors in its distributed memory, which serves both as the main memory and as a part of the checkpoint. The scheme is similar to what is used in RAID (Redundant Array of Inexpensive Disks) systems. As Figure 7.15 shows, in ReVive, memory is organized as parity groups, where a memory block in one node contains the parity bits corresponding to the data bits for the other nodes. Thus, a parity bit P in the memory block holding parity can be computed as P = A XOR B XOR C, where A, B, and C are the corresponding data bits in each of the other memories.

When a memory write occurs in a specific memory block, the corresponding controller sends the XOR of the original parity bits and the newly created parity bits for the memory block that was just written to the memory that holds its parity bits. Assume the new parity bit is C'. Then the controller holding C sends an update U = C XOR C' to the controller with the parity bit. The reader can easily verify that the controller with the parity bits can compute the new parity bit simply as $P' = P$ XOR U. Using this distributed parity scheme, ReVive can completely recover from single-bit transient faults in its memory system and from the complete loss of a node's entire memory.

Logging Writes

ReVive must keep track of the first write to any memory block since the last checkpoint was created. When it detects a write, it copies the old value into a log and writes the new value to memory. Then, on detecting a fault, it can simply use the log to reset the values of all blocks to recreate the checkpoint and restart execution.

Typically, cache-coherent, shared-memory multiprocessors implement a single-writer invalidation protocol, which allows multiple readable copies of a memory

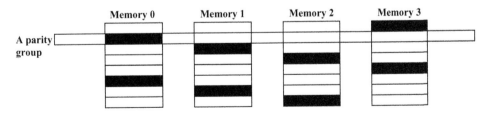

FIGURE 7.15 Parity groups in ReVive's distributed memory protected with parity. Shaded boxes represent parity bits, whereas boxes not shaded represent data bits. Each bit in the shaded parity box is an XOR of the corresponding data bits in the same parity group.

block, but only one node can write the block at any time. To write a block, a node must have the only copy of the block and must be its exclusive owner. Before a block transitions from readable to exclusive, the corresponding node must request an exclusive write access to this block. This makes it easy for a memory controller to detect a node's intention to write a memory block. At that point, the memory controller can save the old value of the memory block in the log.

To only log the first write to block after a checkpoint, ReVive augments its blocks with an additional bit—called the *log bit*—beyond what may be needed by the coherence protocol. The log bit tracks first writes to a block after a checkpoint. If the log bit is set, then ReVive will not copy the block's old value to the log when a write request comes. The log bits are reset when a checkpoint is created.

Global Checkpoint Creation

To create the global checkpoint, ReVive follows a simple two-phase protocol. First, all processors and nodes synchronize and flush all their caches to write all modified data back to memory. Then the processors synchronize again to ensure that all processors have completed writing all their data back to memory. All the log bits are also cleared at this time. The checkpoint is now established, and all processors can resume execution. Because an error may occur during the creation of a checkpoint, ReVive does not discard the old log till all processors complete the two-phase protocol.

7.6.3 SafetyNet: Backward Error Recovery Using Local Checkpoints

Like ReVive, SafetyNet is a backward error recovery scheme targeted to recover from errors in a cache-coherent, shared-memory multiprocessor [20]. SafetyNet's logging scheme is similar to ReVive's. Unlike ReVive, however, SafetyNet generates local checkpoints in a way that allows it to create a globally consistent state, if a roll-back is necessary. In the background, SafetyNet coordinates different local checkpoints to ensure that there always exists a globally consistent checkpointed state that the entire system can roll back to. Once it establishes such a recovery point, it frees up any prior local checkpoints. Because SafetyNet takes local checkpoints, it can create checkpoints much faster than ReVive. The SafetyNet scheme is, however, more complex to implement than ReVive.

Creating Consistent Checkpoints

In SafetyNet, all nodes periodically take local checkpoints at a globally synchronous logical time. The local checkpoint is a combination of the processor's architectural state and a log of first memory writes since the last checkpoint was initiated. To ensure that the local checkpoints reflect a consistent system state, SafetyNet exploits the key insight that, in retrospect, a coherence transaction appears logically atomic once the entire coherence transaction has completed. A transaction's point

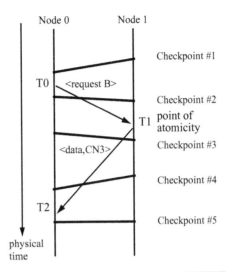

FIGURE 7.16 Example of checkpoint coordination in SafetyNet. Reprinted with permission from Sorin et al. [20]. Copyright © 2002 IEEE.

of atomicity can be established when the previous owner of the requested block processes the request. For example, in Figure 7.16, the point of atomicity could be T1. Node 0 sends a request to the previous owner (Node 1). Node 1 establishes the point of atomicity at T1 and forwards its last checkpoint number (checkpoint #2) to Node 0 as its checkpoint number for the point of atomicity. Node 0, the requestor, does not learn the location of the atomicity point until it receives the response that completes the transaction. To ensure that the system never recovers to the "middle" of a transaction, the requestor does not advance the recovery point until all its outstanding transactions complete successfully. After completion, the transaction appears atomic.

Advancing the Global Recovery Point

At any point of time, SafetyNet maintains a combination of local checkpoints that constitute a consistent global state, which is referred to as the global recovery point. As each node creates local checkpoints, one must advance the global recovery point, so that each node can free up the storage for its prior local checkpoints. Further, any I/O operation can be performed only after a global recovery point is established.

For a checkpoint interval to be fault free, no fault detector in the system should signal an error and every transfer of ownership in that interval must complete successfully. This ensures that the data were transferred fault free to the requestor. Once every component has independently declared that it has received fault free data in response to all its requests in the interval before the checkpoint, the global recovery point can be advanced. At this point, all transactions prior to this checkpoint have had their points of atomicity determined. Then the state for the prior recovery point can be deallocated lazily. One can establish this global recovery point in the background to avoid slowing down the main computation.

The latency to establish the global recovery point depends on the fault detection latency. Some faults that manifest themselves as corrupted messages can be caught immediately through CRC codes, which are often used to protect network messages. Other faults, such as dropped messages, may have to rely on timeout latency that can be many traversals of the interconnect.

7.7   Backward Error Recovery with Fault Detection after I/O Commit

Including I/O devices in the sphere of recovery is a difficult problem and is hard to tackle in hardware. This is because I/O operations can have side effects, which may be hard to control and recover from. For example, an uncached store to an I/O device may trigger the device to send an incorrect message to a terminal, which cannot be recovered from. It is often difficult for a computing system to recover from such an operation.

There are, however, scenarios in which an application can recover even after it commits to an I/O device. For example, databases routinely maintain recovery logs that can be used to recover from incorrect disk operation. These recovery operations are, however, often handled on a per-application basis and in software. Nakamo et al. [13] discuss how disk operations can be rolled back since disk reads are idempotent.

Networking stacks, such as TCP/IP, are allowed to send duplicate messages into the network. This is because if the sender does not receive an acknowledgement, it can send a duplicate copy of the message to the receiver. The same mechanism can be used to send a duplicate message after any operation has been rolled back. Chapter 8 discusses such software-controlled backward error recovery techniques.

7.8   Summary

Fault detection mechanisms either remove any SDC from a system or convert them to DUE. Error recovery mechanisms can reduce or eliminate both the SDC and DUE rates of a system. Broadly, error recovery mechanisms can be classified into forward and backward error recovery mechanisms.

In a forward error recovery, a system can continue executing from its current state after a fault is detected. The forward error recovery schemes usually maintain a redundant, up-to-date error-free state from which the system can continue execution. In contrast, a backward error recovery scheme usually rolls back to a previous error-free state of a system. Such states are referred to as checkpoints and can be generated incrementally or periodically by a system.

There are four options for implementing forward error recovery in hardware: fail-over systems, DMR, TMR, and pair-and-spare. Each of these systems has incrementally smaller downtime compared with the previous one. Fail-over systems typically consist of a mirrored pair of computing slices denoted as the primary and secondary. When the primary slice detects a fault, the secondary slice takes over execution with some state correction (e.g., restarting the failed process).

A DMR system detects faults by comparing outputs from two redundant slices of execution. Further, if an internal error checker, such as a parity checker, fires within one of the slices, then the output comparator can precisely determine the slice that is in error. Then, by copying the correct state from the fault free slice to the faulty slice, the DMR system can recover from the error.

A TMR system extends the concept of DMR by running three redundant slices of execution. The output comparator votes on the output that is correct. If outputs from one of the redundant slices are faulty, then the TMR system freezes that slice but can continue execution in degraded DMR mode with the other two fault free slices. Eventually, the correct state must be copied from the fault free slices to the faulty slice to reintegrate the faulty slice into the TMR system.

Finally, pair-and-spare extends the TMR concept by running four redundant slices. The primary pair is the one that commits to system state. The secondary pair keeps its state up-to-date with the primary pair. When the primary pair detects a fault, it transfers execution to the secondary pair.

Unlike forward error recovery schemes, a backward error recovery scheme rolls back execution to a prior fault free state referred to as a checkpoint. What constitutes a checkpoint depends on the fault detection scheme. Four options are possible: fault detection before register commit, fault detection before memory commit, fault detection before I/O commit, and fault detection after I/O commit. Fault detection after I/O commit is typically achieved in software (not discussed in this chapter). If the fault is detected before registers are committed, then existing techniques that a processor uses to recover from a misspeculation, such as a branch misprediction, can be used to recover from a transient fault.

If the fault is detected after registers are committed, but before memory is committed, then the recovery scheme needs a mechanism to checkpoint the register state. This can be done incrementally through the use of a history buffer that logs all register updates. Alternatively, this can be done periodically by taking snapshots of the register state. Periodic checkpointing may require significant output comparison bandwidth to ensure that the checkpoint is fault free. This can be reduced through a state-compression technique known as fingerprinting.

If the fault is detected after memory is committed, then the system must support a mechanism to roll back memory state. Memory state is typically much larger than register state, so it may be difficult to take a checkpoint of memory state. Instead, systems typically maintain a log of memory values received from the system since the prior register checkpoint. During recovery, the system can then replay execution, starting at the register checkpoint and using the log of memory values as inputs to the execution stream. Distributed memory machines impose an additional problem in which checkpoints must be globally consistent across an entire system. This can be achieved by coordinating checkpoint generation across the individual nodes of a distributed system.

7.9   Historical Anecdote

One of Stratus Technologies' customers had requested an automatic reboot capability in the event of a system hang. This is because the customer had many unattended sites that would benefit from such an automatic reboot. Stratus did implement such a facility through something called a *dead man timer* [10]. The computer automatically rebooted if the dead man timer was not periodically reset by the software running on it. Many of Stratus' customers who were not aware of this dead man timer were not running the appropriate software to reset it. They experienced many mysterious crashes for months. Thus, a capability that Stratus implemented to recover a system from a system hang ended up crashing the system itself.

Stratus did not notice this problem due to a number of issues. The dead man timer was not well documented and was designed in a somewhat ad hoc fashion. It did not record a log and therefore operators had no way of telling if the dead man timer had reset the system. The signal to drive the timer was only 1 bit wide, nonredundant, and unchecked. Consequently, a 1-bit error in the transmission of the signal could also activate the timer mechanism. The timer design was not also subject to usual design reviews that engineering companies normally implement. Once Stratus was notified of the problem, its engineers fixed the problem in a week. Since then, Stratus has imposed strict review and controls to ensure that new features do not cause similar problems.

References

[1] H. Ando, Y. Yoshida, A. Inoue, I. Sugiyama, T. Asakawa, K. Morita, T. Muta, T. Motokurumada, S. Okada, H. Yamashita, Y. Satsukawa, A. Konmoto, R. Yamashita, and H. Sugiyama "A 1.3 GHz Fifth Generation SPARC Microprocessor," in *2003 IEEE Solid State Circuits Conference (ISSCC)*, pp. 1896–1905, 2003.

[2] J. Barlett, W. Bartlett, R. Carr, D. Garcia, J. Gray, R. Horst, R. Jardine, D. Lenoski, and D. Mcguire "Fault Tolerance in Tandem Computer Systems," Technical Report 90.5, Part Number 40666, Hewlett-Packard, May 1990.

[3] W. Bartlett and L. Spainhower, "Commercial Fault Tolerance: A Tale of Two Systems," *IEEE Transactions on Dependable and Secure Computing*, Vol. 1, No. 1, pp. 87–96, January–March 2004.

[4] D. Bernick, B. Bruckert, P. D. Vigna, D. Garcia, R. Jardine, J. Klecka, and J. Smullen, "NonStop® Advanced Architecture," in *Proceedings of the International Conference on Dependable Systems and Networks (DSN)*, pp. 12–21, 2005.

[5] B. Bloom, "Space/Time Trade-offs in Hash Coding with Allowable Errors," *Communications of the ACM*, Vol. 13, No. 7, pp. 422–426, July 1970.

[6] D. Burger and T. M. Austin, "The Simplescalar Tool Set, Version 2.0," Technical Report 1342, Computer Sciences Department, University of Wisconsin–Madison, June 1997.

[7] M. Elnozahy, L. Alvisi, Y.-M. Wang, and D. B. Johnson, "A Survey of Rollback-Recovery Protocols in Message-Passing Systems," Technical Report CMU-CS-99-148, School of Computer Science, Carnegie Mellon University, June 1999.

[8] M. A. Gomaa and T. N. Vijaykumar, "Opportunistic Fault Detection," in *32nd Annual International Symposium on Computer Architecture*, pp. 172–183, 2005.

[9] M. A. Gomaa, C. Scarbrough, T. N. Vijaykumar, and I. Pomeranz, "Transient Fault-Recovery for Chip Multiprocessors," in *30th Annual International Symposium on Computer Architecture*, pp. 96–109, June 2003.

[10] P. A. Green Jr., "Observations From 16 Years at a Fault-Tolerant Computer Company," in *15th Symposium on Reliable Distributed Systems*, pp. 162–164, 1996.

[11] S. Hangal and M. Lam, "Tracking Down Software Bugs Using Automatic Anomaly Detection," in *International Conference on Software Engineering*, ICSE'02, pp. 291–301, May 2002.

[12] S. S. Mukherjee, S. K. Reinhardt, and J. S. Emer, "Incremental Checkpointing in a Multi-Threaded Architecture," United States Patent Application, Filed August 29, 2003.

[13] J. Nakamo, P. Montesinos, K. Gharachorloo, and J. Torrellas, "ReVive I/O: Efficient Handling of I/O in Highly-Available Rollback-Recovery Servers," in *12th Annual International Symposium on High-Performance Computer Architecture (HPCA)*, pp. 200–211, 2006.

[14] E. Perelman, G. Hamerly, M. Van Biesbrouck, T. Sherwood, and B. Calder, "Using SimPoint for Accurate and Efficient Simulation," in *ACM SIGMETRICS, the International Conference on Measurement and Modeling of Computer Systems*, pp. 318–319, June 2003.

[15] M. Prvulovic, Z. Zhang, and J. Torrellas, "ReVive: Cost-Effective Architectural Support for Rollback Recovery in Shared-Memory Multiprocessors," in *29th Annual International Symposium on Computer Architecture (ISCA)*, pp. 111–122, 2002.

[16] P. Racunas, K. Constantinides, S. Manne, and S. S. Mukherjee, "Perturbation-Based Fault Screening," in *13th Annual International High-Performance Computer Architecture (HPCA)*, pp. 169–180, February 2007.

[17] S. K. Reinhardt, S. S. Mukherjee, and J. S. Emer, "Periodic Checkpointing in a Redundantly Multi-Threaded Architecture," United States Patent Application, Filed August 29, 2003.

[18] J. E. Smith and A. R. Pleszkun, "Implementation of Precise Interrupts in Pipelined Processors," in *12th International Symposium on Computer Architecture*, pp. 291–299, 1985.

[19] J. C. Smolens, B. T. Gold, J. Kim, B. Falsafi, J. C. Hoe, and A. G. Nowatzyk, "Fingerprinting: Bounding Soft-Error Detection Latency and Bandwidth," *IEEE Micro*, Vol. 24, No. 6, pp. 22–29, November 2004.

[20] D. J. Sorin, M. M. K. Martin, M. D. Hill, and D. A. Wood, "SafetyNet: Improving the Availability of Shared Memory Multiprocessors with Global Checkpoint/Recovery," in *International Symposium on Computer Architecture (ISCA)*, pp. 123–134, May 2002.

[21] T. N. Vijaykumar, I. Pomeranz, and K. Cheng, "Transient Fault Recovery Using Simultaneous Multithreading," in *29th Annual International Symposium on Computer Architecture*, pp. 87–98, May 2002.

[22] N. J. Wang and S. J. Patel, "ReStore: Symptom-Based Soft Error Detection in Microprocessors," *IEEE Transactions on Dependable and Secure Computing*, Vol. 3, No. 3, pp. 188–201, July–September 2006.

[23] C. Weaver, J. Emer, S. S. Mukherjee, and S. K. Reinhardt, "Techniques to Reduce the Soft Error Rate of a High-Performance Microprocessor," in *31st Annual International Symposium on Computer Architecture (ISCA)*, pp. 264–275, 2004.

CHAPTER

8

Software Detection and Recovery

8.1 Overview

Software techniques to detect transient faults and correct corresponding errors are gaining popularity. Recently, Marathon Technologies introduced its EverRun servers in which fault tolerance is implemented completely in software [15]. Software fault-tolerance schemes typically degrade performance more than the hardware techniques described in previous chapters because of the inherent overhead incurred by software techniques. Further, in some cases, pure software implementations may not have enough visibility into the hardware to provide adequate fault coverage.

Nevertheless, software fault-tolerance schemes can still be attractive for several reasons. Software schemes are cheaper to implement since they do not have to be built into hardware modules. These schemes are more flexible and can be deployed in the field long after the hardware has been in use. Software schemes can typically be run with off-the-shelf hardware and often require little or no modification to existing hardware. For soft errors, recovery time is often not as critical since soft errors occur in days, months, or years, and not in every microsecond. Hence, software implementations of recovery, which may take longer to execute than hardware recovery schemes, are often attractive solutions. The level of protection offered by software schemes can also be adjusted (e.g., traded off with performance) for selected applications.

Like hardware schemes, software fault-tolerance schemes can be classified into fault detection and error recovery techniques. Both software fault detection and software error recovery can be implemented in multiple levels of the software stack. They can be implemented directly in an application, in the OS, or in the virtual machine layer that abstracts the hardware and can run multiple OS (Figure 8.1). Implementing detection and/or recovery in each of these layers has its own advantage and disadvantage. Application-level detection and recovery are easier to implement since the application is often under a programmer's control. A fault in the OS, however, may not be detectable through this mechanism. OS-level detection and recovery may have greater fault coverage but requires changing an OS. It may, however, be difficult to add detection and recovery in a mature OS, such as Windows. Finally, the virtualization layer provides a great place to include fault-tolerance mechanisms since it is closest to the hardware and can provide maximum fault coverage. The complexity of including such mechanisms in the virtualization layer would typically be higher than in an application.

This chapter illustrates the principles of fault detection and error recovery schemes implemented in an application, an OS, and a virtualization layer. Several example implementations are used to illustrate these principles. The first two techniques for fault detection—signatured instruction streams (SIS) and software RMT—can be used to detect faults across an application or in selected regions of an application. Since the OS can be thought of as a specialized application itself, these techniques can also be applied to detect faults in the OS as well. An implementation of a hybrid RMT solution that augments the software RMT implementation with targeted hardware support is also described. Then two implementations of

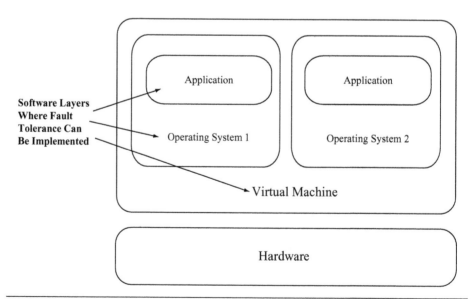

FIGURE 8.1   Implementation options for software fault tolerance.

redundant virtual machine (RVM) that detects faults by running redundant copies of virtual machines and comparing their outputs are discussed.

Like hardware error recovery, software error recovery can also be classified into forward and backward error recoveries. Application-specific recovery can be implemented in various ways. To illustrate the underlying principles of software error recovery, an implementation of forward error recovery done via compiler transformations and two implementations of backward error recovery—one used in databases and the other proposed for parallel shared-memory programs—are described.

Since the OS has better control over the hardware resources, it is often a good place to implement backward error recovery. As discussed in Chapter 7, backward error recovery techniques must deal with the output and input commit problems. The output commit problem arises if a system allows an output from which it cannot recover to exit a domain or a sphere of recovery. The input commit arises if the system rolls back to a previous state or checkpoint but cannot replay the inputs that arrived from outside the sphere of recovery. For an OS, the outputs and inputs are typically related to I/O operations. Since the OS can control I/O devices directly, it is often easier to implement backward error recovery in the OS. This OS-level backward recovery is illustrated using two examples: one that uses a pseudo-device driver and other that checkpoints the state of the OS periodically. Finally, how error recovery can be implemented on top of an RVM is described.

8.2 Fault Detection Using SIS

Fault detection using software checkers has been thoroughly researched in the literature [5]. A software checker will usually check for a violation of a program condition, such as a pointer reference accessing an unmapped region of memory. There are usually two kinds of software checkers: assertion checkers and signature checkers. Assertion checkers typically will assert a property, such as memory bound violation, which can happen due to either a software bug or a hardware fault. These assertions can be application specific, such as ensuring that the value of a computation is within a certain range. Alternatively, these assertions can also be generic, such as ensuring that a program does not dereference a dangling pointer.

Signature checking has typically been used in software fault detection to check control-flow violations. As instructions in a program execute, a *signature* is created for the flow of a group of instructions. These signatures are compared with a set of predetermined signatures that are allowed to be produced in a fault free execution. This section discusses an example of signature-based assertion checking.

These checkers can be implemented purely in software or with some hardware support. Software assertions can be inserted by a programmer, by a compiler, or through a binary translation. Some of these software checkers can add significant overhead to a program's execution. The overhead of software checking can be reduced by performing the checks in a separate coprocessor, often called a watch-dog processor [5], or in a customized hardware engine [8].

SIS [14] is an example of signature checking that uses partial hardware support. SIS detects control-flow violations that could be caused by transient or other kinds of faults. Hence, SIS's coverage is limited to any fault that causes control-flow violation. Nevertheless, SIS can be quite useful in providing partial fault coverage.

SIS detects faults by comparing signatures generated statically at compile time with the corresponding ones generated at run-time. A signature is an encoding of a group of instructions that execute in sequence. Schuette and Shen [14] chose a 16-bit CRC code with the generator polynomial $x^{16} + x^{12} + x^3 + x + 1$. The reader is referred to the section Cyclic Redundancy Check, p. 178, Chapter 5, for a description of CRC codes. A signature mismatch indicates the presence of a fault. To generate a signature, one must statically identify a sequence of instructions for which one can compute the signature. This can be achieved by partitioning a program into blocks or groups of instructions that have one entry point and one or more exit points. Let us call such a block or a group a *node*.

The simplest form of a node is a *basic block* of instructions. A basic block is a sequence of instructions with one entry point and one exit point. The entry point is the first instruction of the basic block. The exit point is the last instruction and consists of a control transfer instruction, such as a branch or a jump. A program can be statically expressed as a combination of basic blocks. Schuette and Shen [14], however, chose a different form of a node, possibly to increase the fault coverage of instructions. Nevertheless, the same principles can be applied to basic blocks as well.

Figure 8.2a shows an example program segment that can be broken up into four nodes, as shown in Figure 8.2b. It should be noted that Node 3 has three exit points and therefore is not a basic block. Figure 8.2c shows how these can be represented in the form of a control-flow graph (CFG). Each node has one entry point but may have more than one exit point. Then, for each node, one computes a signature for the entry point to each exit point within a node. Each signature can be associated with the corresponding exit point of the node. Schuette and Shen [14] showed that such program transformations to embed signatures in a program incurred about a 9%–10% memory overhead.

The run-time system regenerates these signatures for each node as the program executes. These signatures could be monitored purely in software but that may significantly degrade performance. Hence, Schuette and Shen [14] propose the use of a custom monitoring hardware. The monitoring hardware observes these signatures as they are fetched from instruction memory and compares them with the signatures it continuously generates. Because the monitoring hardware runs in parallel to the main processor that executes the program and does not interrupt the executing program, it does not slow down the main processor.

Instruction Number	Label	Operation
1		a ← b
2		b = b + c
3		b = b − d
4		br fifth
5	third	compare a, b
6	second	beq first
7		d = c × d
8		compare d, a
9		bne second
10		f = f xor e
11	first	compare f, b
12		br third
13	fifth	b = a and b
14		br third

(a)

Node	Instruction Number	Label	Operation	Node Entry or Exit?
1	1		a ← b	entry
	2		b = b + c	
	3		b = b − d	
	4		br fifth	
	13	fifth	b = a and b	
	14		br third	exit
2	5	third	compare a, b	entry / exit
3	6	second	beq first	entry / exit
	7		d = c × d	
	8		compare d, a	
	9		bne second	exit
	10		f = f xor e	exit
4	11	first	compare f, b	
	12		br third	

(b)

(c)

FIGURE 8.2 Partitioning a program to create SIS. (a) An example program. (b) How to partition the program into nodes, such that each node has only one entry point. Entry and exit denote entry and exit points, respectively. (c) Representation of (b) in a graphical format. This is also known as the CFG. The arcs/arrows in the graph are labeled with the instruction number (IN) corresponding to the exit condition from the node.

8.3 Fault Detection Using Software RMT

Chapter 6 discussed the concept of RMT and how it can help detect faults by running redundant copies of the same program. This section discusses how RMT can be implemented in software. Unlike cycle-by-cycle Lockstepping, RMT does not require two redundant copies of hardware that are completely cycle synchronized. Instead, RMT compares programmer-visible committed instruction streams from two redundant threads running the same program. Hence, RMT can be implemented purely in software as well since it compares instructions and corresponding results that are visible to software. In contrast, Lockstepping is fundamentally a hardware concept. To detect faults, Lockstepping compares hardware signals that may not be visible to software. Further, in modern speculative processors, it is often hard to keep independent copies of hardware running redundant software in a cycle-synchronized mode. Hence, it is very difficult to implement Lockstepping in software without significant hardware support. Also, unlike software checkers that check individual properties of a program's execution, such as ensuring that the control flow is following the correct path, software RMT typically provides broader fault coverage since it executes redundant copies of the program.

Figure 8.3 shows the sphere of replication of a software RMT system. Recall that the sphere of replication identifies the logical domain protected by the fault detection scheme (see Sphere of Replication, p. 208, Chapter 6). Components within the sphere of replication are either logically or physically replicated. Any fault that occurs within the sphere of replication and propagates to its boundary will be detected by the fault detection scheme corresponding to the sphere of replication. Any outputs leaving the sphere must be checked by an output comparator for errors. Any inputs coming into the sphere of replication must be replicated to the redundant versions at the same instruction boundary to ensure that both versions follow the same execution path. If inputs, such as interrupts, are delivered to the redundant threads at different instruction boundaries, the redundant threads may execute correctly but follow different program execution paths. Hence, the inputs must be delivered at the same instruction boundary in both versions.

Unlike hardware RMT systems that typically use separate hardware contexts, such as registers, address space, and separate program counters, for each of the redundant threads, a software RMT instantiation can implement the redundant versions within the same hardware context. Then the redundant versions would share the same control-flow mechanism, such as the program counter. Consequently, in a software RMT implementation, any control-flow change, such as a branch to a new instruction, must be checked for faults to avoid reducing the fault coverage. In terms of the sphere of replication, hardware RMT systems have the

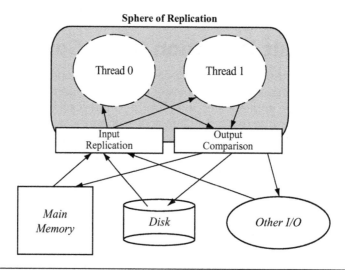

FIGURE 8.3   Sphere of replication of an RMT system. The sphere of replication includes the two thread contexts running redundant code.

program counter inside the sphere, whereas in software RMT implementations the program counter is typically outside the sphere.

The performance degradation from software RMT systems is typically greater than that from hardware RMT systems. This is because the output comparison must be invoked fairly frequently. Since such output comparison is done in software, it slows down the program more than a hardware RMT system would.

This section discusses three implementations of software RMT: error detection by duplicated instructions (EDDI), software-implemented fault tolerance (SWIFT), and Spot. EDDI proposed by Oh et al. [9] implements software RMT on a MIPS instruction set using a compiler. SWIFT proposed by Reis et al. [11] also implements software RMT using a compiler, but for the Itanium architecture. It improves upon EDDI by excluding memory from the sphere of replication and by providing better protection on control transfers. Finally, Spot proposed by Reis et al. [13] implements software RMT using binary translation on an x86 architecture. Unlike EDDI or SWIFT, Spot does not require source code since it operates directly on the binary. None of these software systems, to the best of my knowledge, is used commercially.

8.3.1 Error Detection by Duplicated Instructions

EDDI implements software RMT within a single hardware context [9]. It duplicates instructions within a single thread to create two redundant execution streams. To check for faults, it introduces compare instructions at specific points in a program. The existing architectural register file and memory address space are split in half between the two redundantly executing streams of computation. Consequently, both the register file and the memory are within the sphere of replication. This model will work for cache-coherent, shared-memory multiprocessors as well if the entire shared address space is split among the redundantly executing streams. In this model, I/O resides outside the sphere of replication, so special support may be needed to replicate I/O operations (e.g., DMA from an I/O device to memory). The same issue will arise for interrupts and other forms of asynchronous exceptions.

EDDI Transformation

Figure 8.4 shows an example transformation that the EDDI scheme will perform. Registers R11, R12, R13, and R14 are registers used in the first executing stream, whereas R21, R22, R23, and R24 are the corresponding registers used in the second executing stream. The original program segment shown in Figure 8.4a consists of a load, an add, and a store instruction. Figure 8.4b shows the EDDI transformation. Each instruction is redundantly executed. However, EDDI only checks inputs to the store—registers R11 and R14—for faults. In EDDI, there is a slight subtlety in memory address comparison. Since memory is split into two halves, the corresponding memory addresses in the two redundant versions are not the same. Nevertheless,

LOAD R12 = [R11]

ADD R11 = R12 + R13

(a)

LOAD R12 = [R11]

LOAD R22 = [R21]

ADD R11 = R12 + R13

ADD R21 = R22 + R23

COMPARE R11, R21

(b)

FIGURE 8.4   EDDI transformation. (a) Original program. (b) Program with EDDI transformations.

they could be made to be at a fixed offset from one another. The compare instruction must take this into account. For simplicity, this effect is not shown in Figure 8.4. Also, because the sphere of replication includes main memory, EDDI could have relaxed this constraint and checked only I/O operations through system calls.

EDDI also checks inputs to branch instructions in the same way it checks inputs to store instructions. This guarantees that EDDI has correct inputs to a branch instruction. Nevertheless, the branch instruction could receive incorrect inputs due to a particle strike on the datapath from the register to the branch instruction's execution unit.

EDDI Evaluation

To evaluate the effectiveness of EDDI, Oh et al. [9] performed random bit flips in the code segment to see how many of these bit flips result in an error. This methodology neither reflects the true soft error characteristics of a processor or a machine nor allows computing the AVF. Nevertheless, the methodology gives an idea of how many bit flips injected in a code stream may result in an error. Using this methodology, Oh et al. [9] show that the percentage of undetected incorrect output for eight benchmarks—FFT, Hanoi, Compress, Qsort, Fibonacci, Insert-sort, Matrix multiply, and Shuffle—reduce from 20% to 1.5% on average.

Figure 8.5 shows the performance degradation from introducing extra instructions in the code stream. On a two-way superscalar machine, which in a cycle can execute a maximum of two instructions from either of the two executing streams, the average performance degradation ranges between 36% and 111%. The degradation is not uniformly 100% because in many cases a processor may not be able to issue a maximum of two IPCs from a single execution stream because of interinstruction dependences. In such a case, the redundant instructions can execute in parallel, thereby causing less performance degradation. This parallel execution

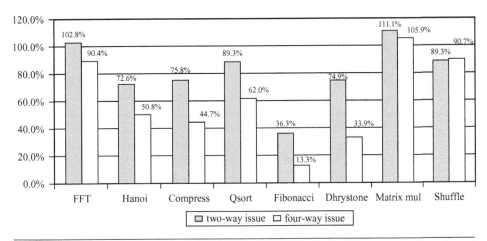

FIGURE 8.5   EDDI execution time overhead in MIPS R4400 (two-way issue) and R10000 (four-way issue) processors. Reprinted with permission from Oh et al. [9]. Copyright © 2002 IEEE.

effect is more pronounced in a four-way issue processor pipeline where the performance degradation ranges from 13.3% to 105.9%.

8.3.2 Software-Implemented Fault Tolerance

In many ways SWIFT is similar to EDDI. SWIFT implements software RMT for an Itanium2 architecture using a single hardware context [11]. Like EDDI, SWIFT splits the registers among the two redundant versions, thereby including the register file in the sphere of replication. Unlike EDDI, however, SWIFT does not duplicate memory since it checks every store for faults, and memory is typically protected with ECC. By placing memory outside the sphere of replication, SWIFT achieves two benefits. First, it reduces system cost. Second, it makes it easier to do input replication for I/O operations. For example, DMA operations can simply transfer I/O data directly to the single copy of memory, thereby making it unnecessary to change how I/O devices or the OS operates.

SWIFT still duplicates the load instruction to ensure that both redundant execution streams have the correct inputs from load instructions. This may cause false DUE since the second redundant load may pick up a different, yet correct, value. This can happen if an I/O device or another processor in a shared-memory multiprocessor changes the loaded value between the two corresponding load operations.

SWIFT provides better control-flow checking than EDDI using an extension of the basic idea of SIS (Fault Detection Using SIS, p. 299). Recall that control-flow changes in a software RMT implementation using a single context may need to be protected since the control-flow mechanism is outside the sphere of replication unlike a hardware RMT implementation. There are two kinds of checks the software must perform. First, one must ensure that the registers used by the control transfer instructions have the correct values. Second, one must ensure that the transfer itself happened correctly. Ensuring the first condition implies checking the register values feeding the control instructions. Both EDDI and SWIFT do this check.

SWIFT uses signatures of predicted hyperblocks, which are extensions of basic blocks, to ensure that the control transfer happens correctly. The mechanism reserves a designated general-purpose register—referred to as the GSR—to hold the signature for the currently executing basic block. These signatures are generated for basic blocks in a manner similar to that for SIS. Before the control transfer happens, the basic block asserts its target using a second register called RTS. After the control transfer takes place, the target decodes its signatures from RTS and compares it to the statically assigned signature of the block it is supposed to be in.

Specifically, let us assume that SIG_CURRENT is the signature of the current basic block. Let us also assume that SIG_TARGET is the signature of the basic block that the control should be transferred to. Before the control transfer in the current basic block, SWIFT would execute the following operation: RTS = SIG_CURRENT XOR SIG_TARGET. Once the branch is taken, the instruction GSR = GSR XOR RTS is executed. It should be noted that GSR contains SIG_CURRENT before this XOR operation is executed. If the control transfer happens correctly, GSR would become SIG_TARGET through this operation (since A XOR A XOR B = B). However, if the control transfer happens incorrectly, then the GSR value will not match the signature of the basic block the control was transferred to. Thus, when SWIFT compares GSR (after the XOR operation) with the statically generated signature for the basic block, it can immediately find the fault.

SWIFT's performance and AVF reduction capability are described later in this chapter in the section CRAFT: A Hybrid RMT Implementation, p. 310.

8.3.3 Configurable Transient Fault Detection via Dynamic Binary Translation

In a software implementation, it is often easier to trade off reliability for performance because the software can be changed or adopted to specific needs. Reis et al. [13] used this observation to implement a software RMT scheme called *Spot* in which they could adjust the level of reliability required by a user. Spot uses the general principles of SWIFT—as described in the previous subsection—but implements software RMT using binary translation for an x86 architecture. Subsequently, the binary translation mechanism, an evaluation of Spot, and how Spot can be used to modulate the level of reliability a user may want are described.

Fault Detection via Binary Translation

Spot implements software RMT using the Pin dynamic instrumentation framework [4]. Pin allows users to insert code snippets into an existing binary. The code snippets for Spot are the reliability transformations to introduce redundant and check instructions. Introducing new code snippets in an existing binary can be challenging. For example, the new binary must change any branch address affected by the introduction of new code.

It is also difficult to statically handle many other challenging issues, such as variable-length instructions, mixed code and data, statically unknown indirect jump targets, dynamically generated code, and dynamically loaded library. Hence, Pin uses a dynamic instrumentation framework in which Pin combines the original binary and new code snippets dynamically at run-time. To reduce the cost of binary translation, Pin uses a software code cache that stores the most recently executed transformed code. Unlike SWIFT, Spot does not do signature checking because it does not have the program's CFG. Instead, it checks the source registers feeding a branch.

Evaluation

Unlike EDDI or SWIFT, Spot computes the AVF of the architectural register file to guide reliability modulation (the reader is referred to Chapter 3 and Chapter 4 for AVF analysis techniques). Current processors do not provide hardware hooks to allow a program to estimate its AVF at run-time, but researchers are exploring the area [16]. Consequently, it is difficult to compute the AVF at run-time as the program executes. Instead, Reis et al. [13] do SFI into the architectural register file as a program runs (see Computing AVFs Using SFI into an RTL Model, p. 146, Chapter 4). The fault is injected into a random bit at a random point in time as the program executes.

If the program runs to completion, then the fault was masked and the corresponding bit into which the fault was injected was un-ACE (unnecessary for ACE). If the program produces the wrong output, if the program does not run to completion, or if the program's exit code indicates an error, then the corresponding bit for that fault injection is classified as ACE. Then dividing the total number of ACE instances by the total number of injected faults gives the average SDC AVF of the register file. Similarly, the DUE AVF is computed as the ratio of ACE and un-ACE instances detected by Spot and the total number of injected faults.

This fault injection study with 15 SPEC benchmarks shows that the SDC AVF for the register file reduces from 20.2% to 6.5%, resulting in three-fold increase in SDC coverage. The difference—20.2% – 6.5%—13.7% is the true DUE AVF. The actual DUE AVF goes up slightly more because of false DUE effects (in which faults in un-ACE bits are flagged as errors).

Figure 8.6 shows the performance degradation experienced by SPEC integer benchmarks when Spot implements the highest level of reliability it can offer. On average, Spots slow down execution four-fold. In the figure, duplication is the overhead from duplicating the instruction streams to create the redundant execution streams, whereas detection shows the overhead for instructions to check for errors. Basepin is Pin's overhead to implement the dynamic binary translation. As

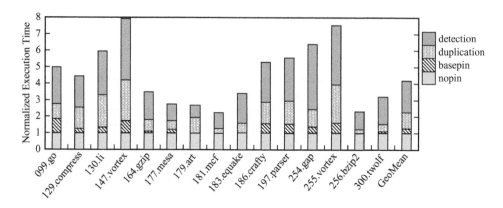

FIGURE 8.6 Execution time increase from Spot's software RMT implementation using the dynamic binary translator Pin. Reprinted with permission from Reis et al. [13]. Copyright © 2006 IEEE.

is obvious from the figure, both detection and duplication add significant overhead to Spot's implementation. Spot's high-performance degradation may be tolerable when a system cannot complete a program because of frequent soft errors. If Spot can be combined with an error recovery mechanism (discussed later in this chapter), then it can let a program proceed to completion. Spot's performance degradation can, however, be reduced if the system requires reduced levels of reliability. The next subsection discusses how to trade off reliability for performance.

Trading off Reliability for Performance

One way to reduce the performance overhead from Spot's software RMT implementation is to incur greater numbers of SDC errors. To do this effectively, one can protect structures with higher AVF or bit counts and potentially remove the protection on structures with lower AVFs. For example, in an x86 architecture, the registers EBP and ESP are typically highly susceptible to transient faults. These registers are used primarily as pointers loading from and storing to memory. Hence, faults in these registers will likely cause a segmentation fault.

Spot allows each architectural register to be either protected or not protected via the binary translation mechanism. If a register is unprotected, instructions to duplicate the operation that uses the register and instructions that detect a fault in that register are both eliminated, thereby improving performance of the application. Given that the x86 architecture has eight registers, one can create 2^8 or 256 combinations. For example, one of the 256 combinations could have protection on the registers EBP and ESP but no protection on the other six registers.

Figure 8.7 shows the performance of each of the 256 combinations against their SDC AVF. The left side of the graph contains data points that have EBP protected, causing greater execution time, and the right side of the graph typically

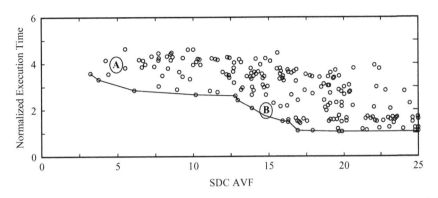

FIGURE 8.7 Trading off performance for reliability for the SPEC integer benchmark gap. Points A and B are discussed in the text. Reprinted with permission from Reis et al. [13] Copyright © 2006 IEEE.

has no protection for EBP. As this graph shows, there is performance-reliability trade-off that software can exploit. For example, a user may choose to get a two-fold degradation in performance for an SDC AVF of 15% (point B in Figure 8.7), instead of a four-fold degradation in performance with an SDC AVF of 5% (point A in Figure 8.7).

Computing AVFs to make this trade-off can, however, be challenging. Obtaining the AVFs for architectural register files requires numerous fault injection experiments on a per-application basis, which can be tedious. A programmer typically has visibility into architectural structures only. Microarchitectural structures are not exposed to the programmer, and current hardware does not provide any way to compute what the AVFs of different structures could be. To make matters more complex, AVFs change across structures, so minimizing the AVF of a register file alone may not mean that the SDC rate of a processor running an application has been lowered. This remains a future research area in computer architecture.

8.4 Fault Detection Using Hybrid RMT

Software fault detection schemes typically have two limitations. First, the extra duplication and/or checks add overhead and degrade performance. Both SIS and software RMT incur significant performance degradation.

Second, software often does not have visibility into certain structures in the hardware and therefore is often unable to protect such structures. For example, a software RMT implementation compares input operands for stores. Then the store can be executed. At this point, the store retires and sends its data to the memory system. Usually, in a modern dynamically scheduled processor, a store resides in a store buffer. Once the store retires, it is eventually evicted from the store buffer, and its data are committed to the memory system. A software RMT implementation that creates the redundant contexts within the same hardware thread cannot, however, protect the datapath from the store buffer to the memory system because it does not have visibility into this part of the machine.

Hence, researchers have examined hybrid RMT schemes that combine the best of software and hardware RMT schemes. Software RMT schemes are typically cheaper to implement since they do not need extra silicon area. Hardware RMT schemes provide better fault coverage and lower performance degradation. Compiler-assisted fault tolerance (CRAFT) is one such hybrid RMT scheme, which is described in this section.

8.4.1 CRAFT: A Hybrid RMT Implementation

CRAFT augments SWIFT's implementation of software RMT with hardware support that improves both fault coverage and performance [12]. CRAFT introduces two changes in hardware: one for store checks and one for load value replication (Table 8-1). As shown in Table 8-1, with no fault detection, the stores and loads execute as is. SWIFT inserts extra instructions to duplicate the instruction stream and then compares the inputs to both store and load instructions to ensure that instructions receive the correct values. Once a value is loaded using the original instruction, SWIFT copies the loaded value to the redundant stream's corresponding register.

Although SWIFT ensures that the registers that are inputs to stores are fault free, it cannot protect the store instruction itself or the datapath from the registers to the branch. For example, if the store instruction is corrupted in the store buffer after the store receives its inputs, then SWIFT cannot detect the fault. Further, the extra check instructions themselves add overhead and degrade performance. To reduce this overhead, Reis et al. [12] propose to introduce new store instructions in the architecture and a corresponding *checking store buffer (CSB)* that will handle the store checks. This creates two flavors of store instructions: the original ones that

TABLE 8-1 ■ Comparison of CRAFT with Other Techniques. Rn = [Rm] Denotes a Load Operation That Loads Register Rn with a Value from Memory Location Rm. Similarly, [Rn] = Rm Denotes a Store Operation That Stores the Value in Register Rm to the Memory Location Rn

Instruction Type	No Fault Detection	SWIFT	CRAFT: CSB	CRAFT: LVQ	CRAFT: CSB + LVQ
Store	[R11] = R12	br FaultDet, if R11 != R21 br FaultDet, if R12 != R22 [R11] = R12	[R11] = R12 Duplicate: [R21] = R22	Same as SWIFT	Same as CRAFT: CSB
Load	R11 = [R12]	br FaultDet, if R21 != R22 R11 = [R12] R21 = R11	Same as SWIFT	R11 = [R12] Duplicate: R21 = [R22]	Same as CRAFT: LVQ

cannot commit their data to memory and the duplicate ones that will commit their values to memory after the stores are checked for faults. CRAFT introduces both flavors of instructions—possibly back-to-back—as shown in Table 8-1. When both the original and duplicate stores enter the CSB, the CSB will compare their inputs to ensure that they are fault free and then commit the value of the original store. The CSB must be implemented carefully to ensure that both redundant streams can make forward progress. This can be usually solved by reserving a few CSB entries for each stream.

SWIFT is also partially vulnerable in its load check and duplication sequence (Table 8-1). The load instruction can still experience a transient fault between the address check (branch instruction shown in table) and the address consumption by the load instruction. Also, there is a vulnerability to transient faults between the original load (R11 = [R12]) and the move operation (R21 = R11). That is, if R11 is corrupted after the load operation, then R21 will receive the corrupted value, and both redundant streams will end up using the same faulty value, which will go undetected.

To avoid these problems associated with a load operation, CRAFT provides the option of using a hardware LVQ as described in Load Value Queue, p. 235, Chapter 6. The LVQ is a structure that allows the primary redundant stream to forward load values to the duplicate stream using a queue of load addresses and values. Unlike SWIFT, CRAFT will issue two load instructions: one is the original one and the other is the duplicate one. The duplicate one will pick up its load value from the LVQ instead of the cache. It should be noted that if both redundant loads read their values from the cache, then there is the possibility of a false error. This is because the value in the cache could be changed by an I/O device or another processor before the second redundant load has the opportunity to read it from the cache. SWIFT avoids this problem by only issuing one load and copying the loaded value to the redundant stream's register. CRAFT not only avoids this using the LVQ but also provides greater fault coverage since the load path is now completely protected.

8.4.2 CRAFT Evaluation

Figure 8.8 shows a comparison of AVF from the different schemes. NOFT has no fault tolerance. SWIFT is described in the previous subsection. CRAFT has three versions: one with CSB, one with LVQ, and one with both the CSB and the LVQ. CRAFT:CSB uses SWIFT's load replication scheme and CRAFT:LVQ uses SWIFT's store checking scheme. SDC is broken into two pieces: dSDC is the one that definitely causes an incorrect output but without crashing the program. pSDC is a probable SDC that causes system hangs or other program crashes without producing an output. $AVF_{SDC} = AVF_{dSDC} + AVF_{pSDC}$. (Total) $AVF = AVF_{dSDC} + AVF_{pSDC} + AVF_{DUE}$.

Both the performance and AVF were measured on a simulated Itanium2 processor using benchmarks drawn from SPEC CPUINT2000, SPEC CPUFP2000, SPEC

System	Integer Register File (GR)				Predicate Register File (PR)				Instruction Fetch Buffer (IFB)			
	AVF	AVF$_{dSDC}$	AVF$_{pSDC}$	AVF$_{DUE}$	AVF	AVF$_{dSDC}$	AVF$_{pSDC}$	AVF$_{DUE}$	AVF	AVF$_{dSDC}$	AVF$_{pSDC}$	AVF$_{DUE}$
No FT	18.65%	7.89%	10.76%	0.00%	1.58 %	0.55%	1.03%	0.00%	8.64%	4.48%	4.16%	0.00%
SWIFT	26.78%	0.13%	1.69%	24.96%	3.95%	0.03%	0.01%	3.91%	19.17%	0.65%	0.53%	17.99%
CRAFT:CSB	23.80%	0.09%	0.41%	23.30%	3.49%	0.01%	0.01%	3.47%	19.82%	0.02%	0.23%	19.57%
CRAFT:LVQ	25.14%	0.12%	1.69%	23.33%	3.52%	0.04%	0.02%	3.19%	16.07%	0.72%	1.18%	14.17%
CRAFT:CSB+LVQ	22.95%	0.04%	1.39%	21.52%	2.68%	0.01%	0.02%	2.65%	14.97%	0.01%	1.05%	13.91%

FIGURE 8.8   SDC and DUE AVF of three structures with different fault detection options. Reprinted with permission from Reis et al. [12]. Copyright © 2005 IEEE.

CPUINT95, and MediaBench suites. To measure the AVF, the authors injected faults randomly into three structures—integer register file, predicate register file, and instruction fetch buffer—in a detailed timing simulator of the Itanium2 processor. Each faulty simulation was run until all effects of the fault manifested in an architectural state or until the fault was masked. Once all effects of a fault manifest in an architectural state, the authors only needed to run the functional (or architectural) simulation, thereby improving simulation speed and accuracy of AVF numbers by running programs to completion, which allowed the authors to precisely determine if the faulty bit was ACE or un-ACE.

As expected, SWIFT reduces the SDC AVF for the three structures 19-fold, which is a significant improvement. CRAFT:CSB decreases the AVF more than 2.5-fold over SWIFT. The introduction of the LVQ, however, makes CRAFT's AVF worse because the loads in CRAFT:LVQ could experience a segmentation fault before the load address is checked. In contrast, SWIFT checks the fault before the load is executed in the program, thereby avoiding this problem. CRAFT could avoid this problem if a mechanism to delay raising the segmentation fault until the load addresses are compared for faults is provided (e.g., see Mechanism to Propagate Error Information, p. 197, Chapter 5).

Reis et al. [12] also introduced a new metric called MWTF to measure the profitability of the different fault detection schemes. MWTF is similar to the metric MITF, as discussed in Exposure Reduction via Pipeline Squash, p. 270, Chapter 7. MITF measures the average number of instructions committed between two errors. MITF, however, does not apply in this case because the number of instructions of each fault detection scheme is different from that of the original non-fault-detecting scheme. However, the total work done by each program under each fault detection scheme is still the same. Hence, MWTF is a more appropriate term. Like MITF, MWTF is proportional to the ratio of performance and AVF. Performance is expressed as the inverse of execution time. Thus, a scheme could be better than another one if its overall MWTF is higher than that of the other.

Figure 8.9 shows the MWTF for SDC and dSDC errors for the four fault detection schemes. As expected, CRAFT:CSB has the highest MWTF and is the most profitable

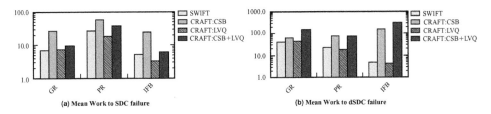

(a) Mean Work to SDC failure (b) Mean Work to dSDC failure

FIGURE 8.9 Normalized MWTF for three structures and the four fault detection techniques. GR = general purpose registers, PR = predicate registers, IFB = instruction fetch buffer. Reprinted with permission from Reis et al. [12]. Copyright © 2005 IEEE.

scheme. CRAFT with LVQ loses out because of the increased AVF experienced due to the faulting loads. However, if only dSDCs—that is, SDC errors that always result in only an output mismatch with no other clue that a fault has occurred—are considered, then CRAFT with LVQ schemes do better.

8.5 Fault Detection Using RVMs

This section examines the implementation of fault detection in the virtual machine layer (Figure 8.1). In the presence of a virtual machine software layer, application accesses to I/O devices and physical memory go through another level of indirection. Instead of accessing and managing the hardware resources directly, the OS implicitly calls the virtual machine layer software that is inserted between the hardware layer and the OS. The virtual machine layer—often referred to as the virtual machine monitor (VMM)—manages these accesses on behalf of the OS. Because the OS does not have direct access to the hardware devices, one can run multiple copies of the same OS or multiple OSs on a single VMM.

In recent years, VMMs have become increasingly popular because of two reasons. First, they provide fault isolation. In some OSs, such as Microsoft Windows, an application crash or a device driver error may bring down the entire OS. With an underlying VMM, such a crash would only bring down the offending virtual machine but not the entire system. Second, in many data centers, machines are underutilized because they often run dedicated applications, such as a file server, an e-mail server. The availability of spare compute power and virtual machine software allows data center managers to consolidate multiple server software—each running on its private copy of the OS—on a single physical machine.

Popular VMMs, also sometimes called hypervisors, include VMWare's ESX server, Microsoft's Viridian server, and the freely available Xen software. Many companies, such as Marathon Technologies, have built their own private virtual machine layer to implement fault tolerance. Certain virtual machines, such as Xeon, require special OS support to run efficiently and therefore cannot run unmodified commodity OSs. This style of virtualization is known as paravirtualization.

Implementing fault detection using a pair of RVMs uses the same principles as RMT described in Chapter 6. The sphere of replication includes the two RVMs running identical copies of applications and OSs (Figure 8.10). The output comparison is done at I/O requests from the application and OS to the VMM. The VMM synchronizes both copies of the virtual machine and sends the I/O request out to the I/O devices. To provide storage redundancy, the same I/O request could be sent to multiple identical disks as well.

Input replication can be tricky since both redundant copies must receive I/O interrupts and responses at the same exact instruction. To facilitate input replication, execution is typically broken up into multiple epochs. An epoch can be a sequence of a preset number of committed instructions. At the end of an epoch, I/O interrupts are replicated and delivered to the RVMundant virtual machines, so that both virtual machines process them at the same instruction boundary.

Like other software fault detection schemes, reads and writes from special machine registers, such as the cycle counter register or time of day register, must be handled with care since the RVMs can read them at different times. Such registers can be treated as being outside the sphere of replication. Writes to such registers must be synchronized and compared by the VMM. Reads from these registers can specify an epoch boundary. The VMM can replicate these reads to the RVMs.

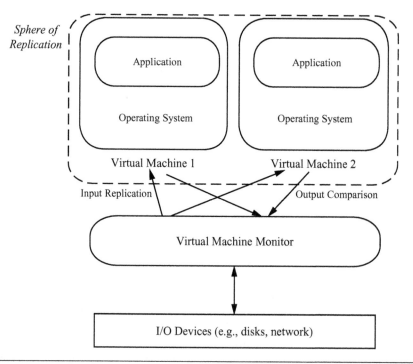

FIGURE 8.10 Redundant virtual machines.

The commercially available Marathon EverRun server uses similar principles of RVMs to implement fault detection completely in software [15]. The EverRun machine also allows recovery in case one of the virtual machines hangs. The EverRun machine accomplishes this by copying the state of one virtual machine to another and then transparently restarting the entire EverRun server. Bressoud and Schneider [1] created a similar fault detection scheme using RVMs.

8.6 Application-Level Recovery

Software fault detection techniques discussed so far can reduce the SDC component of the SER and often replace it with DUE. Once a fault is detected, an error recovery scheme needs to be triggered if the system wants to reduce its DUE rate. Error recovery can be implemented either in hardware or in software. Chapter 7 discussed various hardware error recovery schemes. The rest of this chapter discusses software error recovery mechanisms.

Software error recovery schemes, like software fault detection schemes, can be implemented in three places: in the application, in the OS, or in the VMM. Application-level recovery allows an application programmer to implement recovery without changing the underlying OS. Database applications have traditionally implemented their own recovery schemes to "undo" failed transactions. OS-level recovery requires changing the OS itself (or at least some device drivers), which can be more difficult to implement, test, and deploy. A virtual machine can run several commodity OSs simultaneously and can be a convenient spot to implement both fault detection and software error recovery since it does not require any change to the OSs themselves.

This section discusses three flavors of application-level error recovery: forward error recovery using triplication and arithmetic codes, log-based backward error recovery in a database system using logs, and a checkpoint-based backward error recovery scheme for shared-memory parallel programs. Applications themselves can, of course, implement their own recovery scheme using application-specific knowledge and custom implementations. A complete coverage of all such techniques is beyond the scope of this book. Nevertheless, the three schemes described in this section will illustrate some of the challenges faced by application-level recovery schemes. The basic principles of error recovery, such as output and input commit problems, are similar to what has been discussed for hardware implementations of backward error recovery in Chapter 7.

8.6.1 Forward Error Recovery Using Software RMT and AN Codes for Fault Detection

As the name suggests, a forward error recovery scheme continues to execute—that is, move forward—on encountering an error. In an application, such a forward recovery can be implemented by using three redundant elements: the first two for fault detection and the third for recovery. One possibility is to implement three redundant streams of instructions and use majority voting to decide what stream is in error. Alternatively, one can use two redundant streams augmented with an additional fault detection capability to decide the copy in error. These schemes are described by Reis et al. [10].

The first scheme of Reis et al.—called SWIFT-R—intertwines three copies of a program and adds majority voting before critical instructions (Figure 8.11). Figure 8.11 shows an example of how this can be implemented. Registers R1n denote the original set of registers, registers R2n denote the second set, and registers R3n denote the third set. As Figure 8.11 shows, before a load (R13 = [R14]), majority voting ensures that one has the correct value in R14. The add instruction is triplicated. Finally, before a store operation is performed ([R11] = R12), both its operands are validated through majority voting. On average, Reis et al. found that SWIFT-R degrades performance by about 200% using this triplication mechanism. However, it improves the SDC AVF of the architecture register file about 10-fold.

To reduce the performance degradation from triplicating every instruction, Reis et al. also explored the idea of using only two redundant streams. To facilitate the recovery, the second stream is encoded as an AN code (see AN Codes, p. 182, Chapter 5), where $A = 3$ is a constant that multiplies the basic operand N found in the first redundant stream. Errors in the AN-coded stream can be detected by dividing the operands in this stream by A. If the modulus is nonzero, then this stream had an error. Alternatively, if operands in the first non-AN-coded stream when multiplied by A do not match with the AN-coded stream, but the AN operand is divisible by A, then the first stream must be in error. Once the error is decoded, the state of the correct stream can be copied to the faulty stream and execution restarted. This AN-coding mechanism significantly improves the performance over SWIFT-R and degrades the performance on average by only 36%. However, this AN-coding

Original Code	Fault Tolerant Version
	majority(R14, R24, R34)
R13 = [R14]	R13 = [R14]
	R23 = R23
	R33 = R33
R11 = R12 + R13	R11 = R12 + R13
	R21 = R22 + R23
	R31 = R32 + R33
	majority(R11, R21, R31)
	majority(R12, R22, R32)
[R11] = R12	[R11] = R12

FIGURE 8.11   SWIFT-R triplication and validation.

mechanism reduces the SDC AVF of the architectural register file by only around 50%. This AN-coding scheme has lower fault coverage than SWIFT-R because AN-coded values cannot propagate through logical operations, such as OR and AND. Also, multiplying an operand N by A may cause an overflow. In cases where the compiler cannot guarantee an overflow, it cannot fully protect the operands.

8.6.2 Log-Based Backward Error Recovery in Database Systems

Unlike the somewhat generic forward error recovery technique described in the previous section, the error recovery technique in this section is customized to a database program. A database is an application that stores information or data in a systematic way. It allows queries that can search, sort, read, write, update, or perform other operations on the data stored in the database. Databases form a very important class of application across the globe today and are used by almost every major corporation. Companies store information, such as payroll, finances, employee information, in such databases. Consequently, databases often become mission-critical applications for many corporations.

To avoid data loss, databases have traditionally used their own error recovery schemes. Many companies, such as Hewlett-Packard's Tandem division, sold fault-tolerant computers with their own custom databases to enhance the level of reliability seen by a customer. Databases can get corrupted due to both a hardware fault and a software malfunction. The error recovery schemes for databases are constructed in such a way that they can withstand failures in almost any part of the computer system, including disks. This includes recovering from transient faults in processor chips, chipsets, disk controllers, or any other silicon in the system itself.

To guard against data corruption, commercially available commodity databases typically implement their own error recovery scheme in software through the use of a "log." Database logs typically contain the history of all online sessions, tables, and contexts used in the database. These are captured as a sequence of log records that can be used to restart the database from a consistent state and recreate the appropriate state the database should be in. The log is typically duplicated to protect it against faults.

The rest of this subsection briefly describes how a database log is structured and managed. For more details on databases and database logs, readers are referred to Gray and Reuter's book on transaction processing [3]. There are three key components to consider for a log: sequential files that implement the log, the log anchor, and the log manager. Logs are analogous to hardware implementations of history buffers (see Incremental Checkpointing Using a History Buffer, p. 278, Chapter 7), but the differences between the two are interesting to note.

Log Files

A log consists of multiple sequential files that contain log records (Figure 8.12). Each log file is usually duplexed—possibly on different disks—to avoid a single point of failure. The most recent sequential files that contain the log are kept online. The rest are moved to archival storage. Duplicate copies of each physical file in a log are allocated at a time. As the log starts to fill up, two more physical files are allocated to continue the log. Because the log consists of several duplicated files, its name space must be managed with care.

Log Anchor

The log anchor encapsulates the name space of the log files. The redundant log files use standard file names ending with specific patterns, such as LOGA00000000 and LOGB00000000, which allow easy generation and tracking of the file names. The log anchor contains the prefixes of these filenames (e.g., LOGA and LOGB) and the index of the current log file (to indicate the sequence number of the log file among the successive log files that are created).

The log anchor typically has other fields, such as log sequence number (LSN), for various log records and a log lock. The LSN is the unique index of a log record in the log. The LSN usually consists of a record's file number and relative byte offset of the record within that file. An LSN can be cast as an integer that increases monotonically. The monotonicity property of a log is important to ensure that the log preserves the relative timeline of the records as they are created. The log anchor typically maintains the LSN of the most recently written record, LSN of the next record, etc.

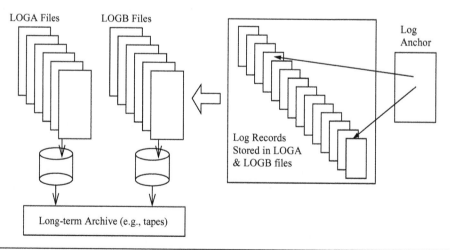

FIGURE 8.12 Structure of a database log.

The log anchor also controls concurrent accesses to the end of the log using an exclusive semaphore called the log lock. Updates to sequential files happen at the end of the file. Hence, accesses to the end of the log by multiple processes must be controlled with a lock or a semaphore (or a similar synchronization mechanism). Fortunately, updates happen only to the end of the log since intermediate records are never updated once they are written. Access to this log lock could become a performance bottleneck. Hence, the log lock must be implemented carefully.

Log Manager

The log manager is a process or a demon that manages the log file and provides access to the log anchors and log records. In the absence of any error, the log manager simply writes log records. However, when an application, a system, or a database transaction reports an error, the log is used to clean up the state. To return each logged object to its most recent consistent state, a database transaction manager process would typically read the log records via the log manager, "undo" the effect of any fault by traversing backward in time, and then "redo" the operations to bring the database back to its most recent consistent state.

8.6.3 Checkpoint-Based Backward Error Recovery for Shared-Memory Programs

Checkpoint-based error recovery is another backward error recovery scheme. Unlike log-based backward error recovery, checkpoint-based schemes must periodically save their state in a checkpoint; hence, the application's performance may suffer if checkpoints are taken frequently. A log-based recovery updates the log incrementally, so it incurs relatively lower performance degradation for the application compared to checkpointing. However, in the log-based error recovery, to create the consistent state of the application, one must traverse the log and recreate the application's state incrementally. Hence, the error recovery itself is slower in log-based backward error recovery than in checkpoint-based recovery, which can simply copy the checkpoint to the application's state and continue execution.

This section discusses an example of application-level checkpointing proposed by Bronevetsky et al. [2]. Although the discussion will be centered around shared-memory programs, the same basic principles can be applied to message-passing parallel programs as well. First, a brief overview of the technique is given. Then, two important components of the program, saving state and avoiding deadlocks for synchronization primitives, are discussed.

Overview

In the method of Bronevetsky et al., an executable with checkpointing capabilities must be created. To achieve this, an application programmer must annotate a shared-memory parallel program with potentialCheckpoint() calls in places where it may be safe to take a checkpoint (not shown in figure). The authors' preprocessor then instruments the original application source code with code necessary to do the checkpointing. The preprocessor avoids taking checkpoints at some of the potentialCheckpoint() calls where it deems it unnecessary. Then the code runs through the normal compilation procedure and creates the corresponding executable.

When this executable is run, there is a coordination layer to ensure that system-wide checkpoints are taken at the correct point without introducing deadlocks (Figure 8.13). The running program makes calls to the OpenMP library that implements the shared-memory primitives. OpenMP is a current standard to write shared-memory parallel programs. These shared-memory primitives in turn run on native shared-memory hardware.

The checkpointing protocol itself is done in three steps. (1) Each thread calls a barrier. (2) Each thread saves its private state. Thread 0 also saves the shared state. (3) Each thread calls a second barrier. The recovery algorithm works in three phases as well. (1) Every thread calls a barrier. (2) All threads restore their private variables. Thread 0 restores the shared variables as well. (3) Every thread calls a barrier. Execution can restart after step 3.

A barrier operation—both for checkpointing and for normal operations—must ensure that all threads globally commit their memory values, so that all threads have the same consistent view of memory after emerging from the barrier. The checkpoint barriers must also not interfere with normal barriers in a shared-memory

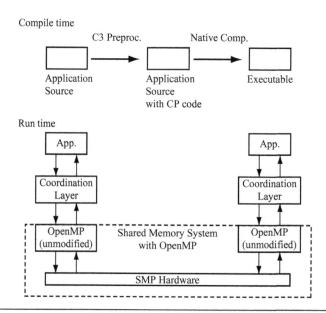

FIGURE 8.13 Overview of compile- and run-time methodologies used in checkpoint-based backward error recovery proposed by Bronevetsky et al.

program. This is because one of the threads could enter a checkpointing barrier—waiting for other threads to do the same—whereas another thread could enter a normal program barrier—waiting for other threads to do the same. This would cause a deadlock. How to avoid such situations is discussed below.

Saving State

Hardware typically does not have visibility into an application's data structures and hence must interpret the entire register file or the memory as an architectural state. In contrast, an application can precisely identify what constitutes its state, which typically consists of the global variables, the local variables, the dynamically allocated data called the heap, and a stack of function calls and returns. An application's checkpoint must consist of these data structures.

The global and local variables are easily identifiable in an application and can be saved at checkpoint time. To tackle the heap, Bronevetsky et al. [2] created their own heap library to keep track of any dynamically allocated data structures, so that at checkpoint time, the application can easily identify how much of the heap has been allocated and must be saved. Instead of keeping track of the stack by using implementation-dependent hardware registers, the authors use their own implementation of the call stack. For every function call that can lead to the potentialCheckpoint() call, they push the function call name into a separate pc_stack. For every function return, they pop the stack. This allowed the authors to precisely keep track of the stack. Variables local to a function call are similarly saved in a separated local variable stack corresponding to every function call push.

Avoiding Deadlocks in Synchronization Constructs

Shared-memory parallel programs usually offer two synchronization primitives—barrier and locks—both of which can cause deadlocks during checkpoint creation. A barrier is a global synchronization primitive. All threads must reach the barrier and usually commit any outstanding memory operation before any thread can exit the barrier. A lock and unlock pair is typically used by a single thread to create a critical section in which the thread can have exclusive access to specific data structures guarded by the lock. When a thread holds a lock, other threads requesting the lock must wait until the thread holding the lock releases it through an unlock operation. Message-passing programs do not have barriers or locks, so deadlocks associated with locks and barriers will not arise in message-passing programs.

Figure 8.14 shows examples of deadlock scenarios for a barrier and for a lock operation. A deadlock could occur if one of the threads (Thread 0 in Figure 8.14a) enters a checkpoint call site, whereas the other threads wait for a normal program barrier. Thread 0 will never enter the normal program barrier because it waits for Thread 1 and Thread 2 to also enter the global checkpoint creation phase. Thread 1 and Thread 2 wait for Thread 0 to enter the normal program barrier instead. This causes a deadlock.

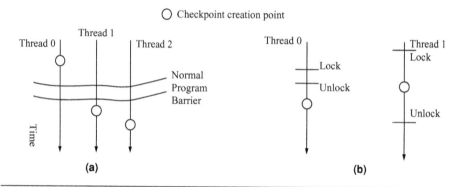

FIGURE 8.14 Deadlock scenarios for barrier (a) and lock (b) operations when global checkpoints must be taken.

Similarly, Figure 8.14b shows a deadlock scenario with a lock operation. Thread 1 grabs a lock and enters its critical section. Then it decides to take a checkpoint and waits for Thread 0 to enter the checkpoint phase as well. Thread 0, on the other hand, tries to acquire the lock Thread 1 is holding, fails, and waits for Thread 1 to release the lock. Thread 1 will not release the lock till Thread 0 enters the global checkpoint phase. Thread 0 will not enter the global checkpoint phase, since it waits for Thread 1 to release the lock. This causes a deadlock.

The solutions to both deadlock problems are simple. For the barriers, Thread 1 and Thread 2 must be forced to take a checkpoint as soon as they hit a normal program barrier, and some other thread (Thread 0) in this case has initiated a checkpoint operation. This can be implemented through simple Boolean flags implemented via shared memory. Similarly, a flag can be associated with each lock. Before a thread initiates a checkpoint, it sets the flag corresponding to all locks it is holding to TRUE. Then it releases all its locks. When a different thread acquires the lock, it checks the value of the flag associated with the lock. If it is TRUE, then it releases the lock and takes a checkpoint. Eventually, the thread that originally held the lock and initiated the checkpoint re-requests the lock, acquires it, and continues with normal operation.

8.7 OS-Level and VMM-Level Recoveries

Recovery in the OS or the VMM is very appealing. One can imagine running multiple applications, such as an editor and a playing audio in the background, when one's machine crashes. When the system recovers from this error, it would be very nice to have both the editing window with unsaved changes visible and the song replayed exactly from where it was left off. This requires extensive checkpointing and recovery mechanisms in either the OS or the VMM. Commercial systems are yet to fully adapt these techniques.

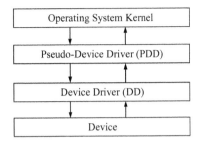

FIGURE 8.15 The pseudo-device driver (PDD) software layer.

An interesting OS-level approach to handle the output commit problem has been proposed by Masubuchi et al. [6]. Disk output requests are redirected to a pseudo-device driver rather than to the device driver (Figure 8.15). The pseudo-device driver blocks outputs from any process until the next checkpoint. Nakano et al. [7] further optimized this proposal by observing that disk I/O and many network I/O operations are idempotent and can be replayed even if the output has already been committed once before. This is because disk I/O is naturally idempotent. Network I/O is not idempotent by itself, but TCP/IP network protocol allows sending and receiving of the same packet with the same sequence number multiple times. The receiver discards any redundant copies of the same packet.

8.8 Summary

Software fault-tolerance techniques are gaining popularity. Recently, Marathon Technologies introduced software fault tolerance in its EverRun servers. Software fault detection and error recovery schemes can be implemented in various layers, such as in an application, in an OS, and in a VMM. Faults can be detected in an application using SIS or software RMT. In SIS, a program precomputes signatures corresponding to a group of instructions at compile time. By comparing these signatures to the ones generated at run-time, one can detect if the program flow executed correctly.

In contrast, software RMT runs two redundant versions of the same program through a single hardware context. Unlike hardware RMT implementations where the program counter is inside the sphere of replication, software RMT implementations share the same program counter among the redundant versions. Hence, the program counter is outside the sphere of replication and must be protected to guarantee appropriate fault coverage. Such software RMT schemes can be implemented either via a compiler or via binary translation. Software RMT schemes can also be augmented with specific hardware support, such as an LVQ or a CSB, to reduce the performance degradation incurred by a software RMT implementation. Similarly, faults can also be detected by running a pair of RVMs that check each other via the VMM.

Like hardware error recovery, software error recovery schemes can also be grouped into forward and backward error recovery schemes. In a software forward error recovery scheme, one can maintain three redundant versions of a program in a single hardware context. Alternatively, one can maintain two redundant versions but add software checks, such as AN codes, to detect faults in individual versions. On detecting a fault, the software copies the state of the faulty version to the correction version and resumes execution.

Software backward error recovery, like hardware schemes, can be based either on logs or on checkpoints. Log-based backward error recovery schemes, typically implemented in databases, maintain a log of transactions that are rolled back when a fault is detected. In contrast, a software checkpointing scheme periodically saves the state of an application or a system to which the application or the system can roll back on detecting a fault. Such recovery schemes can be implemented in an application, an OS, or a VMM.

References

[1] T. C. Bressoud and F. B. Schneider, "Hypervisor-Based Fault Tolerance," *ACM Transactions on Computer Systems*, Vol. 14, No. 1, pp. 80–107, February 1996.

[2] G. Bronevetsky, D. Marques, K. Pingali, P. Szwed, and M. Schulz, "Application-Level Checkpointing for Shared Memory Programs," in *11th International Conference on Architectural Support for Programming Languages and Operating Systems (ASPLOS)*, pp. 235–247, October 2004.

[3] J. Gray and A. Reuter, *Transaction Processing: Concepts and Techniques*, Morgan Kaufmann Publishers, 1993.

[4] C.-K. Luk, R. Cohn, R. Muth, H. Patil, A. Klauser, G. Lowney, S. Wallace, V. J. Reddi, and K. Hazelwood, "Pin: Building Customized Program Analysis Tools with Dynamic Instrumentation," in *ACM SIGPLAN Conference on Programming Language Design and Implementation*, pp. 190–200, June 2005.

[5] A. Mahmood and E. J. McCluskey, "Concurrent Error Detection Using Watchdog Processors—A Survey," *IEEE Transactions on Computers*, Vol. 37, No. 2, pp. 160–174, February 1988.

[6] Y. Masubuchi, S. Hoshina, T. Shimada, H. Hirayama, and N. Kato, "Fault Recovery Mechanism for Multiprocessor Servers," in *27th International Symposium on Fault-Tolerant Computing*, pp. 184–193, 1997.

[7] J. Nakano, P. Montesinos, K. Gharachorloo, and J. Torrellas, "ReViveI/O: Efficient Handling of I/O in Highly-Available Rollback-Recovery Servers," in *12th International Symposium on High-Performance Computer Architecture (HPCA)*, pp. 200–211, 2006.

[8] N. Nakka, Z. Kalbarczyk, R. K. Iyer, and J. Xu, "An Architectural Framework for Providing Reliability and Security Support," in *International Conference on Dependable Systems and Networks (DSN)*, pp. 585–594, 2004.

[9] N. Oh, P. P. Shirvani, and E. J. McCluskey, "Error Detection by Duplicated Instructions in Super-Scalar Processors," *IEEE Transactions on Reliability*, Vol. 51, No. 1, pp. 63–75, March 2002.

[10] G. A. Reis, J. Chang, and D. I. August, "Automatic Instruction-Level Software-Only Recovery," *IEEE Micro*, Vol. 27, No. 1, pp. 36–47, January 2007.

[11] G. A. Reis, J. Chang, N. Vachharajani, R. Rangan, and D. I. August, "SWIFT: Software Implemented Fault Tolerance," in *3rd International Symposium on Code Generation and Optimization (CGO)*, pp. 243–254, March 2005.

[12] G. A. Reis, J. Chang, N. Vachharajani, R. Rangan, D. I. August, and S. S. Mukherjee, "Design and Evaluation of Hybrid Fault-Detection Systems," in *32nd International Symposium on Computer Architecture (ISCA)*, pp. 148–159, June 2005.

[13] G. A. Reis, J. Chang, D. I. August, R. Cohn, and S. S. Mukherjee, "Configurable Transient Fault Detection via Dynamic Binary Translation," in *2nd Workshop on Architectural Reliability (WAR)*, December 2006.

[14] M. A. Schuette and J. P. Shen, "Processor Control Flow Monitoring Using Signatured Instruction Streams," *IEEE Transactions on Computers*, Vol. C-36, No. 3, pp. 264–276, March 1987.

[15] G. Tremblay, P. Leveille, J. McCollum, M. J. Pratt, and T. Bissett, "Fault Resilient/Fault Tolerant Computing," European Patent Application Number 04254117.7, filed July 9th, 2004

[16] K. R. Walcott, G. Humphreys, and S. Gurumurthi, "Dynamic Prediction of Architectural Vulnerability from Microarchitectural State," in *International Symposium on Computer Architecture (ISCA)*, pp. 516–527, San Diego, California, June 2007.

Index

A

Accelerated alpha particle tests, 62–63

Accelerated neutron tests, 63–66
    monoenergetic neutron beam, 64
    proton beam, 65
    white neutron beam, 63

ACE. *See* Architecturally correct execution

Active load address buffer (ALAB), 234–235

Alpha ISA, 152

Alpha particle, 20
    accelerated tests, 62–63
    architectural fault models for, 30–31
    contamination, 2
    impact on circuit elements, 45
    interaction with silicon crystals, 26
    soft errors due to, 63

Alpha radiation, 21

AMD's Opteron^{TM} processor, 133, 187

AN codes, 182–183, 315–316

Application-level recovery, 315

Architectural ACE bits, 90

Architectural ACE *versus* un-ACE paths, 91

Architectural derating factor, 80

Architecturally correct execution:
    instruction per cycle (IPC) of, 101
    principles, 90
    types of, 90

Architecturally correct execution analysis:
    and fault injection, comparison of, 147–149

    using point-of-strike fault model, 106
    using propagated fault model, 114–117

Architectural un-ACE bits:
    dynamically dead instructions, 95
    logical masking, 96
    NOP instructions, 94
    performance-enhancing operations, 94
    predicated false instructions, 95

Architectural vulnerability factor:
    algorithm, data structures for, 106
    basics, 80
    of bit, 81
    of branch commit table, 100
    of CAM arrays, 135, 143–144
    DUE and SDC, 86
    of hardware structure, 96–97
    of Itanium$^{®}$2 execution unit, 113–114
    of Itanium$^{®}$2 instruction queue, 109–113
    of latches, 148
    of RAM arrays, 123, 141–142
    from SoftArch's evaluation, 114–117

Architectural vulnerability factor computation:
    using ACE analysis, 104–105
    using Little's law, 98–101
    using performance model, 101–105
    using SFI, 146

Arithmetic codes. *See* AN codes; Residue codes

AR-SMT, 239

Assertion checkers, 299

AVF. *See* Architectural vulnerability factor